THE ANALYTIC SPIRIT

Henry Guerlac

The ANALYTIC SPIRIT

ESSAYS IN THE HISTORY OF SCIENCE

In Honor of Henry Guerlac

EDITED BY
HARRY WOOLF

Cornell University Press
Ithaca and London

First published 1981 by Cornell University Press.
Published in the United Kingdom by Cornell University Press Ltd., Ely House, 37 Dover Street, London WIX 4HQ

International Standard Book Number 0-8014-1350-8
Library of Congress Catalog Card Number 80-69841
Printed in the United States of America

Librarians: Library of Congress cataloging information appears on the last page of the book.

CONTENTS

III

SCIENTIFIC INSTITUTIONS

IV

STRUCTURE AND FUNCTION IN EARLY MODERN SCIENCE

V

THE REVOLUTIONARY ASPECTS OF MODERN PHYSICS

VI

THE SUPERNOVA OF 1054

PREFACE

A traditional honor accorded a great teacher is to present the public with some of the consequences of his achievement—the work of his students. This volume of essays is meant to do just that, though naturally not all those whom Henry Guerlac taught directly could be given space here, much less the larger number of scholars who came to know his influence.

There are moments in the history of thought when the philosophical urge—that perpetual inquiry into the meaning of existence—and the encyclopedic quest—that powerful, parallel desire to set out the order of and establish the connections among the parts of knowledge—merge to broaden and deepen the river of human consciousness. The eighteenth-century response to the new science with respect to these matters was of this very kind, a confluence marked, in spite of the apparent contradiction, by both a rational and an empirical exegesis of man and nature unequaled in the earlier history of Western intelligence except perhaps for the Greek Miracle. The elevation of materialism and skepticism, the attack on religion, and the early flowering of utilitarianism were direct consequences of this condition, as were the rise of toleration, political liberalism, and the spread and popularization of recondite knowledge. The expansion of the new learning into unexpected places increased almost in direct ratio to the growing success of science and technology in altering the material conditions of life, and perhaps more important, to their growing promise of a progression of improvement and amelioration without end. If the foundation of these achievements was not itself a product of the eighteenth century, but had been formed, in part, by the great men whose genius had lent primary color to the previous epoch, no matter, for the derivative spectrum of the Enlightenment was no less interesting, and certainly no less significant in its social and historical consequences.

Into this fruitfully complicated and fluid epoch which, as one wit has put it, set out to denature God and to deify nature, the skillful scholarship and the superb pedagogy of Henry Guerlac has penetrated to produce enlightenment of its own. His *Essays and Papers in the History of Modern Science* (Johns Hopkins University Press, 1977), recently assembled, are witness enough to over thirty years of scholarship, with the bibliographical addendum below testifying to ongoing accomplishment and vitality. Into this same intellectual arena his students have followed him to create clarity of their own, as this volume, for the most part, and their publications elsewhere reveal. And even when their own scholarly bent has inclined them toward a historical territory somewhat distant from this terrain, such as occurs in Parts V and VI of this book, the values of the master persist. Old pedagogy has become new practice in a pattern that is an essential element in the continuing quest for historical understanding.

The spirit which animates our friend and teacher to whom this Festschrift is dedicated, and which I believe has entered into the essays in this volume, just as it has infused the lives of its authors, can best be epitomized, perhaps, by a passage from a text that Henry Guerlac brought to our attention as students and taught us to read carefully. The *Discours Préliminaire* that d'Alembert wrote to elucidate the purpose of the *Encyclopédie* affected our style and helped to set our goals. There D'Alembert said:

> If one reflects somewhat upon the connection that discoveries have with one another, it is readily apparent that the sciences and the arts are mutually supporting, and that consequently there is a chain that binds them together. But, if it is often difficult to reduce each particular science or art to a small number of rules or general notions, it is no less difficult to encompass the infinitely varied branches of human knowledge in a truly unified system.
>
> The first step which lies before us in our endeavor is to examine, if we may be permitted to use this term, the genealogy and the filiation of the parts of our knowledge, the causes that brought the various branches of our knowledge into being, and the characteristics that distinguished them. In short, we must go back to the origin and generation of our ideas.[1]

This we have tried to do.

HARRY WOOLF

Institute for Advanced Study
Princeton, New Jersey

1. Jean le Rond d'Alembert, *Preliminary Discourse to the Encyclopedia of Diderot*, translated by Richard N. Schwab (Indianapolis: Bobbs-Merrill, 1963), p. 5, quoted with the permission of the publisher.

BIBLIOGRAPHICAL NOTE

HENRY GUERLAC'S *Essays and Papers in the History of Modern Science*, published by the Johns Hopkins University Press in 1977, includes on pages 513–514 a bibliography of his other books and papers. The following items have been published more recently:

"Amicus Plato and Other Friends," *Journal of the History of Ideas,* 39 (1978), 627–633.

"Some Areas for Further Newtonian Studies," *History of Science,* 17 (1979), 75–101.

"The Lavoisier Papers—A Checkered History," *Archives Internationales d'Histoire des Sciences,* 29 (1979), 95–100.

Several other endeavors are under way: an edition and translation of the Lavoisier–Laplace *Mémoire sur la chaleur*, a book entitled *Newton on the Continent*, and a critical edition of Newton's *Opticks*.

I

THE CHEMICAL REVOLUTION

THE ORIGINS OF LAVOISIER'S THEORY OF THE GASEOUS STATE

J. B. GOUGH

Until quite recently, historians of science have assumed that Antoine Laurent Lavoisier's chemical revolution began in 1772 with the signal experiments on the increase in weight of burning phosphorus and sulphur. In traditional accounts of Lavoisier's career in chemistry, the problem of combustion has usually played the central and unifying role, both at the beginning of the chemical revolution and for some time afterwards. Lavoisier's most important contribution to chemistry was generally thought to be his new theory of combustion.

Recent research, however, has revealed that Lavoisier began his career in chemistry as early as 1766, six years before his famous experiments on burning phosphorus and sulphur, and that the problem central to the beginning as well as to the whole of the subsequent revolution in chemistry was not combustion, but the nature of vapors and the air and their relation to one another—in other words, the problem of the gaseous state.[1] The purpose of this study is to examine the origins of Lavoisier's theory of the gaseous state of matter.

In May 1766, Lavoisier read a treatise on the elements written by an obscure scientist and academician of Berlin, Johann Theodore Eller.[2] Eller

I should like to thank Professor Henry Guerlac of Cornell University, under whose guidance and direction the bulk of this paper was written.

1. See J. B. Gough, "Lavoisier's Early Career in Science: An Examination of Some New Evidence," *British Journal for the History of Science,* 4 (1968), 52–57; and Robert Siegfried, "Lavoisier's View of the Gaseous State and Its Early Applications to Pneumatic Chemistry," *Isis,* 63 (1972), 59–78.

2. "Dissertation sur les élémens ou premiers principes des corps & c." and "Seconde dissertation sur les élémens &c.," *Histoire de l'Académie Royale des Sciences et Belles Lettres* (Berlin), 2 (1746), 3–48. Lavoisier took notes on Eller's article (MSS, Académie des Sciences, Lavoisier, 1670 Bm); the full text of these notes has been published in Siegfried, "Lavoisier's View," p. 62.

had placed a quantity of water in the emptied recipient of a vacuum apparatus and, upon heating the liquid, noted that it changed into an elastic fluid, capable of sustaining the mercury in a barometer. Eller thought that the water had actually been transmuted into air through the agency of elemental fire.

When considered in light of eighteenth-century theories concerning the nature of vapors and the air, this idea is not as absurd as it might at first appear. Most scientists of the mid-eighteenth century believed that the only elastic fluid in nature was the air and that vapors were solutions of liquid substances in the atmosphere.[3] Vapors were not widely believed to be innately expansible; if they existed in an expansible state, it was only because they shared temporarily the elastic properties of the air in which they were dissolved. For this reason the experiment on evaporation *in vacuo* was something of an anomaly. A vapor without air, like a salt solution without water, would be an obvious absurdity. Nils Wallerius, who was perhaps the first to observe the phenomenon, confessed that he could not explain it at all.[4] The Abbé Jean-François Nollet, who did extensive experiments on the subject, decided that a subtle air—immune to barometic detection—had invaded the walls of his vacuum recipient and dissolved the water within;[5] and Charles Le Roy, who was one of the earliest and most articulate advocates of the solution theory of vapors, thought that air contained in the water was released *in vacuo* to dissolve the very substance in which it had itself only moments before been dissolved.[6]

After reading Eller's essay on the elements, Lavoisier was apparently inspired to write one of his own.[7] I have dealt at some length with the details of this document in a previous article,[8] and we need not be concerned with them here. Significant to this study, however, are two questions that Lavoisier posed toward the end of his brief essay—questions obviously provoked by Eller's experiment on evaporation *in vacuo*. Lavoisier first asked himself whether vapors enter into a state of expansion because they are dissolved in air or because they are dissolved in the matter of fire (*le*

3. See S. A. Dyment, "Some Eighteenth-Century Ideas Concerning the Aqueous Vapor and Evaporation," *Annals of Science,* 2 (1937), 465–473; and Maurice Crosland, "The Development of the Concept of the Gaseous State as a Third State of Matter," *Proceedings of the 10th International Congress of the History of Science,* II (Ithaca, 1962 [1964]), 851–854.

4. Nils Wallerius, "De l'évaporation dans le vuide," *Collection académique,* 11 (1772), 166–169. (This is a French translation of an article written in Latin for the Swedish Academy in the 1730's.)

5. [L'abbé Jean-François] Nollet, "Recherches sur les causes du bouillement des liquides," Memoires de l'Ac*adémie Royale des Sciences,* 1748, pp. 57–104. This publication abbreviated *Mém. Acad. Sci.*

6. Charles Le Roy, "Evaporation," *Encyclopédie,* VI (Paris, 1756), 123–130.

7. MSS, Académie des Sciences, Lavoiser, 1670 Bn. The full text can be found in Siegfried, "Lavoisier's View," p. 46.

8. J. B. Gough, "Lavoisier's Early Career."

fluide igné) which (according to widely accepted eighteenth-century theories) also had an elastic nature. If the latter were the case, then evaporation *in vacuo* would no longer be an anomaly since heat, unlike air, was thought to be easily capable of traversing the walls of vacuum recipients. Secondly, Lavoisier asked whether, in fact, the air might not also be a fluid in expansion. In other words, was the air a vapor-like compound of some particular substance united to the matter of fire?

In 1960, René Fric, then editor of the Lavoisier correspondence, brought to light another much longer and more detailed treatise on the elements written by Lavoisier probably in the summer of 1772.[9] This document has been named by Henry Guerlac the "Système sur les élémens,"[10] and it gives an extraordinary insight into Lavoisier's thoughts on the eve of his famous course of experiments on the nature of combustion. The "Système" can, in fact, be regarded as a long, remarkably detailed answer to the two questions Lavoisier had asked himself in the treatise on the elements of May 1776. By August 1772, he had come to the conclusion that vapors were elastic compounds of a particular substance united with the matter of fire. According to this theory, any body in the state of vapor would share the elastic properties of elemental fire; it would tend to expand without limit in exactly the same manner as the air. Physically a vapor would be almost indistinguishable from the air; hence Eller's mistaken identification of water vapor with air. Later, Lavoisier would refer to vaporous fluids as "aëriform"—a word that he (or perhaps his colleague, Jean-Baptiste-Michel Bucquet) invented to give expression to the new theory.[11] Obviously such a word was not needed before Lavoisier when only the air was thought to be aëriform.

As for the second question in the treatise of May 1766—the question concerning the nature of the air—Lavoisier had concluded by August 1772 that the ordinary air of the atmosphere was a factitious compound of a particular (and as yet unknown) fluid with the matter of fire. In fact, the air was merely a vapor with a boiling point far below the temperatures of ordinary experience. If vapors were aëriform, then by the same token the air was "vaporiform." In other words, the two things, air and vapor, were really only one.

This identification of the air with vapors is the very essence of Lavoisier's

9. René Fric, "Contribution à l'étude de l'évolution des idées de Lavoisier sur la nature de l'air et sur la calcination des métaux," *Archives internationales d'histoire des sciences,* 12 (1959), 137–138. (This volume was actually published in 1960.)

10. Henry Guerlac, *Lavoisier—The Crucial Year* (Ithaca, N.Y.: Cornell University Press, 1961), pp. 99 ff.

11. Bucquet and Lavoisier both began using the word in 1777. See Lavoisier, *Oeuvres,* ed J. B. Dumas and Edward Grimaux (Paris, 1862), II, 179, from an essay entitled "Expériences sur la réspiration des animaux &c.," which was read 3 May 1777 and published in 1780. Bucquet used the word in the title of a book, *Mémoire sur la manière dont les animaux sont affectés par différens fluides aériformes, méphetiques* & c. (Paris, 1778), which was first given as a paper before the Societé Royale de Médecine, 27 January 1777.

theory of the gaseous state, and it was not only the initial step in the chemical revolution, it was the most important—for it underlay all the major subsequent steps from the early experiments on combustion to the new nomenclature of 1787. The significance of Lavoisier's theory of the gaseous state is a topic that I shall reserve for discussion in another place. Here I shall be concerned solely with how it was that Lavoisier arrived at his theory of the gaseous state in the six years between 1766 and 1772—how, in other words, he was able to answer the two crucial questions of May 1766.

COOLING BY EVAPORATION

If Eller's experiments on the evaporation of water in a vacuum had merely suggested to Lavoisier the possibility that vapors might owe their existence to heat instead of air, then the experiments on cooling by evaporation furnished the decisive proof of it. Lavoisier dedicated nearly half his treatise, the "Système sur les élémens," to an explanation of this phenomenon in terms of a theory which said that fire was absorbed by evaporating fluids and fixed in them by an act of chemical union.

The phenomenon of cooling by evaporation played such a significant role in the early development of Lavoisier's chemical and physical ideas, that it is important to establish just exactly how Lavoisier first became aware of it. But, before attacking this complex and difficult question, it will be helpful first to review some of the background of the problem.

Cullen and Baumé

The cooling effect of evaporating liquids had been described several times before Lavoisier raised the question whether vapors were dissolutions in fire or in air, but despite the relative abundance of descriptions of the phenomenon that had been given, its existence was not very widely known. Only two series of detailed experiments had been performed which demonstrated that cooling was a fact common to all evaporating liquids, and both of them were published in such obscure places that they did not receive the attention that they rightfully deserved.

The earliest observations on cooling by evaporation as a general phenomenon were made by William Cullen, the Scottish physician, chemist, and teacher (most notably of Joseph Black), who published the results of his experiments in the now classic memoir, "Of the Cold Produced by Evaporating Fluids &c." Cullen's memoir first appeared in the second volume of the Edinburgh Philosophical Society's *Essays and Observations,* which volume, unlike the first of that series, was never published in a French translation. Although a German edition appeared not long after its initial publication in English, and a French manuscript version was probably prepared by Pierre Demours (who translated the first volume of the Edinburgh

Society's memoirs), neither could have had a very wide circulation.[12] As a result, Cullen's researches remained (with one remarkable exception which will be discussed further on) largely neglected in France until Lavoisier brought them to the attention of his colleagues in connection with his own work on the nature of gases and their relation to vapors, heat, and combustion.

In 1757, the year after the publication of Cullen's essay, Antoine Baumé, a master apothecary and former student of Claude-Joseph Geoffroy (Geoffroy the younger), read before the Academy of Sciences two memoirs entitled "Sur le refroidissement que les liqueurs produissent en s'évaporant," in which he described a series of experiments remarkably similar to those performed by Cullen. Because Baumé was not a member of the Academy of Sciences (he was not elected to that body until 1772), the publication of his essays was reserved for the collection popularly called *Mémoires des savans étrangers.*[13] The volume of that collection containing Baumeé's essays was not to appear in print until 1768, over a decade after he had first delivered them to the Academy. The Academy's dilatory and uncertain procedure with regard to the publications of nonmembers was too much for Baumé to bear. In a belated effort to establish the priority of his own researches, Baumé hastened to publish large extracts from them in a book, the *Dissertation sur l'aether*, which was printed in Paris, probably in the late months of 1757.[14] Baumé's precipitate publication did not, however, produce quite the results he had intended. The *Dissertation* was of limited interest, and the experiments on evaporation, having only an incidental relation to the main topic of the book, were not likely to have been of much concern to those few specialists who had occasion to consult it. As we shall see, what little notice Baumé's experiments did achieve came in the form of a vicious accusation that he had pillaged them from Cullen.

Dortous de Mairan and His Chinese Refrigerator

A number of observations of cooling by evaporation earlier than those described by Cullen and Baumé had appeared in the late seventeenth and

12. "Of the Cold Produced by Evaporating Fluids and of Some Other Means of Producing Cold," *Essays and Observations, Physical and Literary, Read before a Society in Edinburgh* & c. (Edinburgh) 12 (1756), 145–156. See Guerlac, *The Crucial Year*, p. 14, note 20. The German title was *NeueVersuche und Bemerkungen aus der Arztneykunst and ubringen Gelehrsamkeit einer Gesellschaft zu Edinburg vorgelesen* & c., II (Altenburg, 1757). Pierre Demours (1702–1795), doctor, translator of the *Philosophical Transactions*, academician, and bibliographer, published the French edition of the first volume of the *Essays and Observations*, in Paris in 1759.

13. The memoirs were read 22 January and 21 May 1757. *Mémoires des savans étrangers,* V (Paris, 1768), V, 405–425 and 425–440.

14. *Dissertation sur l'aether dans laquelle on examine les différens produits du mélange de l'esprit de vin avec les acides minéraux* (Paris, 1757). Baumé's book was most likely published after May, when he delivered the second of his papers before the Academy.

early eighteenth centuries, but for the most part they dealt only with water, and in every case the observer had failed to associate the cooling phenomenon specifically with evaporation. Indeed, Guillaume Amontons,[15] who was probably the first to describe this kind of observation (in 1687), thought that the evaporating liquid had only diminished the elastic force of the air in his air thermometer. The fact that cooling had taken place evidently escaped his notice entirely. G. W. Richmann, on the other hand, noticed the cooling, but missed the importance of evaporation.[16] In a series of experiments, published in 1750 in the memoirs of the St. Petersburg Academy, he concluded that the wet bulb of his thermometer had attracted "frigorific particles" from the surrounding atmosphere.

The best known observations on cooling by evaporation were those described by J.-J. Dortous de Mairan in the fourth edition of his prize winning essay, *Dissertation sur la glace* (published in 1749).[17] It was the remembrance of Mairan's experiments which inspired both Cullen and Baumé to investigate the phenomenon in greater detail, using liquids other than water.

Mairan was an amateur of Chinese culture and particularly of Chinese science. In the 1730's he carried on a rather extensive correspondence with Dominique Parrenin, one of the most able and intelligent of the Jesuit missionaries to the court of the early Manchu emperors and one of the few who were allowed to remain in Peking even during the reign of the anti-Christian monarch, Yung Cheng.[18] Mairan later published his portion of the correspondence, consisting mostly of questions on Chinese astronomy and language. Father Parrenin's agent in France, Jean-Baptiste Du Halde, published the replies.[19]

It was perhaps through his well-publicized interest in Chinese science, or even through Parrenin himself, that Mairan encountered Charles-Robert Godehue, a director of the French India Company who later became governor of that company at a most inopportune moment in the history of Anglo-French colonial rivalry in India.[20] Before going to India, Godehue

15. "Moyen de substituer commodement l'action du feu à la force des hommes et des chevaux," *Mem. Acad. Sci.,* 1699, 112–126.

16. G. W. Richmann, "Tentamen legem evaporationis aquae calidae in aere frigidiori constantis temperiei definiendi," *Novi commentarii Academie Scientiarum Imperialis Petropolitane,* 1 (1747–1748 [1750]), 198–205.

17. The additions and revisions that Mairan made in the fourth edition are so extensive that it might well be regarded as an entirely new book.

18. Parrenin (1665–1741) arrived in China in 1698. He was a sinologist of some note, a translator from and into the Chinese language, a diplomat and an amateur of science.

19. *Lettres de M. De Mairan au R. P. Parrenin, Missionnaire de la Compagnie de Jésus à Pékin,* &c. (Paris, 1759). *Lettres édifiantes et curieuses écrites des missions étrangers par quelques missionnaires de la Compagnie de Jésus,* 18 vols., ed. Jean-Baptiste Du Halde (Paris, 1711–1743).

20. Godehue concluded a peace treaty with Thomas Saunders and the British consul in Madras on 11 January 1755. The terms of the treaty gave Britain the upper hand in Indian affairs, and some disappointed Frenchmen regarded Godehue as a traitor.

had spent several years in Canton, and, like many travelers to the East, he had encountered there a refrigerating device which operated on the principle that evaporation produces cooling.[21] Such devices, consisting either of porous terra-cotta jars which allowed water to seep through, evaporate, and thus cool the remainder of their contents, or of simple tin cups surrounded by pieces of moistened linen, had been described many times in scattered travelers' accounts; but until Godehue actually delivered one (of the terra-cotta variety) into the hands of Mairan, no one had ever been able to confirm scientifically the efficacy of these instruments.

Mairan described the design and manner of employing the device Godehue had given him as well as several experiments made with a thermometer to test the extent of its ability to refrigerate liquids. He was able to achieve a cooling of about 2° Réaumur (4 ½° F) below the temperature of the water at the beginning of his experiment and about 6° Réaumur (13 ½° F) below the temperature of the air at the end. In addition, Mairan performed experiments using moistened linen wrapped around the bulbs of his thermometers to increase and prolong the cooling effects. This experimental arrangement might well have been suggested by the linen-wrapped cooling device, which seems to have been common in Persia and India, and which, like the terra-cotta instrument, had attracted the attention of a great many European travelers.[22]

Unfortunately, Mairan's explanation of the phenomenon was not much of an improvement over those of his predecessors, Richmann and Amontons. Just as the rapid freezing of super-cooled water occurs only when it is stirred, so, Mairan thought, the cooling produced by evaporating water occurs only because its particles are agitated. Cooling was a result of the motion of the particles, not of their evaporation *per se*. Also, Mairan confined his observations to a single substance—water. Only after Cullen and Baumé discovered that the cold which had been attributed to ether was not merely a psychological reaction, but a real—and, indeed, an extreme—reduction in temperature, was the cooling produced by the evaporation of water understood to be a particular instance of a general phenomenon common to all liquids.

Augustin Roux and Lavoisier's Awareness of Cullen

Mairan's experiments with Godehue's Cantonese refrigerating device had demonstrated that cooling by evaporation was a phenomenon which had practical as well as theoretical implications. In the temperate regions of

21. Several such voyagers are mentioned by Mairan. A more detailed discussion of them is to be found in the book by Augustin Roux which will be treated below.

22. [Jean] Chardin, *Voyages . . . en Perse et autres lieux de l'Orient,* V (Rouen, 1723), p. 45, describes a device employed in Persia in which the two methods were combined. A porous terra-cotta jar was surrounded by moistened linen.

France, ice collected in the winter and stored away from the heat in large underground "glacières" was used in the torrid summer months to refresh beverages.[23] This method, however, had two annoying disadvantages; ice was expensive, and when the winter season had been mild its supply was unreliable. The possibility of using the cooling effects of evaporation as a more efficient and less expensive way to freshen potable liquids seems to have first been suggested in 1753 by M. Baux, an obscure physician and amateur of science from Nîmes who addressed a short note on the subject to René Antoine Ferchault de Réaumur.[24] Three years later (in 1756), the Abbé Nollet, apparently unaware of Baux's communication, delivered a public lecture in which he proposed replacing ice by chemical means of cooling (that is, using the negative heat of dissolution of saltpeter or sal ammoniac).[25] It was probably with the intention of supplementing this lecture on chemical cooling with a section on cooling by evaporation that Nollet decided, in the spring of 1757, to investigate some of the experiments that Baumé had just presented to the Academy.[26] The Abbé delivered the results of his researches in the form of a "pli cacheté," a sealed note, which when finally opened before the Academy on 19 August 1772 caused Lavoisier to leave the room abruptly to fetch his "Système sur les élémens," a large portion of which concerned the phenomenon of cooling by evaporation and its explanation in terms of a theory of fixed fire.[27]

By far the most complete investigation of the practical aspects of cooling by evaporation was the one conducted in 1758 by Augustin Roux, a teacher, physician, encyclopedist, publisher of dictionaries, editor of the *Journal de Médecine*, and translator from the English.[28] Roux's first publication, a

23. See "Glacière" in *Dictionnaire universel de commerce, d'histoire naturelle et des arts et métiers,* III (Copenhagen, 1761). On the occasional rareness and expense of ice, see "Extrait du mémoire de M. l'Abbe Nollet, lu dans la dernière Assemblée publique de l'Académie des Sciences, intitulé; Recherches sur les moyens de suppléer à l'usage de la Glace dans les temps & dan les lieux où elle manque," *Mercure,* January 1757, pp. 103–113.

E. J. F. Barbier, an eighteenth-century Parisian lawyer, recorded in his diary for February 1724, "It hasn't been below freezing for two days. In order to have ice this summer, Mr. Omberal took the trouble to put under seal the ice houses of several beverage vendors [presumably to halt or restrict sale], and he will tax ice." *Journal d'un bourgeois de Paris sous le régne de Louis XV,* ed. Phillipe Bernard (Paris, 1963), p. 76. Unless the climate of northern France was considerably cooler in the eighteenth century than it is now, such occurrences must have been fairly common.

24. *Histoire de l'Académie Royale des Sciences* (1753), p. 79.

25. "Extrait du mémoire de M. l'Abbe Nollet," *Mercure,* January 1757.

26. "The Abbe Nollet began a reading of a paper concerning experiments on the cold produced by the evaporation of liquids, relative to the memoir by Mr. *Baumé,*" Procès-verbaux, 16 March 1757, fol. 223. (The italics are those of the secretary.) Nollet was chosen (along with P.-J. Macquer) to report on Baumé's second memoir. This report was read 16 July 1757 (Procès-verbaux, fol. 91).

27. The details of this incident were first reported by Henry Guerlac, "A Lost Memoir of Lavoisier," *Isis,* 47 (1956), 211–216.

28. Biographical information on Roux (1726–1776) can be found in [Darcet], "Eloge de M. Roux," *Journal de médecine,* 47 (January 1777), 3–20, and in [Deleyre], *Eloge de M. Roux* (Amsterdam, 1777).

book dealing with the artificial cooling of beverages, appeared anonymously in 1758, without any indication of its publisher or the place of publication. It was a small, almost pamphlet-sized work, bearing the incongruously long, but typically eighteenth century title, *Recherches historiques et critiques, sur les différens moyens qu'on a employés jusqu'à présent pour refroidir les liqueurs, où l'on en indique un connu de temps immémorial & practiqué dans la plus grande partie de l'Univers par lequel il est facile sans null dépense, & avec un soin très-léger, de se procurer dans les plus grandes chaleurs de l'été, des boissons très fraîches.*[29]

Although Roux discussed at some length chemical means of producing artificial cold, the greater portion of his book was devoted to cooling by evaporation. Roux assiduously collected about half a dozen references to travelers' accounts of the linen-wrapped and terra-cotta cooling devices of the East; he described Mairan's experiments and translated Richmann's essay on the cooling effects of evaporating water. The most important section of Roux's book, however, concerned the experiments performed by Cullen and Baumé. Roux not only summarized and examined critically the results of their researches, he also translated the whole of Cullen's treatise and reproduced, for the sake of comparison, the relevant sections of Baumé's *Dissertation*.

Both Roux and his publisher refused to affix their names to the *Recherches,* because, in addition to attempting to introduce the practice of artificially cooling beverages, the book was also written to advance an ill-founded and vicious charge of plagiarism against Baumé. The substance of Roux's allegation was based upon the astonishing similarities between the experiments which Cullen and Baumé had performed. Both claimed to have begun their investigations with the accidental observation of the extreme cooling effect produced by evaporating ether; both, upon noticing this remarkable phenomenon, were reminded of Mairan's experiments on evaporating water, and both decided, therefore, to try to experiment using a variety of liquids. The most significant coincidence, however, was the fact that Cullen and Baumé both attempted the experiment in a vacuum and discovered to their surprise that the absence of air, far from preventing liquids from evaporating, rather, allowed them to do so all the more readily.[30] These remarkable similarities between the two sets of experiments

29. Both of the above mentioned biographers cite Roux as the author of the anonymous pamphlet. Barbier, *Dictionnaire des anonymes,* gives Paris as the place of publication. Roux also wrote the article, "Refroidissement," for the *Encyclopédie* XIII, pp. 902–906, in which he summarized the major points of his book.

30. Roux's awareness of Cullen's memoir requires some explanation. According to his biographer Alexandre Deleyre, Roux learned English in only six months in order to help with the translation of the *Philosophical Transactions of the Royal Society*. The director of this project was Pierre Demours, who, as we have seen, began but did not complete the publication of a French translation of the second volume of the Edinburgh Philosophical Society's *Essays and Observations,* where Cullen's memoir was first published. It was almost certainly as

together with the order of their appearance (Cullen's memoir was published in 1756, Baumé's *Dissertation,* a year later) were sufficient to convince Roux that Baumé had, in fact, stolen his ideas from Cullen. Whether this charge had any validity is a dubious question which need not concern us in this discussion. The important point is that Roux, in presenting the evidence to support his case against Baumé, published the only French translation of Cullen's essay ever made, and at the same time provided an invaluable source book on cooling.

Lavoisier knew of Cullen's experiments at least as early as February 1775, when he referred to him in a little-known essay on "elastic fluids."[31] The source of Lavoisier's knowledge of Cullen was almost certainly Roux. Lavoisier could not read English and it is doubtful that, in 1775, Madame Lavoisier could do so either. Roux's translation was, in effect, the only source available. The most telling piece of evidence, however, is the fact that Roux's pamphlet-sized volume was listed among the items in Lavoisier's library at the time of its confiscation in 1794.[32] The book is now extraordinarily rare and must have been uncommon enough even at the time Lavoisier acquired it.[33] When and how Lavoisier first became aware of Roux's *Recherches* and whether it also provided him with the references to the experiments on cooling which appear in the "Système" are questions which must be reserved for later when we shall investigate the specific origins of the "Système" and of the ideas which it contains.

Demours' assistant that Roux encountered Cullen's classic essay, noticed its similarity to Baumé's researches, and decided to publish a translation for the benefit of French-speaking readers.

The hostility between Baumé and Roux apparently lasted a long time and descended to a personal level. In the second edition of his *Manuel de chymie* (Paris, 1766), avertissement, p. xi, Baumé made the following comment: "I have added many instructive notes in response to the minute criticism which Mr. Roux published in his *Journal de médecine* for March 1764. I dare to flatter myself that he will be satisfied with the clarifications that he will find in my responses, and that he will at last cease to tire me with his badly made objections, with unreliable quotations, or with interpretations far removed from the true sense of my statements."

31. "Delelasticite et de la formation des fluides elastiques," MSS, Académice des sciences, Lavoisier, 171. For text, see J. B. Gough, "Nouvelle contribution à l'étude de l'évolution des idées de Lavoisier sur la nature de l'air et sur la calcination des métaux," *Archives internationales d'histoire des sciences,* 22 (1969) 267–275. Lavoisier also referred to Cullen in a published memoir, "De la combinaison de la matière du feu avec les fluides évaporables et de la formation des fluides élatiques aériformes," *Mém. Acad. Sci.* (1777 [1780]), p. 424 (*Oeuvres,* II, p. 416). The incorrect dating of Cullen's memoir (1775 instead of 1755) was apparently a printer's error. Lavoisier's manuscript copy cites the correct date (MSS, Académie des Sciences, Lavoisier, 1315, fol. 6).

32. MSS, Bibliothèque de l'Arsenal, Paris, 6496, Tome X, 6°, fol. 170.

33. The book is not listed in the British Museum Catalogue. I have been able to find it in only one of the Paris libraries, the Bibliothèque Nationale. As far as I have been able to determine, the work has been cited by no modern author. References to it in the contemporary literature are rare. Shortly after the appearance of Roux's book, at least two reviews of it were published, one in the *Journal des sçavans,* another in the *Mémoires de Trévoux.*

TURGOT AND VAPORIZATION

Lavoisier had known for some time—perhaps as early as 1763 when he took a course in chemistry from G.-F. Rouelle—that the air existed in a "fixed" form in most animal and vegetable and even in some mineral substances. The experiments on "fixed air," as it was commonly called, were widely known in eighteenth-century France through the numerous popularizers of the experiments of the Reverend Stephen Hales.[34] I have shown elsewhere that Lavoisier was directly acquainted with Hales's work as early as 1768.[35]

There is, however, a serious conceptual difficulty associated with the phenomenon of fixed air. In the first place, the pneumatic sciences had so stressed the mechanical properties of the atmosphere that air had very nearly come to be defined in terms of its unique sort of elasticity, an elasticity which, unlike that of the spring, seemed to possess no limits of expansion or contraction. Under these circumstances it was difficult to conceive of the air existing in any form other than its expansible one; yet fixed air was a reality too common to be denied, except perhaps by those few who had never had the good fortune to witness a repetition of Hales's experiments or to read his book. The air which existed in chemical compositions was generally admitted to be unelastic. If it were merely compressed (which was the only conceivable alternative) its force, as calculated by some of the more mathematically inclined chemists of the eighteenth century, would have been sufficient either to rend apart the constituents of the substances in which it was "confined" or to cause a disastrous explosion upon its release.[36] Experiments had shown some bodies to contain as much as six hundred times their volume of air.[37]

This dichotomy between the theory of the air's elasticity and the reality of its fixation constitutes what might conveniently be termed "the paradox of fixed air." Part of the problem in resolving this paradox stemmed from the fact that the experience of fixed air was entirely limited to its evolution, its

34. Henry Guerlac, "The Continental Reputation of Stephen Hales," *Archives internationales d'histoire des sciences,* 15 (1951), 393–404.

35. See J. B. Gough, "Lavoisier's Early Career," pp. 55–56.

36. Stephen Hales himself broached this question in the preface to his treatise (see Hales's *Statical Essays,* 4th ed. [London, 1769], p. viii). Also Père Laurent Béraut in his *Dissertation sur la cause de l'augmentation de poids que certaines matières acquièrent dans leurs calcinations* (Bordeaux, 1747), pp. 22–23, made calculations of this sort in order to deny that the air was the cause of the increase in weight of metallic calxes. Other examples exist in Brossier de Sauvage, *Dissertation où l'on recherche comment l'air suivant ses différent qualités agit sur le corps humain* (Bordeaux, 1753), p. 18; and [Robert de] Limbourg, *Dissertation . . . Quelle est l'influence de l'air sur les végétaux?* (Bordeaux, 1758), p. 7; and finally as an objection to Lavoisier, Delafollie, an obscure chemist from Rouen, "Questions précises concernant le système de l'air fixe," *Observations sur la physique, sur l'histoire naturelle & sur les arts & métiers,* 5 (1775), 60–62.

37. A figure cited by Lavoisier in the "Système" (see Guerlac, *The Crucial Year,* 223).

coming out of bodies. The reverse process, fixation of the air, was a phenomenon which had not yet been observed, or, if observed, not recognized as such. Thus, while the circumstances under which air arose were familiar and well defined (heating, putrifaction, and effervescence), it was impossible to specify exactly how the air had come to exist there in the first place. It was all very well for Rouelle and his disciples to explain the fixation of air in terms of elastic "aggregations" of air particles, "breaking up" under the "force" of "chemical affinity," but they could cite no specific examples.[38] The understanding of the ways in which air became fixed was limited to theoretical imaginings constructed with, as Lavoisier expressed it, "vain words," that is to say, words which were detached from any concrete experience.[39]

When Lavoisier, in writing the "Système," raised this question which we have termed "the paradox of fixed air," he began to outline a solution to it in terms of the air being, not an innately elastic element, but a composition of a "particular fluid" united to the matter of fire; in other words, the air was only a vapor, after the concept of vapors which he had just presented at some length in the previous sections of the "Système."

> But how can air exist in bodies? How can this fluid, capable of such a terrible expansion, fix itself in [bodies and there] a solid and occupy a space six hundred times less than that it occupies in the atmosphere.
>
> The solution to this problem is related to a singular theory that I will attempt to make clear. It is that the air that we breathe is not a simple entity. It is a particular fluid combined with matter of fire. . . .[40]

> Mais comment lair existe-t'-il dans les Corps Comment Ce fluide Susceptible dune Si terrible expansion peut il Se fixer dans [les corps et y] un solide et y occuper un espace six cent fois moindre quil n'occupoit dans l'atmosphere. Comment concevoir que le meme corps puisse exister dans deux etats Si differens.
>
> Le Solution de ce problem tient a une theorie Singuliere que je vais essayer de faire entendre cest que lair que nous respirons nest point un etre simple Cest un fluide particulier combine avec la matiere du feu. . . .[40]

38. One example of the fixation of air was known. G. F. Venel, one of Rouelle's most ingenious disciples, demonstrated that "air" (actually carbon dioxide) could be "fixed" artificially in water, just as Priestley showed some twenty years later in his *Directions.* Venel's article, "Mémoire sur l'analyse des eaux de selters ou de seltz" (in *Mémoires des savans étrangers,* II [Paris, 1755], 53–59 and 80–112), also contains one of the best expositions of (what perhaps can be called) the "aggregation theory" of the air's elasticity. Simply stated, this theory supposes that air is elastic in the "aggregate" and unelastic when the individual particles are separated from one another by the force of chemical affinity. Venel's theory almost certainly comes from Rouelle.

39. In an early essay on hygrometry, dated 1768, Lavoisier complained about the inability of chemists to specify how air became fixed in bodies. "Chemistry, when consulted on these various subjects, will answer with empty words about rapports, analogies to effervescence . . . which convey no idea, and have no other effect than to accustom the mind to content itself with words." *Oeuvres* III, 450.

40. Guerlac, *The Crucial Year,* appendix, pp. 222–223.

This "singular theory" and (as he described it further on in the same passage) "completely new idea" was nothing of the kind. Lavoisier had already encountered the idea that the air was a factitious composition in Eller's essay on the elements.[41] Eller also believed that air was a composition of a particular fluid with elemental fire, although Eller, unlike Lavoisier, specified that the particular fluid involved was water. Lavoisier at this point was unsure what the fluid was. Later, of course, he was to discover that it was oxygen. In addition, there is very strong evidence that Lavoisier had seen at least one other theory of this kind, in an anonymous treatise entitled "Expansibilité," published in 1756 in the sixth volume of the *Encyclopédie.*[42] Moreover, it seems nearly certain, on the basis of this evidence, that it was here Lavoisier first discovered the arguments and experimental support on which he based much of his theory of the air as well as the very words in which he gave that theory expression.

The author of this lengthy and little known article on "expansibility" was A.-R.-J. Turgot, the economist, intendent, and minister of finance, who, like many of his illustrious contemporaries, was also an enthusiastic amateur of science.[43] Unfortunately, it is impossible to determine exactly how or why Turgot came to write this important contribution to the *Encyclopédie.* Of his scientific activities little is known apart from the facts that he studied chemistry with Rouelle,[44] and that, in several letters written to M.-J.-A.-N. Caritat Condorcet in 1771, he argued in favor of a theory of combustion which bore a striking resemblance to the one Lavoisier was to advancê only a year later.[45] It was probably through his association with the circle of friends of Madame Marie Du Deffend that Turgot met Denis Diderot and Jean d'Alembert and was invited to participate in the production of the *Encyclopédie.*[46] Altogether, he wrote only five articles, "Etymologie," "Existence," "Expansibilité," "Foire," and "Fondations." In 1759, when the *Encyclopédie* was temporarily suppressed, Turgot withdrew from the project entirely, in spite of Diderot's strongest supplications to continue and his promise to guard carefully the secret of Turgot's collaboration.[47] The essay

41. For a discussion of the details of Eller's theory, see Siegfried, "Lavoisier's View," pp. 61–64.

42. (Paris, 1757), pp. 274–285. For another discussion of this article in the context of French Stahlism, see Martin Fichman, "French Stahlism and Chemical Studies of Air, 1750–1799," *Ambix,* 18 (1971), 117–122.

43. For Turgot's authorship of the article, "Expansibilité," see Condorcet *Vie de M. Turgot* (London, 1786), pp. 22–23.

44. Rhoda Rappaport, "G. F. Rouelle: An Eighteenth-Century Chemist and Teacher," *Chymia,* 6 (1960), 76.

45. Henry Guerlac has treated this incident in *The Crucial Year,* pp. 146–155.

46. See Gustave Schelle, *Oeuvres de Turgot et documents le concernant* (Paris, 1913), I, notice historique, xxxi.

47. Ibid., p. 594. Diderot sent him a list of titles. He wrote, "Choose the articles that you find agreeable. You can be sure that I shall guard the secret most inviolably and that you will not be exposed to any inconvenience that would give you pause. Some articles please,

on expansibility was the only article Turgot wrote for the *Encyclopédie* on a scientific subject and, indeed, the only one of that kind he was ever to write. Its consequences, as we shall see, were more far-reaching than he ever knew.

The words *expansibilité* and *expansible* were both neologisms, coined probably by Turgot himself. The English equivalents had appeared somewhat earlier, at the turn of the century, when Nehemiah Grew had employed them in a vague reference to the properties of the atoms which compose fluids.[48] Although Turgot might well have borrowed them from the English, which he was capable of reading perfectly well,[49] it seems just as likely that he formed them himself. either from the Latin root, *expandere,*[50] or from the word, *expansion,* which had appeared previously in French, primarily in medical contexts.[51]

Turgot's intention in introducing this new expression was to make clear the distinction between the limited elasticity of the spring, which he called *ressort* (springiness), and the unlimited elasticity of the air, which he termed *expansibilité.* The confusion between these two entirely different things had resulted in some authors' taking the expression "springiness" too literally when it was applied to the air. The air, it was thought, was actually composed of tiny springs. To avoid this kind of thinking, Turgot proposed that "elasticity" be a generic expression, divided into two separate classes, "springiness" and "expansibility."[52] The word *expansible* was officially admitted by the French Academy into the French language in 1762; *expansibilité* followed in 1798. Its definition was taken from the opening paragraphs of Turgot's article.[53]

Monsieur, some articles!" Turgot, nominally a theologian of the Sorbonne, was perhaps frightened of being subjected to the same treatment that had been given to the Abbé de Prades only a few years before by the enemies of the *Encyclopédie.* In any case, he could not afford to jeopardize his governmental career by associating himself with a project that was, in principle, illegal.

48. *Cosmologica Sacra: or a Discourse of the Universe as It Is the Creature and Kingdom of God* &c. (London, 1701), bk. 1, chap. 3, par. 19, p. 14. Grew used "expansible" on the same page. The *O.E.D.* also cites Boyle's *History of the Air* (1692) for the use of the word "expansible."

49. Turgot translated at least one book from the English.

50. This is the etymology for the word expansion given in most English dictionaries. *Expandere* is a post-Augustan neologism, meaning to spread out.

51. Antoine Furetière's dictionary (1727) lists *Expansion* as a "Terme de Médecine" and defines it as an "allongement," as, for example, in a cancerous growth.

52. The word *élasticité* was itself of fairly recent origin, having been coined by Jean Pecquet (1622–1674) in his *Dissertatio anatomica* (1651). The word *élastique* appeared in Furetière's dictionary in 1690. *Elasticité first appeared in the Dictionnaire de Trévoux* in 1732. Both words were accepted by the French Academy in 1740 (third edition of their *Dictionnaire*). Ironically, Pecquet intended the term specifically to describe the expansive properties of the air, but the word came quite rapidly to be used also in the sense of "springiness" or the limited elasticity of solid bodies.

53. Fifth and sixth editions of the Academy's *Dictionnaire* respectively. The Academy's definition of *expansibilité* reads, "Quality by which fluid bodies tend to occupy a larger space" (Qualité par laquelle les corps fluides tendent à occuper un plus grand éspace). Turgot's

> Expansibility is a property of the air . . . it is a property of all bodies in a state of vapor. . . . Thus spirit of wine [i.e., Alcohol], mercury, the most concentrated acids, and a great number of liquids that differ in their nature and specific gravity can cease being incompressible and acquire the property of extending themselves as the air does in all directions without limits, of sustaining the [mercury in a] barometer and of overcoming enormous resistances and weights. . . .
>
> Many solid bodies even, after having been liquified by heat, are also capable of acquiring the state of vapor and of *expansibility,* if one heats them further: such as sulfur, cinnabar, which is still heavier than sulfur, and many other bodies. There are, indeed, very few [things] which, if one persists in augmenting the heat, will not finally become expansible, either altogether or in part.[54]

> L'expansibilité appartient à l'air . . . elle appartient aussi à tous les corps dans l'état de vapeur . . . ainsi l'esprit-de-vin, le mercure, les acides les plus pésens, & un très-grand nombre de liquids très-différens par leur nature & par leur gravité spécifique, peuvent cesser d'être incompressibles, acquérir la propriété de s'étendre comme l'air en tout sens sans bornes, de soûtenir le barometre & de vaincre des résistances & poids énormes. . . . Plusieurs corps solides mêmes, après avoir été liquifiés par la chaleur, sont susceptibles d'acquérir aussi l'état de vapeur & *d'expansibilité,* si l'on pousse la chaleur plus loin: tels sont le souphre, le cinnabre plus pésant encore que le souphre, & beaucop d'autres corps. Il en est même très peu qui, si on augment toûjours la chaleur, ne deviennet à la fin expansibles, soit en tout, soit en partie. . . .[54]

Turgot's attribution of elasticity to vapors was based entirely on those experiments which others had performed on evaporation *in vacuo.* Turgot apparently regarded these demonstrations as decisive proof of vapors' "expansibility." There was no question of water becoming air nor of air being introduced surreptitiously into the vacuum, either through the walls of the glass recipient, as Nollet had thought, or dissolved in the liquid, as Le Roy believed. Turgot did not bother to refute these explanations or even to mention them. He merely enunciated his principle that vapors were expansible and offered the experiments *in vacuo* as evidence.

According to Turgot, all "expansibility" was caused by heat, a subtle fluid which entered between the constituent parts of substances and, by some unknown mechanical action, weakened the attractive forces which held them together. When heated, bodies first increased in volume, then melted, and finally expanded into vapors. The idea of matter existing in three states is one of the most important implications of Turgot's doctrine of expansibility. "Thus water applied successively to all degrees of known temperature, passes successively through three states: solid body, liquid, and

---

somewhat longer definition states that expansibility is the "property of certain fluids by which they tend unceasingly to occupy a larger space" (Propriété de certains fluides par laquelle ils tendent sans cesse à occuper un éspace plus grand).

54. Turgot, "Expansibilité," p. 276.

vapor or expansible body."[55] (Ainsi, l'eau appliquée successivement à tous les degres de température connus, passe successivement par les trois états: de corps solide, de liquide, et de vapeur ou corps expansible.) This opinion was not by any means the common one. Bucquet and P. J. Macquer, for example, thought that water existed in only two states: liquid and solid.[56] The vaporous form of matter was no more a separate state of its existence than salt water was a separate state of salt. In his courses of chemistry, Rouelle taught that elasticity was "so peculiar to the air that only it can be properly described as elastic,"[57] and d'Alembert in an article on "Fluidité" wrote, "the only fluid bodies that are not liquid are *fire* and *air*."[58]

Turgot's idea of the air followed quite naturally from his doctrine of the three states of matter. The air, too, owed its expansibility to fire; it was, in fact, only a vapor which in theory could be reduced to a liquid or even a solid, providing of course that some means could be discovered to deprive it of its heat. The arguments that Turgot marshaled in favor of this position were much more convincing than Eller's experiments could possibly have been. First, since air had weight, it was not exempted from the law of universal gravitation; therefore, it followed that some extraneous repelling force was interposed between the molecules composing it. Second, air expanded when heated; if part of its "expansibility" was caused by heat, why not then all of it? Finally, and most significantly, there were circumstances in which the air was altogether unelastic. Here, Turgot cited without mentioning his name the experiments of Stephen Hales.

> *Everyday experience reveals to scientists an air that is in no way expansible; it is this air that chemists have showed [to exist] in an infinity of solid and liquid bodies, [an air] which has contracted with their elements a veritable union, which enters as an essential principle in the combination of many mixes [i.e., compounds] and which disengages from them, either through decomposition . . . or by the violence of fire.[59]
>
> (L'expérience met tous les jours sous les yeux des Physiciens de l'air qui n'est pas en aucune manière expansible; c'est cet air que les Chymistes ont démontre dans une infinité de corps, soit liquides, soit solides qui a contracté avec leurs élémens une véritable union, qui entre comme un principe essential dans la combinaison de plusieurs mixtes, & qui s'en dégage, ou par des décompositions . . . ou par la violence du feu.)[59]

If elasticity were an immutable property of air, then "fixed air" would be a paradox. But the air is not, in fact, immutably elastic, any more than steam is. Turgot argued that the expansible fluid Hales had found to exist in solid

55. Ibid., 277.
56. See [Jean-Baptiste-Michel] Bucquet, *Introduction à l'étude des corps naturels tirés du règne minéral* (Paris, 1771), p. 64. P. J. Macquer, *Dictionnaire de chimie* (Paris, 1766), I, 365.
57. Rouelle, *Cours de chimie,* MSS, Bibliothéque de l'Institut de France, Paris, 6035, fol. 32.
58. *Encyclopédie,* VI, 892.
59. Turgot, "Expansibilité," p. 278.

bodies was in exactly the same condition as water of crystallization; it remained solid above its melting or even boiling temperature because it was united by a strong chemical force to another body. Lavoisier would later employ this same line of reasoning, and he would extend it further, in light of the experiments on cooling by evaporation, to "fixed fire," the forerunner of his fixed caloric or latent heat.[60]

Vaporization

In spite of the proofs established by the experiments on evaporation in the vacuum, Turgot was unable to reject entirely the "solution theory" of vapors. How was it possible, assuming that vaporous substances were made expansible only by heat, to account for their existence in the atmosphere on the coldest of days? How could one explain on the basis of such an hypothesis the evaporation of ice and snow?

Turgot attempted to solve this problem by supposing that there were two entirely different ways in which a liquid could become a vapor—either through chemical combination with the air, or through the mechanical action of an intense heat. In the first case, the transformation into vapors took place slowly and only at the surface of the liquid, or at those portions of the surface which were in direct contact with the air. The resulting vapor was not in itself expansible; rather, it owed that property to the air in which it was dissolved. In the second case, the particles of the liquid were rent asunder by the intense agitation of concentrated heat. Rapid evaporation (boiling) took place even in the body of the liquid, not only at its surface, and the vapor thus formed was itself expansible so long as it was either maintained above its boiling temperature or conserved in the vacuum, away from the chemical activity of the air. A vapor in this condition behaved exactly as the air. It was, as Lavoisier would later term it, an "aëriform fluid."

To distinguish these two kinds of evaporation, Turgot coined yet another expression—*vaporiser,* to vaporize, and its substantive form, *vaporisation.* Expansible vapors, for example the air or steam, were liquids which had been vaporized; humidity on the other hand was only evaporated.

The word *vaporisation* and its cognates appear about thirty times in Turgot's article and always italicized to distinguish them as neologisms. Unfortunately, the differences between the ideas of evaporation and vaporisation are treated rather lightly in the article itself.[61] However, in the *errata,*

60. René Fric, "Contribution." This analogy between fixed fire, fixed air and water of crystallization is a theme which is constant in Lavoisier's early speculations on the elements. In the "Essay sur la nature de l'air," dated 15 April 1773, the analogy is drawn very clearly. Both Turgot and Lavoisier had been students of Rouelle, and it was from him that this notion originally came.

61. In the article the distinction between evaporation and vaporization is made only in passing, and it would be unlikely that any but the most attentive readers would perceive it,

published in the following (seventh) volume of the *Encyclopédie,* Turgot attempted to remedy this deficiency:

> My design in substituting throughout this article the word *vaporization* for *evaporation,* used in this sense by some scientists, was not at all to put a new word in the place of an old one, but to prevent comfounding under a single denomination two very different phenomena."[62]

> (Mon dessein en substituant dans tout cet article le mot *vaporisation* à celui d'*évaporation,* employé dans ce sens par quelques physiciens, n'étoit nullement de mettre un mot nouveau à la place d'un ancien, mais de ne pas confondre sous une seule dénomination deux phénomènes très différens.)[62]

After this short introduction, Turgot went on to describe in some detail the qualities which distinguished the phenomena in question. Years later, in a letter written to Condorcet, Turgot cited these *errata* and cautioned that his article was not comprehensible without them.[63]

Turgot's Influence

The theory of the air and vapors which Lavoisier presented in the "Système" was very largely founded on Turgot's new concepts of expansibility and vaporization. Lavoisier, too, believed that the air was only a vapor and that vapors, at least in the vacuum, were capable of "performing the office of air."[64] Although neither Turgot nor Lavoisier used the term "gas" in describing these expansible fluids, the modern reader can readily see that that is what, in effect, they were.

The influence of Turgot on Lavoisier is evident not only in the similarity of their ideas. In the "Système," Lavoisier used the expression, "fluid in vaporization" (*fluide en vaporisation*) to describe the condition of water

---

especially since there was a printer's error that made Turgot's line of reasoning nearly unintelligible. "On the other hand, one can say that water thus elevated and sustained in the air by the simple way of *vaporization* ["evaporation" is the word that Turgot meant to place here; see errata VII, p. 1028], that is to say by chemical union with the air, is not strictly speaking expansible except by the expansibility belonging to the air, and can be subjugated to the same law, from which [fact] one need not necessarily conclude that water made expansible by *vaporization* proper . . . doesn't follow different laws."

(De l'autre coté, on peut dire que l'eau ainsi élevée et soûtenue dans l'air par la simple voie de *vaporisation* ["évaporation" is the word that Turgot meant to place here; see *errata* VII, p. 1028], c'est-à-dire, par l'union chimique de l'air n'est à proprement parler expansible que par l'*espansibilité* propre de l'air, et peut être assujettie à la même loi, sans qu'on puisse rigor eusement en conclure, que l'eau, devenue expansible par la *vaporisation* proprement dite et par l'action de la chaleur . . . ne suivroit pas des lois différentes.) Turgot, "Expansibilité," p. 281. As we shall see, Lavoisier would make somewhat the same distinction, however, altered in light of the experiments on cooling by evaporation.

62. *Encyclopédié,* VII, 1028.

63. *Correspondance inédite de Condorcet et de Turgot, 1770–79,* ed. Charles Henry (Paris, 1883), p. 112. (From a letter dated 27 November 1772.)

64. "The combination of water and the matter of fire is durable and . . . this fluid in [the state of] vaporization behaves exactly like air." Fric, "Contribution," 144.

heated in a vacuum. This expression could only have come from Turgot's article.[65]

In April 1773, Lavoisier enunciated for the first time his doctrine that matter existed in three states, solid, liquid,[66] and "a third state which I shall call the state of expansion [of vaporization] or of vapors" (un troisieme etat que je nommerai etat dexpansion [de vaporisation] ou de vapeurs).[67] (This docrtine had been contained only implicitly in the "Système.") By at least February 1775, Lavoisier began to make clear a distinction between evaporation and vaporization drawn along the same lines as Turgot's differentiation, but altered somewhat to account for the phenomenon of cooling.[68] In the latter months of 1774, Lavoisier began to employ the expression, "state of expansibility"[69] and the following spring, in using this term for the first time in print, Lavoisier cited Turgot's anonymous article as its source, saying: "This word is today employed by physicists and chemists, after a modern author fixed the meaning in a very lengthy article which is filled with the newest and most far-reaching views, and which carries everywhere the imprint of genius." (Ce mot est aujourd'hui consacré par les physiciens & par les chymistes, depuis qu'un auteur moderne en a fixé le sens dans un article très étendue, rempli des vues les plus vastes & les plus neuves, & qui portent par-tout l'empreinte du génie.)[70] Far from being an act of ingratiation, as has been suggested,[71] this brief reference to Turgot's article, in spite of its flattering remark about "genius," was hardly an adequate expression

65. It was rarely used by others—see note 73.

66. Lavoisier used the word "fluide" instead of "liquide." Fric, "Contribution," p. 147.

67. Idem. The whole sentence is worth reproducing, because it is the first time that Lavoisier explicitly states the concept of the three states of matter. "All bodies in nature present themselves to us in three different states. Some are solid like stones, earths, salts, and metals. Others are fluid like water, mercury, spirit of wine; and others finally are in a . . . third state that I shall call the state of expansion [of vaporization] or of vapors, such as water when one heats it above the boiling point. The same body can pass successively through each of these three states, and in order to make this phenomenon operate it is necessary only to combine it with a greater or lesser quantity of the matter of fire."

68. In a MS entitled, "Delasticite et de la formation des fluides elastiques." See J. B. Gough, "Nouvelle contribution," pp. 271–275.

69. In an essay entitled, "Dela nature des acides dela nature des differens airs," MSS, Académie des Sciences, Lavoisier, Régistre de Laboratoire III, fol. 52. See M. Berthelot, *La revolution chimque, Lavoisier* (Paris: Albert Blanchard, 1964), pp. 261–262, for transcription.

70. *Observations sur la physique*, 5 (1775), 429.

71. Henry Guerlac has written that Lavoisier's reference to the article on expansibility was "probably a gesture of ingratiation rather than a scientific judgment on Lavoisier's part, for when he wrote in the spring of 1775 Turgot was Controller-General of Finance and had just organized the Régie des Poudres to which he appointed Lavoisier in June." *The Crucial Year*, p. 147, note 2.

The evidence, however, is otherwise. First, Turgot's article was, in fact, a brilliant piece of scientific reasoning. Second, it is not at all clear how Lavoisier could have known that Turgot was the author. Turgot was apparently very discrete about his relations with the *Encyclopédie*. Diderot, when asking him to write more articles assured him that he would guard "le secret le plus inviolable." When Turgot wrote Condorcet recommending that he read the article on expansibility, he did not even mention that he had been its author (see note 63). The only contem-

of the real debt Lavoisier owed Turgot.[72] In all likelihood, Lavoisier did not know who had written the article on expansibility. If he had, he might have been more careful to indicate exactly just how much he had gained from it.

Before Lavoisier began to publish his theories concerning the nature of vapor and air, the word *vaporisation* was only very rarely used—several times by Venel in isolated contexts and at least once by Diderot, who had seen Turgot's article to press.[73] Lavoisier, however, employed the word hundreds of times in the course of his career, and it was primarily through his incessant repetition of it that it became incorporated into the language of science. Not surprisingly, the first occurrence of the word in English was in a translation of the *Traité élémentaire de chimie.*[74] Lavoisier's English translators, Thomas Henry and Henry Kerr, usually gave "evaporation" as the nearest equivalent to *vaporisation* thus obscuring the important distinction that Lavoisier was attempting to make.[75] However, in the third book of the *Traité,* Lavoisier contrasted *évaporation* with *vaporisation* and Kerr was obliged in some manner to follow suit. He did so by changing the "s" to a "z"; and he placed the word in italics, much as Turgot had done some thirty years before, to distinguish it as a new word.[76]

It is significant that the three earliest users of "vaporization" and "vaporize" listed in the *Oxford English Dictionary* were all chemists who, most

---

porary reference that I have been able to find for Turgot's authorship of any of the articles in the *Encyclopédie* is contained in Condorcet's anonymous *Vie de Monsiuer Turgot* (London, 1786), written shortly after Turgot's death.

72. In the version of the essay published in the *Mémoires* of the Academy of Sciences (1775 [1778]), p. 520, Lavoisier expunged the phrase referring to "genius."

73. For example, in his article, "Operations chimiques," *Encyclopédie,* XI, 500, Venel talks of vaporization as an operation of chemistry differing from evaporation in that its object is to reduce to vapors rather than to concentrate, crystallize, or desiccate. Denis Diderot in an item entitled "Expériences intéressantes," from the Grimm correspondence, commented on the experiments performed on the diamond chez d'Arat in August 1771: "What is, then, this precious stone, this diamond which is so admired? A drop of congealed water like another drop of water, with this single difference, that a moderate heat suffices to vaporize the one and a violent heat is needed to vaporize the other." *Oeuvres complètes,* IX (Paris, 1875), 461.

74. *Elements of Chemistry in a New Systematic Order Containing All the Modern Discoveries,* trans. Henry Kerr (Edinburgh, 1790).

75. Thomas Henry, the translator of Lavoisier's *Opuscules,* also translated a collection of Lavoisier's signal memoirs, *Essays on the Effects Produced by Various Processes on Atmospheric Air; with a Particular View to the Investigation of the Constitution of Acids* (London, 1783). In the ninth essay, "On the Combination of the Matter of Fire with Evaporable Fluids; and on the Formation of Elastic Aeriform Fluids," the word "vaporize" and its cognates appear twenty-four times in the original French essay (*Mem. Acad. Sci.,* 1777 [1780], pp. 420–432). In each instance, Henry replaced the word with evaporate, evaporation, rapid boiling, or some such equivalent.

76. This occurs in the section dealing with the operations of chemistry. "But the evaporation which takes place from a fluid kept continually boiling, is quite different in its nature, and in it the evaporation produced by the action of the air is exceedingly inconsiderable in comparison with that which is occasioned by caloric. This latter species may be termed *vaporization* rather than *evaporation.*" *Elements of Chemistry* p. 376. In the original French text, the italicization of the words does not, of course, occur.

probably, were familiar with the untranslated works of Lavoisier and of his students and followers: James Smithson, an eminent biologist and chemist who spent a large part of his life in France, used "vaporize" in a treatise written in 1803 on some compounds of zinc.[77] "Vaporization" was first used by an English author in 1799 in an anonymous review of several volumes of the *Annales de chimie* and again in 1807 in an article written by Humphry Davy for the *Philosophical Transactions.*[78] Not mentioned by the *O.E.D.* are Joseph Black, the son of a Bordeaux wine merchant, and his student John Robison who, in the published version of Black's lectures (1803), spelled it "vaporisation:—with an "s" instead of a "z".[79]

In French historical dictionaries the earliest users mentioned are usually Lavoisier's late contemporaries and immediate successors—Marie Riche de Prony (1790), Jean Baptiste Biot (1807), Jacques Thénard (1813), R.-J. Haüy, Georges Cuvier, and François Raspail.[80] Ferdinand Brunot, in his multi-volume *Histoire de la langue française* cites the first occurrence of *vaporisation* in Lavoisier's *Traité* (1789), although, in fact, Lavoisier had used it in published works dating from 1777 and in manuscript essays as early as 1772 (namely, the "Système").[81]

In the nineteenth century, the words *vaporisation* and *vaporiser* (and their English equivalents) passed from the language of chemistry into the language of physics. In 1822 an article entitled "Vaporisation" appeared in the section of the *Encyclopédie méthodique* devoted to the physics.[82] In 1835 the French Academy accepted the word along with most of Lavoisier's

77. "The reguline zinc, vaporized by the heat, rises from the crucible as a metallic gas, and is, while in this state converted to a calx." *Philosophical Transactions of the Royal Society,* 92 (1803), 26.

78. The anonymous article appeared in the *Monthly Review,* 30 (Supplement for September-December 1799), 560. The review concerned the volumes of the *Annales de chimie* for the year 1799, and the word "vaporization" appears in a discussion of Hassenfraz's report on means of obtaining antimony from its ores. For the citation of Davy's use of the word "vaporization," consult the *Oxford English Dictionary.*

79. Like Lavoisier and Turgot, Black, too, made a distinction between what he called "natural evaporation" and vaporization. The former takes place only at the surface of a liquid and then only slowly and produces a vapor that is inelastic or only weakly elastic. *Lectures on the Elements of Chemistry,* ed. John Robison (Edinburgh, 1803), p. 205. It is interesting to note that Black entitled his lectures, *Lectures on the Effects of Heat and Mixture* (see preface, p. lxiv).

80. See *Dictionnaire général de la langue française,* ed. Adolphe Hatzfeld et al. (Delagrave, 1964), for the reference to Prony. The Larousse *Grande dictionnaire universale du XIX siècle* (Paris, 1876) cites Haüy, Cuvier, and Raspail. Littré's *Dictionnaire de la langue française* (1878) lists Biot and Thénard.

81. Ferdinand Brunot, *Histoire de la langue française des origines à 1900,* VI, pt. I, fas. II, 638.

82. *Encyclopédie méthodique* (section dedicated to physics), IV (Paris, 1822), 776–777. Vaporization is defined as, "Passage of a body from the liquid state to the vaporous state, by the action of caloric" (Passage d'un corps de l'état de liquide à l'état de vapeur, par l'action du calorique), p. 776. The article also makes the distinction between evaporation and vaporization along somewhat the same lines as Lavoisier's distinction.

new chemical nomenclature, and in 1902 the *Encyclopaedia Britannica* published its first article on "vaporization," which in subsequent editions has come to replace almost entirely the one on "Evaporation."[83]

Latent Phlogiston

The word "vaporization" today continues to flourish in spite of the fact that the original gounds for its existence—the distinction between two kinds of vapors—were nullified by the researches of John Dalton. Even before Dalton, however, Lavoisier had been moved by a consideration of the phenomenon of cooling by evaporation to obscure the differences between evaporation and vaporization. Since cooling occurred even when evaporation took place in the atmosphere, it was obvious that the phenomenon involved the action of heat and was not simply the result of a chemical solution in the air. Lavoisier thought that evaporation was a twofold process: Liquids first attracted and combined with heat at temperatures below the boiling point, and the resulting, expansible vapors were then absorbed to the point of saturation by the air.[84] Similarly, Lavoisier was obliged by a consideration of this phenomenon to reinterpret Turgot's conception of the role of heat in the production of vapors. To account for the seeming disappearance of a quantity of heat when fluids vaporized or evaporated, Lavoisier supposed that the matter of fire was actually capable of uniting with other bodies in such a way that it no longer manifested itself as heat. This idea was quite obviously modeled after Rouelle's theory of phlogiston as fire in a fixed form;[85] however, instead of limiting this phlogiston or fixed fire to combustible substances, Lavoisier placed it as well in vapors and air. This would later allow him to explain the production of heat and light in the process of combustion without having to assume that the burning substance itself lost phlogiston; rather it was the air, or more specifically the oxygen, which lost it. Lavoisier did not so much deny the existence of phlogiston as relocate it. When he later stated his objections to the doctrine of phlogiston, Lavoisier was in actuality objecting to a name; the thing itself survived "latent" in his theory of caloric.

THE ANSWERS

Buried anonymously in the sixth volume of the *Encyclopédie* under a heading which few readers were likely to find occasion to consult, Turgot's

83. The fifth edition of the Academy's dictionary and the tenth edition of the *Encyclopaedia Brittanica.*

84. See the *Traité élémentaire de chimie* (Paris, 1789), p. 433. The earliest reference to this twofold process of evaporation is found in the "Système sur les élémens" (see *The Crucial Year,* p. 220, note 2, and pp. 221–222).

85. See Guerlac, *The Crucial Year,* p. 32. For an extended discussion of Rouelle's chemical theories, see Rhoda Rappaport, "G. F. Rouelle: An Eighteenth-Century Chemist and Teacher," *Chymia,* 6 (1960), 68–101.

article on "expansibility" was nearly unknown to his contemporaries. References to it in the literature of the period are very few indeed, and the arguments it advanced in favor of the identity of vapors and air appear to have attracted little attention.[86] How then did Lavoisier discover this important but obscure work upon which so many of his own theories and conceptions seem to have been founded? What inspired him to look there for the answers to the questions he had raised concerning the nature of the air and its relation to vapors and heat?

In 1768, just after his election to the Academy of Sciences, Lavoisier began his famous course of experiments pertaining to the transmutation of water into earth.[87] As a kind of introduction to his own researches, Lavoisier wrote a separate, historical memoir in which he examined the theories and experiments of his predecessors. This "first memoir" (as it is called) was probably completed after the "second memoir" (dealing with Lavoisier's own experiments) but certainly before November 1770, when Lavoisier in reading the second memoir at a public meeting of the Academy of Sciences mentioned to his audience that the historical section of his discourse had already been delivered to the Perpetual Secretary of the Academy.[88]

Just about the time Lavoisier was engaged in research on the historical background to the problem of transmutation, there appeared in Avignon the volume of the *Collection académique* containing a selection of essays from the Academy of Sciences at Berlin.[89] This volume included another edition of the second of Eller's essays on the elements to which had been added a

86. It is not strange that Turgot's article should have attracted so little attention. "Expansibility" was a neologism, and one does not usually consult an encyclopedia for an unknown word.

87. "Sur la nature de l'eau et sur les expériences par lesquelles on a prétendu prouver la possibilité de son changement en terre," *Oeuvres,* II, first part, 1–11; second part, 12–28.

88. A copy of the second memoir was delivered to the secretary of the Academy of Sciences 10 May 1769. A note on the manuscript reads, "The present paper, containing 30 pages, was presented to me the 10 May 1769, to be initialed which I did and I returned it immediately, signed Defouchy." MSS, Académie des Sciences, Lavoisier, 1299. In the text of this manuscript, Lavoisier expresses the desire to do an historical introduction on the experiments of others concerning the transmutation of water into earth, but because of the limitations of time, he says that he will reserve his historical introduction for another memoir.

Another copy of the second memoir in the same dossier has written across the top, "Read at the public reopening of the Academy after the long holiday of the late summer and early fall of 14 November 1770." In this copy, a note at the bottom of the first page states that an historical introduction has been written, but, "I haven't the memoir with me. It was returned to Mr. deFouchy." Thus, Lavoisier did not read both his papers on the transmutation of water at the *séance publique* of 10 November 1770, as Maurice Daumas erroneously states; *Lavoisier théoricien et expérimentateur* (Paris, 1955), p. 27.

89. Volume VIII of the *partie étrangère,* entitled *Mémoires de l'Académie Royale de Presses* &c. (Avignon, 1768). A quarto and an octavo edition were published at the same time. In 1770, a duodecimo edition appeared in Paris chez Panckoucke. All references are to the quarto edition.

long, critical introduction by the editor, an obscure Avignon doctor, translator from the Latin, and amateur of science named François Paul.[90]

Judging from his commentary on Eller, Paul was a man of unusual good sense with an extraordinary knowledge of scientific bibliography.[91] In the course of his discussion of Eller's theories, Paul raised exactly the questions which Lavoisier had asked himself after reading Eller's essay two years before—are vapors dissolutions in fire or in air, and is air elastic by its own nature or because of heat? Paul recognized that Eller's presumed transmutation of water into air was nothing but the production of elastic vapors in the vacuum. As evidence of this fact, Paul cited the experiments which had been performed on evaporation in the vacuum, and in particular he cited those of Cullen, which had shown that evaporation was even more pronounced in the absence of air and that cooling took place. In a long footnote, Paul discussed the significance of Cullen's experiments and pointed out to his French-speaking readers where they might find Cullen's article in translation.

> [The experiments], of Mr. *Cullen,* an Edinburgh academician . . . support those of Mr. *Wallerius,* for they establish in the least doubtful manner that evaporating fluids cause the thermometer to descend considerably, but this effect takes place in the vacuum of a pneumatic machine, as it does in the air, and even more perceptibly. One may consult on this matter the curious and learned researches of an anonymous author (whom we believe to be *Roux,* author [i.e., editor] of the *Journal de Médecine*), on the different means which have been used for refrigerating liquids, brochure in duodecimo which appeared in 1758, without indication of place of publication or printer, and in which one finds a translation of the two memoirs of Messrs. Richmann and Cullen.[92]

On the page opposite this footnote, Paul cited yet another work which lent support to the contention that vapors were elastic—Turgot's anonymous article on expansibility.[93]

Lavoisier in 1766 had "read hastily" only a few pages of Eller's memoir

90. François Paul was the brother of a famous French grammarian, Armand-Laurent Paul. He received his medical degree from Montpellier and practiced medicine in Avignon. He translated several works of Hermann Boerhaave.

91. Paul was one of the few persons in France who seems to have been very familiar with the school of British pneumatic chemists. A casual survey of his notes and appendices reveals about a dozen references to Stephen Hales, three references to David MacBride, two references to John Pringle, and a single reference to the work of Joseph Black. In addition, Paul reprinted in an appendix Venel's article on effervences from the fifth volume of the *Encyclopédie.* All of this was sure to have been of interest to Lavoisier, who as we have seen, was already quite concerned with the role of air in chemistry.

92. Paul, *Collection Académique* [*partie etrangère*], VIII, introduction, p. xl.

93. Ibid., p. xli. Another reference to Turgot's anonymous article on expansibility appears in a note on page lxxxi in which Paul describes it as, "One of the finest and most learned articles on physics in this dictionary."

on the elements.[94] In 1769 or 1770 he must have returned to Eller's essay in connection with his own work on transmutation. Eller's experiments were among the most important which had been performed on this subject; Lavoisier treated them at length in the first or historical memoir on the transmutation of water into earth. Lavoisier would also have been interested in other opinions of Eller's work, and Paul's critical essay was readily available to Lavoisier from the time he began his researches. Indeed, Lavoisier may well have been a subscriber to the *Collection académique;* all thirteen volumes of it were in his library.[95] More substantial, although admittedly uncertain, evidence of Lavoisier's having read Paul is the similarity of the arguments they used to counter Eller's alleged transmutation of water. Eller performed a variation of J. B. Van Helmont's famous willow-tree experiments by raising hyacinth bulbs in a measured quantity of earth and noting at the end of the experiment that the increase in weight of the plants was greater than the loss in weight of the earth. Eller argued that the water used to feed the plants had been transmuted into earth. Paul objected that Eller's experiment did not take into account the possibility that his plants may have acquired a portion of their substance from the air. To give authority to his argument, Paul cited passages from the works of Hales and Charles Bonnet showing that plants derive nourishment from the air.[96] In his historical memoir on transmutation, Lavoisier advanced this same objection against Eller and similarly cited Hales and Bonnet to support his position.[97]

Although the evidence is not by any means conclusive, it appears more than likely that Lavoisier read Paul's essay in the course of his research on the transmutation problem, discovered the references to Roux's *Recherches* and Turgot's "Expansibilité," and was thus led to investigate those works for himself. As we have already seen, the fruit of that investigation was to be Lavoisier's "Système sur les élémens."

94. "I rapidly read several pages of it" (j'en ai lu rapidement quelques pages), MSS, Académie des Sciences, Lavoisier, 1670 Bm.

95. MSS, Bibliothèque de l'Arsenal, 6496, Tome X, 6°, fol. 183v.

96. Paul, *Collection Académique* [*partie etrangère*], VIII, p. xliii.

97. *Oeuvres,* II, 6. Lavoisier also mentioned the work of DuHamel and Guettard on the same question.

THE TRIUMPH OF THE ANTIPHLOGISTIANS

CARLETON E. PERRIN

Few of the major conceptual shifts in the history of science rival the chemical revolution for compactness in time and consequent sense of drama. As usually defined, the episode spanned a mere twenty years, from the time of Antoine Laurent Lavoisier's pivotal insights of 1772 into the nature of combustion and the formation of acids until the early 1790's when proponents of the new theory had clearly gained the upper hand, at least in France. Throughout the period there was a sense of excitement pervading chemistry—due intially to the burst of discoveries in pneumatic chemistry and subsequently to the confrontation between competing theories. The period of open confrontation was of even shorter duration, becoming visible only around 1785 when chemists began to choose between opposing camps, rallying to the "old" Stahlist theory or the "new" antiphlogistic theory of Lavoisier and his associates. Although the issues were conceptual and professional, and relatively free of political, moral or theological implications, the ensuing debate produced much heat and some bitterness.

In this essay I propose to examine those years beginning in 1785 when the chemical community was polarized into hostile factions. My emphasis will be upon the skillfully organized campaign strategy of the antiphlogistians and the power shift which they achieved. First, a few remarks relating to developments leading up to 1785 will be necessary as a backdrop to the discussion which follows.

It has been standard practice in accounts of the chemical revolution to juxtapose Lavoisier's personal development of the new theory of chemistry and the conversions to his view, beginning in 1785, of several prominent chemists including Claude-Louis Berthollet, Antoine-François Fourcroy, and Louis-Bernard Guyton de Morveau. This scenario appears to have been inspired by the recollections of Fourcroy himself, who wrote:

> From 1777 to 1785, in spite of Lavoisier's great efforts and his numerous papers, he was quite truly alone in his opinion on the exclusive influence of air in natural processes; while admitting the basis of his experiments and the truth of his results, chemists, witness to their exactitude and their merit, did not yet renounce the existence of phlogiston; and the theory which they followed in their books, their papers, and their demonstrations was still but a more or less forced compromise between that of Stahl and the action of air. For the "bons esprits," for the coolest heads and those most practised in the cultivation of the sciences, there was a sort of neutrality which resisted not the discoveries, but the total overthrow of the old order of ideas; this sensible party, before adopting a total change was waiting for a more decisive victory over the march of nature, and it was only to be found in the time of the decomposition and recomposition of water.[1]

Fourcroy's account and others following his lead have perpetuated a few misleading impressions concerning the reception and spread of the new theory. For one, the scenario exaggerates Lavoisier's degree of isolation in the period prior to 1785. While it is true that no other names were publicly linked with the new theory of which Lavoisier was the principal architect, it is equally true that he seldom worked alone. From his early experiments in the laboratory of Jean-Charles-Philibert Trudaine de Montigny, through the intense collaborations with the chemist Jean-Baptiste-Marie Bucquet and the mathematician and physicist Pierre-Simon de Laplace and the series of researches in the 1780's in which he was assisted by younger scientists including P. Gengembre, J.-B.-M. Meusnier, and J.-H. Hassenfratz, Lavoisier displayed a knack for drawing talented colleagues into his work.[2] In general, they shared not only Lavoisier's equipment, but also his point of view on the interpretation of chemical phenomena.

In addition to these individual collaborations, Lavoisier had quietly gathered around him a group of friends and associates which developed into a new school. The core of this group was made up of colleagues from the Academy of Sciences, drawn especially from the ranks of mathematicians and physicists, who met regularly at Lavoisier's home for discussion and to witness experiments. Once again it is Fourcroy who has left us a vivid impression of these sessions.

> Twice a week at his home he held gatherings to which were invited those men most distinguished in geometry, physics, and chemistry; instructive conversations, exchanges resembling those which had preceeded the establishment of the academies, there became the center of all enlightenment. There were dis-

1. *Encyclopédie méthodique, Chimie,* III (Paris, 1796), 541. My translation.

2. For a concise and reliable account of Lavoisier's scientific career see Henry Guerlac, *Antoine Laurent Lavoisier, Chemist and Revolutionary,* (New York, 1975). Lavoisier's major collaborations are examined by E. McDonald, "The Collaboration of Bucquet and Lavoisier," *Ambix,* 13 (1957), 219–234, and H. Guerlac, "Chemistry as a Branch of Physics: Laplace's Collaboration with Lavoisier," *Historical Studies in the Physical Sciences* (Princeton), 7 (1976), 193–276.

> cussed the opinions of all the most enlightened men of Europe; there were read the most striking and the newest passages from works published by our neighbors; there were theories compared with experiment. . . . I shall never in my life forget the privileged hours which I spent in those erudite exchanges where it was so pleasant for me to be admitted. . . . Among the great advantages of those meetings, that which struck me the most and whose invaluable influence soon made itself felt in the heart of the Academy of Sciences, and subsequently in all the works of physics and chemistry published for twenty years now in France, is the harmony which was established between the manner of reasoning of the geometers and that of the physicists. Precision, severity of the language, of the expressions and of the philosophical method of the former, passed gradually into the minds of the latter.[3]

This emerging school was united by a common spirit of inquiry inspired by the rigor and precision of mathematics and by a commitment to extend such an approach systematically to areas of "experimental physics" including chemistry. "Lavoisier, placed as if at the focus of all this brilliance, missed none of the comparative ideas which were put forth in these scientific sessions, often brought them to bear on the object of his researches of the moment in order to find the means to eliminate difficulties which he found there, to know what others had done or thought on the same subject, and especially to sow little by little among scholars the seed of his own ideas and to confide in them its development."[4] The members of Lavoisier's inner circle witnessed the development of the new chemistry and participated in it. Dress rehersals of Lavoisier's experiments and conclusions were staged before this sympathetic group before presentation to the full assembly of the Academy. Points of dispute between members of the inner circle were resolved by further experiments in which the disputants participated. In short, Fourcroy's reminiscences of his exposure to Lavoisier's inner circle belie the impression of isolation suggested by other passages in his writings. While participants were perhaps not agreed on all points of theory, they shared a methodological commitment to a rigorous and quantitative approach to chemistry, an intimate familiarity with Lavoisier's ideas and experiments, and a receptivity to these ideas which no doubt grew in proportion to their personal contributions.

Familiar accounts of the reception of Lavoisier's theory mislead, too, in portraying the early conversions as more sudden and straightforward than they in fact were. The passage from Fourcroy suggests that chemists reserved judgment until the advent of more conclusive experimental evidence; then the discovery of the composition of water provided the required demonstration of the superiority of Lavoisier's view. It is quite true that the water experiments were an important turning point and may have been instrumental in persuading individuals such as Fourcroy. But closer exam-

3. *Chimie*, p. 425. My translation.
4. Ibid., p. 426. My translation.

ination of the conversions of Fourcroy, Berthollet, and Guyton shows that in each case the abandonment of phlogiston was the culmination of a gradual process extending over several years or more, during which the positions of these individuals moved progressively closer to that of Lavoisier.[5] Fourcroy and Berthollet had been admitted to Lavoisier's inner circle, and the favorable impression which that association made is evident in Fourcroy's remarks quoted above. All that really separated them from Lavoisier in the early 1780's was their retention of phlogiston in the form of the modified theory promoted by Pierre-Joseph Macquer in the 1778 edition of his *Dictionnaire de chymie.*[6] Fourcroy's reconstruction of the attitude of chemists toward the new theory should consequently be regarded with some caution; to a large extent it is a rationalization of his own position in the pre-1785 period. For chemists outside Lavoisier's inner circle the gulf was considerably greater. They were not familiar with Lavoisier's views to the same extent; many of them had difficulty in repeating the experiments of which they had some reports. When word spread of the alleged synthesis of water, the news came not as a decisive development but as a topic of controversy that lasted for several years. The obstacles opposing the new theory were greater for them and their decision was less straightforward than Fourcroy's account—or a number of more recent versions—would allow.

LAUNCHING THE ASSAULT

As the year 1785 opened, Lavoisier could congratulate himself on the great progress of his theory over the preceeding decade. It now encompassed a general theory of combustion, a theory of the composition of acids, and a theory of the nature of heat and the vapor state. The experiments on the composition and decomposition of water had resolved to his satisfaction the major anomalies confronting his theory. Yet the chemical community at large did not appear eager to adopt his views. There was an antiphlogistic theory but as yet no visible antiphlogistic party supporting it.

5. For Guyton's conversion see W. A. Smeaton, "Guyton de Morveau and the Phlogiston Theory," in *Mélanges Alexandre Koyré* (Paris, 1964), I, 522–540; for Berthollet see H. E. Legrand, "The "Conversion" of C.-L. Berthollet to Lavoisier's Chemistry," *Ambix,* 22 (1975), 58–70.

6. Just as Tycho Brahe's theory offered the advantages of the Copernican system without the inconvenience of a moving earth, Macquer's theory offered the advantages of Lavoisier's theory of combustion without the need to abandon the inflammable principle, phlogiston. Macquer equated phlogiston with the matter of light and assumed that the absorption of air by a combustible body in burning is accompanied by a simultaneous release of the matter of light (manifested in rapid combustion as flame). Macquer's theory could account for the quantitative effects of diminution of the air and weight gain of the combustible in precisely the same manner as Lavoisier's. See Macquer, *Dictionnaire de chymie* (Paris, 1778), article "Phlogistique."

Lavoisier perceived that the time was ripe for an open assault on the phlogiston theory.[7] His confidence was bolstered in part by recent experimental successes (on the composition of water) but also by the private support he enjoyed from his inner circle of associates and his awareness that some of his younger chemical colleagues were on the point of breaking with phlogiston. His personal influence had never been greater; he was elected Director of the Academy of Sciences for 1785. His initial assaults on phlogiston were launched within the Academy and took the form of paragraphs added to earlier papers which were given a second reading to that august body in the first half of 1785. The theme of these insertions was the fictitious nature of phlogiston and its retarding effects on the progress of chemistry. "This entity, introduced by Stahl into chemistry, far from having brought light to bear upon it, seems to me to have created an obscure and unintelligible science for those who have not made a highly specialized study of it; it is the *Deus ex machina* of the metaphysicians: an entity which explains everything and which explains nothing, to which one ascribes in turn opposite qualities."[8] The theme presented in this initial insertion on 22 January was repeated on several occasions and blossomed into Lavoisier's famous "Reflections on Phlogiston" which he read to the Academy in the summer of 1785. His initial objective seems to have been to mobilize the private support of his inner circle and to prod those wavering chemists, such as Berthollet and Fourcroy, into a definitive and public renunciation of phlogiston.

The offensive on phlogiston was beginning to bear fruit by the spring of 1785. On 6 April Berthollet read "On Dephlogisticated Marine Acid [chlorine]" in which he explained the generation and the properties of the gas according to Lavoisier's theory.[9] The opinions of Lavoisier and Laplace were espoused with little drama or fanfare; Berthollet's move was probably

7. In his memoir "On Combustion in General," *Oeuvres de Lavoisier* (Paris, 1862–93), II, 225–233, read to the Academy of Sciences on 12 November 1777, Lavoisier had made it clear that his goal was nothing less than a total alternative to Stahl's phlogiston theory. However, he did not resume the direct attack on phlogiston until 1785. Lavoisier's caution in the intervening period was due to the inability of his own theory to account for several important phenomena including the release of inflammable air (hydrogen) when metals dissolve in acids. This and related difficulties were resolved only with the recognition of the compound nature of water.

8. "Considérations générales sur la dissolution des métaux dans les acides," *Oeuvres de Lavoisier,* II, 510. There was an habitual publication lag in the Academy's *Mémoires,* so that the 1782 volume which contained the paper appeared in 1785. Lavoisier often took advantage of this delay to give his papers a second reading with corrections and additions.

9. *Observations sur la physique,* 26 (1785), 321–325. In subsequent footnotes this journal will be abbreviated in conventional fashion as *Obs. phys.,* while in the body of the text it will be called the *Journal de physique* following the eighteenth-century custom. In this paper Berthollet wrote concerning the reactions of iron and zinc with dephlogistocated marine acid (in which no inflammable air is evolved): "it is thus true, as M. Lavoisier has claimed, that these metals contain no inflammable gas. . . . This fact confirms the opinion of M. de la Place who first realized that inflammable gas from metallic solutions was due to the decomposition of water."

no surprise to his academic colleagues who were, no doubt, aware of his close association with Lavoisier. Still his public alignment with Lavoisier must have produced some ripples of uneasiness in academic ranks.

There are signs of growing tension in a related development of early May. A young chemist named Bertrand Pelletier (who probably aspired to a place in the Academy), had submitted a memoir within two days of the submission of Berthollet's paper and on the same subject. An academic commission, consisting of Lavoisier, Jean Darcet, and Louis-Claude Cadet de Gassicourt, was duly appointed to report on Pelletier's paper and, as was frequently the case, Lavoisier drafted the report.[10] While commending Pelletier's ability, he was clearly unhappy with Pelletier's interpretations and could not resist an opportunity to promote his own view. Concerning the properties of dephlogisticated marine acid he wrote: "M. de Fourcroy, who has taken it upon himself in his *Leçons de chimie* to present impartially the phlogistic and antiphlogistic theories, developed this theory in several places in his books; there he too announced that dephlogisticated marine acid was an acid supercharged with vital air (oxygen) and it was to this excess of air that it owed its property of dissolving gold." Lavoisier recommended that prior to publication Pelletier should include citations of several passages from Fourcroy and Berthollet (that is, several passages favoring Lavoisier's view). What gives this incident an interesting twist is the fact that Pelletier was the protégé of Darcet, Lavoisier's co-reporter. Darcet responded to Lavoisier's draft report with a long note replying point by point to his criticisms. He referred pointedly to the *hypothetical* nature of the several theoretical claims Lavoisier was promoting and urged greater reserve. He closed on a conciliatory note: "You see that my remarks, although long, reduce to very little. For the most part they concern a theory on which neither I, nor others more qualified than I, have taken any position; but I am not of the opinion that it be left behind; it is good to speak of it as a theory which is becoming current and of which one of the commissioners is the author. It is thus that I have spoken of it and shall speak of it in public until I am more decided." Awareness of the new theory as a serious threat was growing in the wake of the controversy surrounding the composition of water and—for Parisian scientists—in the light of the open assault Lavoisier was now undertaking. Darcet's reaction was mild compared to others yet to come.

Lavoisier's campaign of 1785 reached a peak with the reading of his "Reflections on Phlogiston" at the Academy's sessions of 28 June and 13 July. It was a masterful polemic which appears to have been directed at his colleagues within the Academy and only secondarily at the general Euro-

10. *Oeuvres de Lavoisier,* IV, 419–423. The text of Lavoisier's report is followed by Darcet's response, pp. 423–425.

pean scientific community. His principal targets were the modified phlogiston theories of Macquer and Antoine Baumé which he compared with the original theory of G. E. Stahl, pointing out that the various proliferations of phlogiston theory were both incompatible with one another and internally inconsistent. Here we find the condemnations of phlogiston which have echoed down through historical accounts of the theory to our own time: phlogiston is a "fatal error in chemistry" which "has considerably retarded its progress." It is a "veritable Proteus which changes its form at any instant." The remainder of the paper presents a coherent account of Lavoisier's own theory, emphasizing that the principal phenomena of chemistry can be accounted for in a simple and adequate way without the need to invoke the existence of phlogiston at all.[11] The heated reaction to Lavoisier's broadside is evident in the testimony of the Dutch scientist Martinus van Marum, visiting in Paris, who was present at the 13 July session: "Then violent objections were made against this [paper], as a result of which the reading was continually interrupted. This, together with the simultaneous efforts of the reader [Lavoisier] and of his opponents to be heard, led to my understanding very little."[12] An open schism had developed within the ranks of the Academy which was ultimately to be resolved only with the demise or deposition of the most stubborn opponents of the new chemistry.

The opposition to his views was anticipated as Lavoisier's closing remarks indicated: "I do not expect that my ideas will be accepted right away; the human mind inclines toward a way of seeing things, and those who have envisioned nature from a certain point of view, during a part of their career, only arrive at new ideas with difficulty.... Meanwhile, I see with great satisfaction that young people who are beginning to study the science without prejudice, that geometers and physicists who have fresh minds concern-

11. *Oeuvres de Lavoisier,* II, 623–655. The intention and impact of this paper have been misread in curiously divergent ways. On the one hand it has often been cited as the "knockout blow" to phlogiston. But here I would point out that it came at the beginning, not the end, of the campaign to establish the new theory and, secondly, that it was less well known and less effective than other vehicles which were used to spread the new theory. It looks more decisive in retrospect than in context. On the other hand, a couple of recent dissertations have criticized the paper as a piece of propaganda attacking outmoded versions of phlogiston while ignoring the most viable alternative to Lavoisier, namely, the British versions equating inflammable air with phlogiston. See B. Langer, "Pneumatic Chemistry, 1772–1789: A Resolution of Conflict" (Ph.D. diss., University of Wisconsin, 1971), pp. 196–198. (A draft dissertation by John Elliott at the London School of Economics dealing with Priestley's research program, which I had the pleasure to read, makes a similar point.) But this ignores the French context in which Lavoisier was working. In France, Macquer's theory had gained a number of adherents *especially* among those younger chemists who appreciated the significance of Lavoisier's experiments and who were the objects of Lavoisier's wooing.

12. Quoted from T. Levere, "Martinus van Marum and the Introduction of Lavoisier's Chemistry into the Netherlands," *Martinus van Marum: Life and Work* (Haarlem, 1969), I, 184.

ing chemical truths, no longer believe in phlogiston in the sense Stahl presented it."[13]

Lavoisier's reference to his growing support indicates the early success of his campaign and the emergence of what was to become the antiphlogistic party. The core of the new party was that group of mathematicians and physicists already adhering to Lavoisier's inner circle who apparently rallied to open support of Lavoisier's position. (In all probability they took part in the Academy's shouting match.) Prominent among them were the mathematicians Laplace and Gaspard Monge who were emerging as the giants of French science. To their numbers were added a few beginners in chemistry who found the new theory appealing and a trickle of converts among established chemists. Berthollet was apparently the first of these; he was joined by Fourcroy, whose conversion was so undramatic that we are unable to date it.[14] The roster of the leaders of the antiphlogistic party was completed with the conversion of the Dijon scientist and parlementarian L.-B. Guyton de Morveau early in 1787 during a visit to Paris. With the emergence of the antiphlogistic party, the efforts on behalf of the new theory became increasingly communal and the activities of its promoters became more and more political.

COUNTERATTACK

Strictly speaking, the opposition to the antiphlogistic party was the entire chemical community. The cool and skeptical reaction of Darcet, for example, was fairly typical of individuals trained in the old way; they varied only in degree. Yet there was initially little reaction to Lavoisier's campaign outside the Academy; there was an interval between presentation of the papers in question and their publication and general circulation. The opposition was to find its rallying point in the *Journal de physique* founded by the Abbé François Rozier. The current editor, the Abbé J. A. Mongez, was engaged as one of the researchers in the scientific expedition organized by J.-F. de La Pérouse for a voyage around the world, from which they were never to return. Prior to departure Mongez handed over editorial responsibilities to Jean-Claude De La Métherie in 1785.

The following January, De La Métherie introduced a new feature into the journal, a preliminary discourse which surveyed interesting developments in

13. *Oeuvres de Lavoisier,* II, 655.

14. See W. A. Smeaton, *Fourcroy, Chemist and Revolutionary* (London, 1962), pp. 13–14. As I have indicated, Fourcroy was already a member of Lavoisier's circle, had adopted his view on a number of issues and was teaching both the antiphlogistic and the phlogistic theory side by side in his courses prior to 1785. It is likely that some of those "young people who are beginning to study the science without prejudice" were students of Fourcroy.

the various sciences and arts over the previous year. He continued this practice until his death in 1817. The first of De La Métherie's annual surveys, in January 1786, reflected the prevailing interests of the scientific community and particularly his own preoccupation. Of the essay's fifty pages, forty-six were given over to the recent developments in chemistry. The article surveyed the implications of the old and the new theory on a variety of topics, including the nature of the elements, combustion, acid theory, metals, and so forth. Toward the end of the essay De La Métherie responded directly to the challenge thrown down by Lavoisier in his opening attack of the previous January. Quoting the passage (cited above) in which phlogiston was condemned as a *Deus ex machina,* he retorted: "Surely anyone who has not made a specialized study of chemistry will not understand any better the combinations of the oxygen principle, or those of water decomposed in one reaction, not decomposed in another analogous one. And it seems to me that by regarding . . . inflammable air as phlogiston, one has a theory just as clear at least as the opposing one, and one conforming better to experiment."[15] De La Métherie adopted the view popular in Britain, and associated especially with Richard Kirwan and Joseph Priestley, which equated phlogiston with inflammable air (hydrogen). His preoccupation with the issue and his opposition to Lavoisier's theory is evident in the initial preliminary discourse; but the tone of his objections was still mild.

De La Métherie rapidly emerged as the principal foe of the new chemistry in France. He enjoyed a unique position of influence as editor of the *Journal de physique,* a monthly with a wide circulation in France and abroad. In his hands the journal became a forum for criticism of the new theory; and he was not reluctant to use his editorial powers to further his cause. There were, of course, the annual preliminary discourses which continued to be dominated by the question in chemistry. They are as much concerned with airing De La Métherie's personal point of view as with keeping the reader informed on the state of the debate; and their tone grew more bitter each year. In his role as editor he corresponded with prominent scientists at home and abroad. He persistently sought contributions for the journal critical of the new theory, and published some of the more caustic passages from his personal correspondence. Finally, he made liberal use of his editorial prerogatives in publishing summaries and critiques of various antiphlogistic works and especially in the notes he frequently appended to articles which he published.

Internationally, De La Métherie was able to solicit the support of such prominent scientists as Priestley in England, Lorenz von Crell in Germany (himself the editor of the *Chemische Annalen*), Felice Fontana in Italy, and

15. *Obs. phys.,* 28 (1786), 52.

Jean Senebier in Geneva. Locally, his active supporters are more difficult to identify. One is certainly Balthazar George Sage, a colleague of Lavoisier's in the Academy of Sciences, who contributed several articles to the journal during this period, including some critical of the new theory. I believe that De La Métherie was also on cordial terms with Jean-Baptiste-Louis Romé de l'Isle, the crystallographer, and the Baron Etienne-Claude de Marivetz. A series of letters published consecutively in one issue of the journal indicate an air of camaraderie among the three men.[16] The first is a letter from Marivetz condemning the new nomenclature proposed by the antiphlogistians (to which De La Métherie added a note in support), the second a letter from Romé to Marivetz with friendly criticism of his theory of the igneous fluid, and the third a cordial reply from Marivetz to Romé encouraging him to publish his letter. The opening of Rome's letter is indicative of the mutual esteem of these men and of their mutual dislike of the new chemistry: "I have just read, Sir and most excellent friend, your VIIth volume of the *Physique de Monde,* which M. De La Métherie had the kindness to lend me. I was waiting impatiently to see the complement to your theory of fire, with which I was most satisfied, except for one point, that I would have liked to see you approach more closely the theory of our modern chemists, I mean those who have not banned Phlogiston, and especially M. De La Métherie who, as you say yourself, is after Euler, the physicist who approaches nearest your principles."[17] To complete this ring of association, it remains only to point out that Romé was a student and friend of Sage.

These were some of the Parisian scientists who were personally acquainted with De La Métherie and lent him moral support in his campaign. Apart from their rejection of the new chemistry, they were united by a mutual hostility toward the academic authority represented by Lavoisier and his friends. Romé, Marivetz, and De La Métherie had all been snubbed by the Academy and, at least in De La Métherie's case the resentment bordered on paranoia.[18] Sage was a member of the Academy but clearly not in favor in Lavoisier's circle.

The men who stood up in favor of phlogiston were a varied group. They were bound together by little more than a common loyalty to Stahl and

16. *Obs. phys.,* 32 (1788), 61–71.

17. Ibid., pp. 64–65.

18. On Romé and De La Métherie see the articles in the *Dictionary of Scientific Biography.* Romé never succeeded in gaining admission to the Academy despite Sage's patronage. Furthermore, he resented the election of the Abbé R.-J. Haüy whose crystallographic theories he regarded as largely derivative from his own. (Haüy was welcomed into Lavoisier's circle.) As for Marivetz, in 1781 he had submitted his *Physique du monde* for the Academy's approval. Joseph-Jérôme Lalande, Jean-Sylvain Bailly, and Lavoisier reported as follows: "This work . . . is a tissue of nonsequiturs and absurdities; in it one finds neither cogency in its ideas, nor knowledge of the most elementary principles of physics; in a word it is the most monstrous assemblage which has ever been presented to the Academy," *Oeuvres de Lavoisier,* IV, 368.

phlogiston, which they often interpreted in divergent ways. A number of them managed to rally around the *Journal de physique,* in spite of their differences, in the face of the offensive by the antiphlogistians. The international contributors were, by and large, individuals of greater stature in the scientific world; their eminence no doubt helped to restore wavering loyalties among the larger silent readership of the journal. De La Métherie's intimate circle was bound together as much by bitterness toward Lavoisier's school as by an intellectual opposition to his theory. On the whole, the defenders of the old view were a less cohesive and poorly organized party. Initially, however, they enjoyed the advantage of the support of the majority of the chemical community and the editorial bias of the *Journal de physique* and Crell's *Chemische Annalen.*

A CAMPAIGN OF PERSUASION

Between 1785 and 1787 the French scientific community effectively split into two hostile camps of which the antiphlogistians were the smaller but better organized party. In the face of stubborn opposition by De La Métherie and his friends and the inertia of chemists in general, the antiphlogistians intensified their campaign of persuasion. They operated on several levels simultaneously and deployed their forces with considerable skill. Most visible were the public efforts of the leaders of the party. They collaborated in a series of major works beginning with the *Méthode de nomenclature chimique* in 1787, a reply to Kirwan's *Essay on Phlogiston* in 1788, and the first issues of a new chemical journal, the *Annales de chimie* in 1789.

On a second level, younger converts to the new theory were pressed into service as a hatchet squad, assigned the task of replying to the various critiques by De La Métherie and others. Such an occasion arose in the spring of 1786 when De La Métherie published a letter by Fontana relating to experiments on the composition of water. De La Métherie added a note critical of the new theory. In June young Pierre-Auguste Adet published a reply to Fonatna's letter, pointing out, among other things that "M. De La Métherie has not quite grasped the state of the question," prompting an indignant reply by the latter.[19]

On yet a third level, the antiphlogistians undertook a letter campaign in which all could participate. In the eighteenth century there was an extensive network of scientific correspondence. A number of prominent and not-so-prominent scientists scattered over various points in Europe attempted to keep one another abreast of recent news and gossip concerning develop-

19. *Obs. phys.*, 28 (1786), 436–441 (Adet's letter; the remark quoted is from p. 440); 442–446 (De La Métherie's reply).

ments in science. An important link in this network was Senebier in Geneva who had a number of correspondents and furthermore often acted as intermediary between contacts in Italy or Germany and the centers of scientific activity in Paris and London. I have found his correspondence of immense value in following the international reception of the antiphlogistic theory. Lavoisier, on the other hand, despite a large correspondence, was not an integral figure in this network. Lavoisier's letters were for the most part businesslike and concerned with his various official duties as a member of the Tax Farm, the Gunpowder Commission, and the Academy of Sciences. He was an extremely busy man who did not have time for newsy letters, nor did he need to participate. As a member of the Academy he was kept informed by that body's network of official correspondents. So his recruits (including Madame Lavoisier) were able to serve him well in this area. The letter campaign, which was intensified in the late 1780's, was directed especially at prominent foreign scientists. A personal contact between one of the antiphlogistians and an individual outside of Paris provided an opportunity to keep the latter informed on the debate, to correct mistaken impressions he may have gathered from other sources and, through the individual nature of the contact, to encourage a favorable rapport with the antiphlogistians.

Finally, the fourth and probably most effective level on which the antiphlogistians promoted their view consisted in intense personal indoctrinations to which visitors to Paris were treated. We have already mentioned one of these visitors, the Dutch scientist van Marum, who was in Paris in the summer of 1785. He met only briefly with Lavoisier, but, following a series of interviews with Monge, Berthollet, and Alexandre-Théophile Vandermonde, his faith in the old theory was shaken. Following his return to Holland he carried out a series of experiments which persuaded him of the validity of the new theory and, by 1787, he brought out a work in Dutch supporting Lavoisier's view.[20] Early in 1786 the young Scot, James Hall, arrived in Paris at the close of a grand tour of the continent. He fared even better than van Marum: "I am personally acquainted with Mr. Lavoisier & have received the greatest civilities from him—I have a standing invitation to dine with him every Monday—he has one of the clearest heads I have ever met with & he writes admirably."[21] Hall stayed on in Paris for several months, perfecting his grasp of the new theory through direct contact with its principal advocates, and returned to Scotland a convert. These are only two early examples of individuals converted in this manner; their numbers increased as the decade wore on. Lavoisier's inner circle made a profound impression upon visitors to Paris, functioning as an almost irrestible instrument of conversion to the new theory.

20. Levere, "Martinus van Marum," pp. 183–189.
21. V. A. Eyles, "The Evolution of a Chemist," *Annals of Science,* 19 (1963), 167.

THE INTENSIFIED CAMPAIGN OF 1787

Once again De La Métherie opened the year with a preliminary discourse in which he gave a précis of the new theory "as they now present it" and an argument in favor of his own version of the phlogiston theory.[22] The following month he published a contribution from Senebier, the first explicit reply to Lavoisier's "Reflections on Phlogiston." Senebier titled his paper "Doubts concerning certain inconveniences attributed by M. Lavoisier to the use of phlogiston,"[23] and in it he replied to Lavoisier's criticisms. In addition, he announced a project which he proposed to undertake: "I believe that it will not be difficult to show that the theory of phlogiston can be as general and as solid as the new doctrine ... and if I should succeed in making such a comparison of the two theories with sufficient impartiality and in an appropriate manner... this work would furnish the necessary materials for the decision of this important controversy."[24] Senebier was a respected member of the scientific community and such a contribution would be widely welcomed. Following the announcement, De La Métherie dramatically increased his correspondence with Senebier to urge him on in the project.[25]

Meanwhile the antiphlogistians were on the verge of securing one of their most important coups, the conversion of Guyton de Morveau. Guyton had been working for several years on a project to reform the nomenclature of chemistry. Early in 1787 he came to Paris to enlist the cooperation of Parisian chemists, and through contact with Lavoisier and his associates was finally persuaded to break definitively with phlogiston. Following his conversion Guyton remained in Paris for several months and collaborated with Lavoisier, Fourcroy, and Berthollet in developing a new system of chemical nomenclature grounded in the new theory. The work proceeded rapidly; on 18 April Lavoisier read a paper to the Academy urging the need for reform and on 2 May Guyton described the proposed system.

The hatchet squad, too, had been in action; in March a reply to De La Métherie's preliminary discourse was published by Adet and Hassenfratz. They found it difficult to recognize Lavoisier's theory from De La Métherie's description of it. "All those who know the modern theory only from what you say of it, will not be able to have a fair idea of it; they will see in it errors instead of truths, they will believe it founded upon hypotheses, rather than on facts, & they will *judge* it without being instructed in it."[26]

22. *Obs. phys.*, 30 (1787), 29–45.
23. Ibid., pp. 93–99.
24. Ibid., p. 99.
25. Over the previous four years, De La Métherie had sent four letters to Senebier; the correspondence jumped to seven letters in nine months following the announcement of Senebier's intention in February 1787, and continued at a high level.
26. *Obs. phys.*, 30 (1787), 215.

Again De La Métherie fired back a reply; he also complained in his correspondence of his treatment at the hands of the antiphlogistians: "our adversaries are becoming ill-humored . . . The leaders have had things written to me they wouldn't have wished to sign themselves."[27]

On 13 June the Academy's commission consisting of Baumé, Cadet, Darcet, and Sage reported on the new nomenclature. It was a curious situation in that one group of the Academy's chemists (those who had developed it) favored the nomenclature, while the rest (those reporting on it) rejected it. After summarizing and criticizing the proposed nomenclature, the commission managed to close on an impartial note: the decision should be left to "the test of time." "And it is in this view that we recommend that the table of the new Chemical Nomenclature, with the associated memoirs be printed and made public under the *privilège* of the Academy, in such a way however that it can not be inferred that it approves or rejects the new theory."[28]

The nomenclature appeared later that summer shortly after a major work for the opposition became available in France: Richard Kirwan's *Essay on Phlogiston.* The near simultaneous appearance of these two works touched off a renewed wave of activity and controversy. The controversy over the nomenclature eclipsed even the continuing debate over the composition of water. On 28 August De La Métherie wrote to Senebier to inform him that the nomenclature was just out, that he had read Kirwan's book and that he had heard from Hassenfratz that Madame Lavoisier was preparing a French translation.[29] In September he published a précis of the nomenclature (based on the Academy's report) and followed this in October with his own detailed critique, the first of many attacks to appear in the journal and elsewhere. Reactions to the nomenclature were not slow in coming. Before the end of September De La Métherie reported to Senebier that the nomenclature was poorly received in England, and from Turin he had heard that the Count des Saluces (Saluzzo) "was alarmed at the harm it could do to the science."[30]

By this time the debate was followed closely in France and abroad. An excellent illustration of the impressions which the battle raging in Paris made upon a scientist in the provinces is contained in a letter from Jean-

27. De La Métherie to Senebier, 9 April 1787. Bibliothèque Publique et Universitaire de Genève (hereafter abbreviated B.P.U.G.), MS Suppl. 1039, fol. 230. De La Métherie informed Senebier that Fontana had also promised a major work relating to the debate and that M. Landriani was supposed to contribute a treatise on fire, complaining: "But in Paris we receive almost nothing from Italy although there are excellent physicists there."

28. *Méthode de nomenclature chimique* (Paris, 1787), p. 251.

29. B.P.U.G., MS Suppl. 1039, fols. 232–233. De La Métherie also revealed a rift with Kirwan. Reading his book had confirmed "what I have been told that he did not like me at all, although I do not believe I have given him any grounds for discontent and have always rendered justice to his talents. He attributes to Mr. Cavendish what I have said concerning the combustion of phosphorus, of sulfur."

30. Ibid., fols. 234–235v.

Huet de Froberville in Orléans to Senebier.[31] He complained that the reform of chemical nomenclature was a task more suited to grammarians and metaphysicians than to geometers. His principal objection was that made by a number of serious critics, namely that the system was founded in hypothesis: "they are constructing a new language which takes as axioms that which is in question." For example, "if MM. Fontana and De La Métherie have proved that the capital experiment on the formation of water is illusory, what then does the alleged hydrogen signify?" "Such then is the theory for which one must abandon that of Stahl and of Macquer, and whose principles one must inculcate exclusively among the young by means of the new language. I find that this is moving rapidly to fraudulence." De Froberville had apparently heard something of the *modus operandi* of the antiphlogistians also: "the dinners and patronage of M. Lavoisier bring him more proselytes than the strength of his reasoning." But his discontent was not directed solely at the antiphlogistians:

> I am especially annoyed that M. De La Métherie, and a few other scientists of merit, instead of uniting and making common cause to follow the doctrine begun by Macquer, wander each on his own path into vague opinions which serve only to confuse the science and to retard its progress. You [Senebier] would be worthy of placing yourself at the head of a party capable of resisting that of the new nomenclators, and of rallying under the standards of the old doctrine, modified according to the view of Macquer, the deserters who are running aimlessly, and notably our dear M. De La Métherie, whose ideas seem to me quite muddled, let it be said between you and me.

De Froberville's remarks reflect a malaise inspired by the growing strength of Lavoisier's party and frustration at the inadequacy of De La Métherie's leadership. They also give further indication of the high esteem which Senebier enjoyed and the pressures which were mounting upon him to complete his judicial comparison of the two theories.

Meanwhile the loss of Guyton de Morveau was keenly felt by the defenders of the old theory. Crell wrote to De La Métherie from Germany: "I mourn the loss of a scientist as respectable as M. de Morveau for our phlogiston; however I do not find the new reasons strong enough to leave our party. . . . The friends of the Stahlist system are greatly indebted to you for standing so firm against the new anti-phlogiston system."[32] Count Marsilio Landriani of Milan, whose travels took him in and out of Paris several times during 1787–88, was already speaking of Guyton's persuasiveness in arguing for the new theory. "I have seen the new chemical nomenclature. M. de Morveau speaks only this language. He has almost convinced me of the nonexistence of phlogiston. You will see the notes for the translation of

31. Ibid., fols. 176–177v. De Froberville to Senebier, 17 October 1787.
32. *Obs. phys.*, 31 (1787), 367.

Kirwan's book done by Mde. Lavoisier. They appear quite strong to me. You must try to obtain a copy as soon as it appears. It is absolutely necessary for you."[33]

The notes to which Landriani refers are, of course, the chapter-by-chapter rebuttals which were added to the French edition of Kirwan's book by Guyton, Lavoisier, Laplace, Monge, Berthollet, and Fourcroy. The publication was delayed for some time, however, and appeared only in the summer of 1788. This work was the second major collaborative effort of the antiphlogistic chiefs. A master touch in its eventual format was the assignment of the final summation to Guyton (like Kirwan and Lavoisier, he was trained as a lawyer) and his listing at the head of the contributors. For Guyton had been named by Kirwan in his book as one of about a dozen internationally prominent scientists who shared his view. It must have come as a blow to Kirwan to find his former ally and correspondent orchestrating the attack.

Both parties continued to campaign vigorously in 1788 and 1789; the tone of the exchanges became increasingly bitter. The antiphlogistians continued to woo potential converts by means of the gentle persuasion of their letters and their hospitality in Paris. For example, both Berthollet and Guyton were working on Senebier; Berthollet encouraged him in his plan to examine the two theories dispassionately. "Not only, Monsieur, do I applaud the stance which you have taken to weigh without attachment to any systematic interpretation the facts and the observations as well as the explanations given of them; but I can assure you that this is the same stance which persuaded me in favor of what you call the pneumatic doctrine."[34]

At the same time Madame Lavoisier was exchanging a series of letters with another prominent Genevan, the naturalist and physicist Horace-Bénédict de Saussure, culminating in de Saussure's admission: "You triumph over my doubts, Madame, at least concerning phlogiston. . . ."[35] The antiphlogistians were less than gentle, however, with their confirmed enemies. Lavoisier's distinguished mathematical colleague Jacques-Antoine-Joseph Cousin was not above leaping into the fray in the manner of some of the younger converts. He published a caustic and sarcastic review in three parts of Sage's *Analyse chimique et concordance des trois règnes* (Paris, 1786). A letter was subsequently addressed to Sage (with the signature *Astrologue*) expressing surprise that in view of what Cousin had said that Sage had not accepted the new nomenclature and abandoned his doctrine. Sage published a series of three letters in De La Métherie's *Journal de physique*, replying with equal sarcasm to "the aristarchus, M. Cousin, cele-

33. Landriani to Senebier, 10 September 1787. B.P.U.G., MS Suppl. 1041, fol. 174.

34. Berthollet to Senebier, 24 March 1788. B.P.U.G., MS Suppl. 1039, fol. 29.

35. Draft of letter from de Saussure to Madame Lavoisier. B.P.U.G., Archives de Saussure 67. Cahier 10, Brouillons divers, commencé le 22 Sept. 1788.

brated Geometer, & Professor in the College Royal de France."[36] The feud of De La Métherie with Lavoisier's disciples, Adet and Hassenfratz, continued, too, in a similar vein.[37]

De La Métherie remained equally active. In the spring of 1788 he undertook a trip to England to see that "free and happy people."[38] The visit also provided an opportunity to revitalize his stance against the new theory through personal contact with those staunch proponents of phlogiston, the Birmingham group of scientists including Priestley. He continued to urge Senebier to bring out his now long-awaited study. On 15 January 1789 he wrote: "I am delighted that you have resumed your work on the comparison of the two doctrines which at the moment divide the science. Your just and impartial eye will make us see things better than we do, we who are predisposed. For I admit to you that the antiphlogistic opinion seems to me more and more indefensible."[39] But behind this optimism lurked the awareness that the battle was going badly, at least in Paris. In the same letter he wrote: "I do not believe that this quarrel can exhaust itself in Paris. The decision is [left?] to you, *MM. les savants étrangers,* who far from the hearth [find yourselves in?] the shelter from the heat generated there. Besides you know the [?] intrigues of our country and all the resources which wealthy and well-placed people can bring into play. All the young people who wish to appear at the Academy and aspire to places affect an alliance with that side because they know that party although much less numerous is the more powerful."[40]

That same month in his preliminary discourse De La Métherie produced a *tour de force.* He included a sketch of the origins and foundations of the new doctrine systematically avoiding *any* reference to Lavoisier's work or even the mention of his name.[41] Each of the key observations relating to the new theory was credited to someone else. For example, concerning the important experiments on reduction of mercury calx without the addition of charcoal he wrote: "It is this famous experiment which, although its author has not been cited, served as the base for the entire new doctrine. . . . Thus one could almost call this doctrine, the system of M. Bayen."[42] Similarly, the

36. *Obs. phys.,* 33 (1788), 478–479; 34 (1789), 66–67 and 138–142. Cousin's review had appeared in the *Gazette de France,* 1788, nos. 49, 50 and 51.

37. *Obs. phys.,* 33 (1788), 384–388.

38. De La Métherie to Senebier, 28 April 1788. B.P.U.G., MS Suppl. 1039, fols. 240–241.

39. B.P.U.G., MS Suppl. 1039, fols. 246–247.

40. Ibid. A few words in this passage are completely obliterated by the torn seal.

41. *Obs. phys.,* 34 (1789), 22–40. Lavoisier's name is, in fact, mentioned only once as the source of a quotation concerning the work of Stephen Hales. Even this quotation (on page 23) is calculated to diminish the originality of Lavoisier's celebrated study of the weight gained by metals in calcination.

42. Ibid., p. 24. Pierre Bayen had published experiments in 1774 showing that mercury calx could be reduced without addition of phlogiston. His results led him to challenge Stahl's phlogiston theory; but he does not appear to have pursued these theoretical implications further. See my "Prelude to Lavoisier's Theory of Calcination: Some Observations on Mercurius Calcinatus per se," *Ambix,* 16 (1969), 140–151.

discovery of the absorption of air in combustion, calcination, and respiration is credited to Jean Rey, Robert Boyle, and Stephen Hales, the determination of the specific heat of vital air (oxygen) to Adair Crawford, and the composition of water to Henry Cavendish. James Watt is said to have discovered that water can be decomposed by iron. De La Métherie seems to have decided that if he could not defeat the antiphlogistians, he would rob them of any originality in the development of their system. De La Métherie's bitterness toward the antiphlogistians and toward Lavoisier in particular was due largely to the disdain with which he felt he was treated. His pet views, which he promoted so vigorously in the journal, were simply ignored by Lavoisier and butchered by the hatchet squad.

SEIZURE OF THE INSTITUTIONAL BASES OF POWER

The offensive of the antiphlogistic party was clearly gaining ground as the 1780's drew to a close. The campaign of persuasion had produced first a trickle then a stream of converts; as the turn of the decade approached, efforts were renewed to turn it into a decisive tide. The leaders sensed that in Paris, at least, victory was at hand on the eve of the great political revolution which shook France in 1789; by the time the country was emerging from the effects of the Reign of Terror in 1794–95, they had consolidated their control on the principal scientific institutions in France. I shall now turn to a brief examination of the key moves in this coup, beginning with the Academy of Sciences.

As 1785 opened, the chemistry class of the Academy consisted of Cadet, Lavoisier, Sage, Baumé, C.-M. Cornette, Berthollet, D.-B. Quatremere d'Isjonval, and Darcet.[43] Although, as we have seen, Berthollet was on the verge of converting, Lavoisier was only one among eight in support of the new theory. Under Lavoisier's directorship in 1785 the Academy was reorganized to introduce two new classes; to keep the overall size of the Academy manageable, each existing class was decreased by one member to six. Lavoisier was able to shuffle around the existing members of the class of chemistry (Darcet and Sage went to natural history and Quatremere to experimental physics) to make the needed cut, to eliminate the *surnuméraire* position and still to open up one place for election. Fourcroy was elected to this open place on 12 May; with his and Berthollet's conversion, the antiphlogistians controlled three out of six places in chemistry by 1786. There was no further opportunity to elect members to the chemistry section until

43. There were normally seven members in the class; Darcet was *surnuméraire,* that is, admitted exceptionally to the Academy until one of the regular places should be vacated. For documents relating to the reorganization of the Academy in 1785 see *Oeuvres de Lavoisier,* IV, 565–614. This needed reform of the Academy is described by Roger Hahn in *The Anatomy of a Scientific Institution: The Paris Academy of Sciences, 1666–1803* (Berkeley, 1971), pp. 97–101.

1792.[44] Pelletier, who had by then converted to the new theory, was chosen. The Academy was abolished in 1793 and when it was reformed as the First Class of the Institut de France in 1795, the chemistry section read like a roll call of leading antiphlogistians.[45] By contrast, Baumé and Sage, unyielding opponents of the new theory, obtained only nonresident positions in chemistry and natural history, respectively; Cadet was absent altogether. In addition, the antiphlogistians had enjoyed strong support from geometry, mechanics, and general physics right from 1785 on. By the time of the creation of the Institut they were clearly the dominating force within the Academy.

In view of De La Métherie's antagonism toward the new theory, the antiphlogistians soon became aware of the need to find an alternative means of disseminating their work. Lavoisier's young convert, Adet, had attempted as early as 1787 to obtain permission to publish a French version of Crell's *Chemische Annalen,* including original works.[46] After some initial setbacks the project finally went ahead late in 1788 with the backing of Lavoisier and his friends. On 23 December 1788, Lavoisier, Guyton, Adet and Hassenfratz asked the Academy to appoint commissioners to examine material for the initial volumes; by 2 May 1789 the first volume was ready for official presentation to the Academy.[47] De La Métherie immediately recognized the threat which the new journal represented: "It is a work directed against the poor Journal de physique, as if the editor had ever refused them space; quite on the contrary we have always done everything . . . to insert their memoirs immediately. They reproach three or four notes; as if all authors and they themselves did not use them."[48]

The founding of the *Annales de chimie* was the last major collaboration of the antiphlogistic party and a successful one which allowed their papers to circulate freely throughout the chemical community without De La Métherie's appended critiques. There was a rivalry between the two publications during the early years following the appearance of the *Annales* and publication in the latter was almost tantamount to siding with the antiphlogistians.

44. Cornette had gone into exile in Italy in 1789, but the Academy was not allowed to elect replacements for emigrés.

45. The members of the chemistry class were Bayen, Guyton, Fourcroy, Berthollet, Pelletier, and N. Vauquelin. Lavoisier himself had been executed during the Terror as a member of the Tax Farm. Bayen, although not active in the antiphlogistic party, rejected phlogiston at an early date on his own. He had been ignored by Lavoisier, and, I believe, his belated election to the Academy was due to the desire of Lavoisier's lieutenants to make amends for their earlier neglect. N. Vauquelin was the former assistant and close associate of Fourcroy. With the exception of Baumé, the nonresident members in chemistry were also converts: A. Seguin, J.-A, Chaptal, F. Chaussier, P.-F. Nicolas, and the Belgian J.-B. van Mons.

46. See Susan Court, "The *Annales de Chimie,* 1789–1815," *Ambix,* 19 (1972), 113–128.

47. Archives de l'Académie des Sciences, Procés-verbaux, 1788, p. 295; and ibid., 1789, p. 124.

48. De La Métherie to Senebier, 6 April 1789. B.P.U.G., MS Suppl. 1039, fols. 248–249.

Another influential publication which the antiphlogistians managed to secure as a vehicle for their views was the chemistry portion of the *Encyclopédie méthodique.* Guyton was the original editor of these volumes, and prior to his conversion he had brought out the first part of volume I and was midway in the second. At that point he inserted a second preface explaining the reasons for his conversion and pointing out corrections which should be made to the earlier portions in the light of the new theory and nomenclature. The second part of the first volume was published in 1789, but the following year administrative commitments forced Guyton to abandon the project and it was handed over to Fourcroy, who continued to employ the antiphlogistic theory. Another of Lavoisier's intimates, Monge, edited the physics volumes; finally, the section on plant physiology was brought out by Senebier in 1791, by which time he, too, had converted. The new theory was firmly entrenched in this major reference work.

Perhaps the most effective coup scored by the antiphlogistians was the virtual domination of science education which they managed to achieve by the mid-1790's. Lavoisier's early collaborator Bucquet was the first to teach the new theory, introducing it into his courses at the Faculty of Medecine in Paris between 1777 and 1779, prior to his untimely death in 1780 at the age of thirty-four. Fourcroy, a student and close friend of Bucquet, continued in his own courses to teach the new theory alongside the old until his conversion, when he abandoned the latter completely. Lavoisier was able to remark as early as 1785 on the ease with which new students pick up the theory, but at that time Fourcroy was practically alone in teaching it systematically in Paris. Still, he was an eloquent lecturer and as Macquer's successor in the famous public lecture course at the Jardin du roi he was able to draw large audiences.[49] Several physicists were among the early teachers of the new theory in the portion of their courses dealing with the nature of gases. Thus, in 1788 we find L. Lefèvre-Gineau, a protégé of Cousin, publicly demonstrating at the Collège Royale experiments on the composition of water; M.-J. Brisson similarly taught the new ideas in his course at the College de Navarre.[50]

The teaching of the new theory received a tremendous boost in 1789 when the third edition of Fourcroy's textbook *Elémens d'histoire naturelle et de chimie* appeared in January. Fourcroy's book enjoyed a great popular-

49. One example of Fourcroy's influence: Young Alexandre Brogniart (nephew of Fourcroy's demonstrator A.-L. Brogniart and seventeen years old at the time) gave a lecture course four years in a row at the Hôtel des Invalides, beginning with a modest three lessons in 1786 and expanding it to a series of twenty lessons by 1789. His explicit model was Fourcroy's course on gases; he wrote: "a distinguished professor of whom I am proud to be a disciple has traced the path I am going to try to follow the best I am able, however difficult it may be. This course is only an attempt at imitation of that of M. Fourcroy." Muséum d'histoire naturelle, MS 2322, A. Brogniart, Cours de Chimie.

50. On Lefèvre-Gineau see *Obs. phys.,* 33 (1788), 457. Concerning Brisson see the article by René Taton in the *Dictionary of Scientific Biography.*

ity, going through several editions and various translations; and this third edition contained an extensive and systematic presentation of the new theory. It was followed within two months by Lavoisier's own *Traité élémentaire de chimie* which introduced a new manner of organizing the content of introductory chemistry. Early in 1790 a third major work, J.-A.-C. Chaptal's *Elémens de chimie,* became available. In addition to any intrinsic merits of the antiphlogistic theory itself, these works profited from the lucidity and the eloquence of their authors, all of whom were extremely capable men. Among the three of them they reached a very large audience and became the models for chemistry courses at the turn of the century.

During the period of the political revolution in France, significant reforms were undertaken affecting science education, and Fourcroy was one of the principal architects of these reforms.[51] There were a series of shortlived institutions, the *écoles révolutionnaires,* created in 1794 for specialized instruction in areas of urgent importance to the republic. In addition, a few surviving institutions for training professionals date from this period, including the Ecole Polytechnique, which was created as the Ecole Centrale des Travaux Publiques in 1794, and of which Monge was the director. If one examines the lists of instructors at these various institutions, a certain group of names crops up again and again: Monge, Laplace, J.-L. Lagrange, R.-J. Haüy, Hassenfratz, Berthollet, Fourcroy, Vauquelin, Pelletier, F. Chaussier, Vandermonde, Guyton. From earlier parts of this paper these names will now be recognized as converts to the antiphlogistic party;[52] their domination of the revolutionary institutions was almost total. Among the prerevolutionary institutions which survived, the Jardin du roi became the Muséum d'histoire naturelle and Fourcroy himself emerged as its dominant figure. The Ecole des Mines was completely reorganized during the revolution and Sage, its former director and founder, was ignored, retaining not even his teaching post. It was still possible for defenders of the old theory to continue to give private courses, as Sage in fact did, but the advantage gained by the antiphlogistians was insuperable.

One last move by the antiphlogistians deserves mention here, the stepped-up letter campaign of 1789–1791, whose goal was to win over leading international authorities in science to the new view—with the assurance that the rest would follow. In this activity Lavoisier himself played a major role. He sent off a number of complimentary copies of his book, the *Traité élémentaire,* to well-known individuals all over Europe and America (including such celebrities as Benjamin Franklin, Alessandro Volta, and Joseph Black). Each of these copies was accompanied by a personal letter from Lavoisier emphasizing the great progress of the new theory in France

51. Hahn, *Scientific Institution,* pp. 274–285.

52. Except Lagrange, who moved from Berlin to Paris in 1788 and was listed by Fourcroy in the *Encyclopédie méthodique, Chimie,* III, 425, as a regular in Lavoisier's circle.

and flatteringly pointing out to the recipient what a milestone *his* support would be.[53] This campaign, together with the proselytizing of foreign nationals converted in Paris, was instrumental in carrying the new chemistry to an international audience.

In the closing months of Lavoisier's life, whatever political concerns and anxieties over his personal safety he may have felt, he was assured of the success of his theory. His only discontent on this score was that the communal promotion of the theory by the antiphlogistic party was so effective and so intimately linked with the theory itself, that it was becoming known as the *French* chemistry and the dimensions of his personal contribution were in danger of being overlooked.

CONCLUSION

The tide of victory was turning in favor of the antiphlogistians as the 1790's opened and before the middle of that decade their triumph was certain. The victory required several years of effort and was the result of a campaign by a party of determined and capable men. The penetration of the theory outside of France was generally slower but displayed similar trends; the scope of this paper does not permit closer examination of those developments, although it is a topic worthy of further study.

Throughout this essay I have spoken of two factions, one phlogistic and the other antiphlogistic. This, of course, is an oversimplification, especially in the case of the former group. There was a wide range of attitudes toward the claims of the new theory ranging from highly sympathetic to hostile. If we gauge the attitudes of individuals by their behavior, their reactions indicate a spectrum of receptivity. There were, first of all, a number of individuals who adopted the new theory with seeming ease, without going through an initial period of opposition. These were predominantly, as Lavoisier himself pointed out, mathematicians and physicists, on the one hand, and beginners in chemistry, on the other. The ease of their conversion has a relatively straightforward explanation: they had never become attached to the old view either by force of habit or by virtue of having contributed to it.

Those educated in the old way—in effect, the entire French chemical community—all resisted to some extent. If we break the spectrum down systematically, we can separate them roughly into several groups of decreas-

53. Here, for example, is an extract from the letter to Volta. "I do not know Monsieur what opinion you have embraced concerning the question which divides chemists relative to the existence of Phlogiston. I hope that you will grant a few moments of attention to the work which I have the honor to send you. You will judge whether one can explain all the phenomena of Chemistry without recourse to a hypothetical substance whose existence is not proved by any direct experiment. I shall regard your suffrage as of great weight in this question." Lavoisier to Volta, 1 January 1791. *Epistolario di Alessandro Volta,* Editione Nazionale (Bologna, 1949–55), III, 95.

ing receptivity: (1) Those chemists who entered the debate at an early stage (initially supporting some version of the phlogiston theory), were converted through interaction with Lavoisier and company, and became champions of the new theory. Fourcroy, Berthollet, and Guyton, among others, fall into this group. (2) Those who entered the debate on the side of phlogiston, fought vigorously, but eventually capitulated to the antiphlogistians. Examples are Kirwan, Black, and Senebier.[54] (3) Those who contributed to modified phlogiston theories in response to the pneumatic discoveries, but died in the early stages of the debate, such as Macquer or the Swedish chemist C. W. Scheele. It remains a matter of conjecture which way they would have turned. (4) Those silent observers, that is, those who did not enter publicly into the debate. They include a variety of types ranging from a few who had a great deal in common with the antiphlogistians, such as the modest Pierre Bayen, through level-headed skeptics in the manner of a Darcet or a Froberville, and on to more hostile individuals. Their very silence makes it difficult to trace their attitudes; one has to look for implicit clues such as adoption of the new nomenclature. With leading authorities advocating the new theory and a new generation of chemists nurtured on it, most of them probably acquiesced; it was either that or face increasing isolation. (5) Finally, those whose opposition seemed to harden as time went on, the die-hard resistance: Baumé who wrote against the new theory in 1798, Sage who was still campaigning against the nomenclature in 1810, and Priestley who died a phlogistian in 1804.

The existence of such divergent attitudes within the French chemical community and abroad raises the tempting question: Why? Any adequate answer to that question would have to take into account conceptual or methodological outlooks which predisposed an individual one way or the other and the reasons given in support of his choice. I have intentionally avoided those aspects in this study in order to focus attention upon another range of factors of a more social or political nature. In particular, I have tried to explore the mechanism by which the new theory was spread, the causal influences at work rather than the arguments used.

It is clear that personal contact was of paramount importance in winning individuals over to the new view. The antiphlogistians had almost total success in converting visitors to Paris to their point of view, although these same individuals had steadfastly supported phlogiston prior to their visits. Other means were less effective; journals and letters brought conflicting opinions and conflicting experimental results from which one could pick and choose. But seeing an experiment well performed had quite a different impact than attempting to repeat that experiment from a sketchy report.

54. Senebier's long-awaited comparison of the two theories was never published, despite De La Métherie's repeated pleas. However, I have located the heavily reworked manuscript of this document and will describe its history in a subsequent article.

Having one's questions and doubts answered on the spot had a force of a greater magnitude than reading a journal article. Most studies of the spread of Lavoisier's theory have emphasized bibliographic indicators—the appearance of translations and original works based on the new theory. These indicators do give a measure of the success of the new theory but in order to determine how the theory arrived, one has to go beyond the bibliographic information to ask who wrote these works, how they became interested. My inquiry has revealed an amazing number of contacts and movements of individuals which I have been able to report here only in part. The antiphlogistic theory spread outward from a core of initial supporters, through master-apprentice relationships and through the stream of visitors turned proselytes. In assessing the triumph of the antiphlogistians one must also bear in mind the personalities involved, the common purpose and cohesiveness of their group, and the wisdom of their strategy. In retrospect, their victory does not appear surprising; they clearly outclassed De La Métherie and his circle. Their ranks included the leading French scientists of the revolutionary and Napoleonic eras in several fields of science.

Personalities and social interactions also influenced the rejection of the new theory by many, especially the die-hard opposition. We have seen indication of the animosity between the circles surrounding Lavoisier and De La Métherie. I believe that the continued hostility of individuals such as Baumé and Sage was the result of frustration at being pushed aside, a resentment which found its outlet in diatribes against the new theory and its proponents. This brings us to a reaction that runs deeper than the clash of individual personalities or even the social stratification in French science. The antiphlogistians were largely men trained in physics and mathematics; Lavoisier himself had closer ties with these traditions than with mainstream chemistry. Traditional chemists were well aware of the intrusion of outsiders into their preserve, and they were quick to impugn the efforts of the "geometers." The controversy involved not merely a collision of methodologies but an invasion of professional territory. The outcome was the establishment not only of a new orthodoxy but also of a new set of authorities in French chemistry.

LAMARCK'S CHEMISTRY: THE CHEMICAL REVOLUTION REJECTED

LESLIE J. BURLINGAME

ANTOINE LAURENT LAVOISIER'S important contributions to the chemical revolution have been studied in detail by Henry Guerlac and other historians of science. In his efforts to promote the new chemistry, Lavoisier was joined by his wife and the converts C.-L. Berthollet, A.-F. Fourcroy, L.-B. Guyton de Morveau, G. Monge, and P.-S. de Laplace to form what Guerlac has called the "anti-phlogistic task force."[1] The group launched a multi-pronged campaign between 1785 and 1789 which succeeded in convincing most scientists and intellectuals of the significance and correctness of the new chemistry. There were a number of scientists, however, who remained unconvinced, Joseph Priestley and a group of Germanic chemists being the best known. Less well known is the group in France which included J.-C. De La Métherie, A. Baumé, B. G. Sage, J. F. Demachy, C. Opoix, A. G. Monnet, and Lamarck (P.-J. Macquer is not included in this list because he died in 1784). Their very opposition can tell us something more about the nature of the chemical revolution. This paper concentrates on the resistance to Lavoisier's theories by one important member of that group: Jean-Baptiste Lamarck.

Lamarck's chemistry has been ignored almost completely because it did not acknowledge the chemical revolution. In the full-scale studies of Lamarck's thought by Alpheus S. Packard and Marcel Landrieu, it was

I thank Henry Guerlac for originally inspiring my interest in Lamarck. The National Science Foundation generously supported my research from 1974 to 1976. I also thank the Radcliffe Institute, where I was a Fellow (1974–1976), for its support of my work. Some of the material in this paper has been published in "The Importance of Lamarck's Chemistry for his Theories of Nature and Evolution or Transformism," *Actes du XIIIe Congrès international d'histoire des sciences, 1971* (Moscow, 1974), section IX, pp. 92–97.

1. "Lavoisier," *Dictionary of Scientific Biography,* 8 (New York: Scribner's, 1974), 80.

minimized and treated in an ahistorical way. These two authors, and others, dismiss it as an unfortunate case of wild speculation.[2] They pretend it has no relationship to the rest of Lamarck's work or to anything else happening at the time. Charles C. Gillispie, and more recently, Gabriel Gohau have been among the very few to suggest the need for a reexamination of Lamarck's chemistry because of its importance for his other work, particularly his theory of evolution.[3] Gillispie has gone one step further and stressed the connections between Lamarck's chemistry and a more general phenomenon which he termed Jacobin science, and which Guerlac has called anti-Newtonian science. Maurice Crosland has characterized this phenomenon as the persistence of an older natural history tradition in a field which had received a new physico-mathematical orientation. Roger Hahn has recently suggested that this attitude toward science was part of a broader pattern of anti-corporate sentiments.[4]

Despite these suggestions of the significance of Lamarck's chemistry, there has been no detailed analysis of it. Lamarck's chemistry was an essential part of his view of nature, the basis of his theories of geology and evolution. The methodology used by Lamarck in his chemistry was similar to that employed in other areas, such as botany, meteorology, invertebrate zoology, geology, and evolutionary theory.

As we examine Lamarck's work in this area, we will notice his increasing hostility toward the established French scientific community. Richard W. Burkhardt has suggested that Lamarck's evolutionary theories were ignored not only because they depended on a pre-Lavoisierian chemistry, but also

2. Alpheus S. Packard, *Lamarck the Founder of Evolution* (New York: Longmans, Green, 1901), pp. 83–88. Marcel Landrieu, *Lamarck le fondateur du transformisme* (Paris: Société zoologique de France, 1909), pp. 150–155, 161–165. Emile Guyénot, *Les sciences de la vie aux XVIIe et XVIIIe siècles: l'idée d'évolution* (Paris: Albin Michel, 1957), p. 414. Albert V. Carozzi, "Lamarck's Theory of the Earth: *Hydrogéologie*," *Isis*, 55 (1964), 304–305.

3. Charles C. Gillispie, "The *Encyclopédie* and the Jacobin Philosophy of Science: A Study in Ideas and Consequences," in *Critical Problems in the History of Science*, ed. Marshall Clagett (Madison, Wisc.: University of Wisconsin Press, 1959), pp. 255–289; "The Formation of Lamarck's Evolutionary Theory," *Archives internationales d'histoire des sciences*, 9 (1956), 323–338; "Lamarck and Darwin in the History of Science," in *Forerunners of Darwin*, ed. Bentley Glass et al. (Baltimore, Md.: Johns Hopkins Press, 1959), pp. 265–291. Gabriel Gohau, "Le cadre minéral de l'évolution Lamarckienne," in *Colloque international "Lamarck,"* ed. J. Schiller (Paris: A. Blanchard, 1971), pp. 105–133. James R. Partington, *A History of Chemistry* (London: Macmillan, 1962), III, 490, has a very brief discussion of Lamarck's chemistry in his section on the reception of Lavoisier's. See also criticism of Gillispie's views by John C. Greene, *The Death of Adam* (Ames, Iowa: Iowa State University Press, 1959), p. 359, n. 63.

4. Gillispie, "The *Encyclopédie*"; Henry Guerlac, "The Anatomy of Vandalism," unpublished paper presented to the History of Science Society, December 1954, and "Comment on the paper by C. C. Gillispie," in Clagett, ed., *Critical Problems*, pp. 318–320; Maurice Crosland, "The Development of Chemistry in the Eighteenth Century," *Studies on Voltaire and the Eighteenth Century*, 24 (1963), 369–441; Roger Hahn, *The Anatomy of a Scientific Institution: The Paris Academy of Sciences, 1666–1803* (Berkeley: University of California Press, 1971), p. 325.

because Lamarck did not supply the evidence he could have to support them. This sin of omission occurred because Lamarck believed that his ideas would be ignored no matter what he said; his conviction, according to Burkhardt, was a direct result of the hostile reception his chemistry received.[5]

In attempting to understand the origins of Lamarck's chemical theories and methodology, it is instructive to compare his education with that of Lavoisier, one year his elder.[6] Both began their formal education at age eleven, but while Lavoisier was receiving the best scientific training available in France at the Collège Mazarin and an introduction to quantitative techniques from N.-L. De Lacaille, Lamarck was being prepared for the priesthood at a provincial Jesuit school. Lamarck left school at age fifteen to pursue a military career. By the age of twenty-three, he had become only a self-taught amateur botanist. During this same period, Lavoisier was completing his scientific education, studying with J.-E. Guettard and G.-F. Rouelle, and launching his own career as an experimental geologist-mineralogist and then chemist. "The crucial year" (1772) saw Lavoisier making the breakthrough that was to lay the foundation for his new chemistry, and Lamarck just beginning medical school in Paris.

Lamarck's introduction to chemistry was probably through pharmacy courses. At some point in the middle of the decade Lamarck began attending lectures on botany at the Jardin du roi. He probably also went to the chemistry lectures of H. M. Rouelle, the younger brother of Lavoisier's teacher, who lectured in his own right at the Jardin from 1768 to 1779. Rhoda Rappaport has said that the younger Rouelle learned most of his chemistry from his older brother, and there are definite resemblances between many of G.-F. Rouelle's theories and Lamarck's.[7] Lamarck is probably indebted to the elder Rouelle for many aspects of his modified Stahlianism, including his general views on the nature of matter and chemical substances; the four-element-instrument theory, especially the dual roles of fire and air; and the anti-Newtonian theory of colors based on phlogiston theory.

5. Richard W. Burkhardt, "Lamarck, Evolution, and the Politics of Science," *Journal of the History of Biology,* 3 (1970), 280–284, 287–291.

6. Background information from: Guerlac, "Lavoisier," pp. 67–74, and his "A Note on Lavoisier's Scientific Education," *Isis,* 47 (1956), 211–216; Leslie J. Burlingame, "Lamarck," *Dictionary of Scientific Biography,* 7 (1973), 584–587, and Franck Bourdier, "Esquisse d'une chronologie de la vie de Lamarck" (with the collaboration of Michel Orliac), unpublished memorandum, 3e Section, Ecole Pratique des Hautes Etudes, 22 June 1971, pp. 4–7.

7. Rhoda Rappaport, "G.-F. Rouelle: An Eighteenth-Century Chemist and Teacher," *Chymia,* 6 (1960), 75–76, speaks of the difficulties in determining who attended public lectures at the Jardin du roi. She discusses G.-F. Rouelle's chemical theories in "Rouelle and Stahl—The Phlogistic Revolution in France," *Chymia,* 7 (1961), 73–83. Both articles contain a number of references to H. M. Rouelle, whom she treats separately in a short *Dictionary of Scientific Biography* article.

Crosland has made the useful distinction between the natural history tradition and the physico-mathematical tradition in eighteenth-century French science. He places Rouelle firmly within the former.[8] Lamarck also belonged in the natural history tradition, and remained within it for his entire life. In the years in which Lamarck was absorbing the natural history approach to chemistry, Lavoisier was creating the chemical revolution by developing a physico-mathematical approach to chemistry, influenced in the late 1770's and early 1780's by his collaboration with Laplace.[9] Rouelle and Lamarck were willing to accept Stephen Hales's proof that air was a constituent element. Rouelle died before English pneumatic chemistry became crucial to Lavoisier's work, but Lamarck, following in the same natural history tradition, never understood the importance of the pneumatic chemists or of exact quantitative experiments with "airs."

Lamarck certainly did not owe his natural history orientation entirely to Rouelle. His early interest in botany had continued to develop, and botany in this period was considered the most advanced field of natural history. At the same time that he began his first work in chemistry, *Recherches sur les causes des principaux faits physiques* (1776), he started the draft of his *Flore françoise*. Georges Louis Leclerc Buffon was so impressed by his criticism of Linnaeus in 1777 that he arranged for its publication at the Imprimerie Royale (1779) and then engineered Lamarck's election to the Academy of Sciences as an adjunct member in botany. In published texts from the *Flore françoise,* Lamarck showed that he had absorbed much of Buffon's natural history orientation, as well as some of his ideas relating to chemistry.

Lamarck began his *Recherches* in 1776, and finished early in 1780. He had tried to include some of his chemical views in the "Discours préliminaire" of the *Flore*, but they were removed during the drafting by R.-J. Haüy.[10] Presumably Lamarck had been pursuing his self-education in chemistry after attending H. M. Rouelle's course and this rebuke may have stimulated his reading, because the *Recherches* shows a knowledge of, if not appreciation for, much of the work of his day, including that of Buffon, Lavoisier, Benjamin Franklin, Baumé, Macquer, Guyton de Morveau, Hales, Priestley, Joseph Black, and others.

Lamarck said that he submitted the *Recherches* to the Academy in 1780, one year after he had become a member.[11] The entry in the "Procès-verbaux" for 22 April 1780, the date Lamarck gives for the submission of

8. Crosland, "Development of Chemistry," pp. 393–402.

9. Henry Guerlac, "Laplace's Collaboration with Lavoiser," *Actes du XIIe Congrès international d'histoire des sciences, 1968* (Paris, 1971), vol. III-B, 31–36; Crosland, "Development of Chemistry," pp. 371, 388–393, 398–411.

10. Bourdier, "Chronologie," pp. 5–6; Lamarck, *Recherches* (Paris; Maradan, 1794), II, 366.

11. "Avertissement," *Recherches,* I, vii.

his *Recherches,* is rather puzzling. It says that Lamarck "read the preliminary discourse of a work on the variations of the atmosphere," and that A.-D. Fougeroux De Bondaroy, M.-J. Brisson, and N. Desmarets were named as commissioners to report on it; on 3 May, Mathieu Tillet and Macquer were added to the committee.[12]

Lamarck had read a paper to the Academy in 1777 entitled "Mémoire sur differens phénomènes de l'atmosphère," which dealt with a number of issues taken up in the *Recherches.* The manuscript of the supposedly missing unpublished paper has revisions in Lamarck's hand showing that he intended to include it in the *Recherches,* and it must have been part of his 1780 submission to the Academy.[13] He later decided to develop the material further in a book which would be a sequel to the *Recherches,* and so he withdrew this section.[14] The version which was eventually published was in the hands of the Academy by 3 May 1781, when Condorcet wrote: "The resumé of the principles of citizen Lamarck has been deposited with the work which contains their development and initialed by me."[15] The commission never issued a report. In the "Avertissement" to the *Recherches,* Lamarck accused them of stalling and ultimately dropping the whole matter; he says that he eventually had Condorcet initial it so that he could claim priority for his ideas.[16] Given the composition of the commission, Lamarck's work could only have received a negative reaction, not because of its adherence to phlogiston theory (Lavoisier's ideas were not yet accepted), but because of its speculative, qualitative, nonexperimental natural history orientation. Many of his ideas would have seemed old-fashioned in 1780 to such men as Macquer. The commissioners must have wished to avoid public embarrassment of a fellow academician. The result of this lack of action by the Academy was extreme bitterness on Lamarck's part, as seen in the "Avertissement," when he finally published the work in 1794, after the suppression of the Academy. He dedicated his work to the French people and the Revolution, hoping for public vindication of his ideas during a period of hostility toward "established science."[17]

12. "Procès-verbaux," Académie des Sciences, entries for 22 April and 3 May 1780, Paris, Archives, Académie des Sciences.

13. MS 755-1, Muséum national d'histoire naturelle de Paris. Landrieu, whose work is the only secondary source which discusses Lamarck's meteorology in any detail (pp. 133–149), refers to the memoir as missing and relies on a summary by Louis Cotte, "Sentiment de M. le Chevalier de la Mark," *Mémoires sur la météorologie* (Paris: Imprimerie Royale, 1788), I, 205–215. A comparison between the manuscript and Cotte's summary proves that the manuscript is the lost memoir.

14. Lamarck, *Recherches,* I, 17–18.

15. Ibid., II, 400.

16. Ibid., I, vii–viii. It does not appear that Lamarck changed the text after 1781; it contains no references to works published later than 1780. He did add some footnotes attacking the new chemistry and defending his views.

17. "Au peuple Français," *Recherches,* I, v–vi.

Although the *Recherches* was not published until 1794, Lamarck had managed to print a brief summary of his chemistry earlier. It appeared at the end of the 1786 article "Classe" in his *Dictionnaire de botanique.*[18] Lamarck's *Recherches* might have fared better if it had been published in 1780. By 1794, however, all the important scientists in France had accepted the new chemistry, and Lamarck's work could only be seen as hopelessly out of date and not even worth attacking.

Lamarck was not one to give up easily. He tried a new approach in 1796, when he published his *Réfutation de la théorie pneumatique ou de la nouvelle doctrine des chimistes modernes,* which was a direct attack on the new chemistry. He chose Fourcroy's *Philosophie chimique* (1792), an extremely influential presentation of the new ideas, as the target "because his book appeared to present a description of the principles of the pneumatic theory [that is] more complete, more precise, and better developed than any other work."[19] Lamarck reprinted Fourcroy's work in entirety on the left-hand pages and countered each point on the opposite pages with criticisms and his alternative theories. Landrieu suggests that Lamarck chose a polemical form of presentation in order to force his opponents at least to discuss his ideas, but they did not respond.[20]

Following that failure, in 1797 he began to read a number of memoirs on the same subject at the Institut. The response was largely to ignore him, and he therefore stopped reading in the middle of the fourth memoir. Lamarck became almost paranoid on the subject, thinking that there was some plot to keep his ideas in obscurity.[21] The memoirs he had written, including the few read at the Institut, were collected and published as the *Mémoires de physique et d'histoire naturelle* in 1797.[22]

This book was his last full-scale work on chemistry. In 1799, he published two memoirs in the *Journal de physique, de chimie, et d'histoire naturelle,* which had begun as François Rozier's *Journal* but had been taken over in 1785 by De La Métherie, who was very hostile to the new chemistry.[23] One of the memoirs was entitled "Mémoire sur la matière du feu considérée

18. Ibid., II, 34–36
19. *Réfutation,* (Paris: Agasse) p. 66.
20. Landrieu, pp. 150–151.
21. *Mémoires de physique et d'histoire naturelle* (Paris: Agasse, 1797), pp. 408–410.
22. See Landrieu, p. 453, for complete title; at the beginning of the first four memoirs, Lamarck indicated the date they were read to the Institut. On page 2 of the *Mémoires,* Lamarck stated that he had started to publish them, as he read them, under a different title. We cite this title in full, as it was not indicated in Landrieu's bibliography and has been found in several libraries; except for the title page, it is identical to the one listed by Landrieu: *Mémoires présentant les bases d'une nouvelle théorie physique et chimique, fondée sur la considération des molécules essentielles des composés, et sur celle des trois états principaux du feu dans la nature; servant en outre de développement à l'ouvrage intitulé: Réfutation de la théorie pneumatique. Lus à la première classe de l'Institut National, dans ses séances ordinaires.*
23. Partington, *History of Chemistry,* III, 494–495.

comme instrument chimique dans les analyses." The other dealt with the matter of sound.[24] Both memoirs were reprinted at the end of Lamarck's *Hydrogéologie* (1802). That work proper also devoted a large section (chapter four) to his chemistry as it related to his geological theories.

Lamarck never abandoned his chemical theories. In addition to appearing in the *Hydrogéologie,* they are also to be found in his major evolutionary works. They play a very prominent role in the *Recherches sur l'organisation des corps vivans* (1802), and are also present in the *Philosophie zoologique* (1809), and the *Histoire naturelle des animaux sans vertèbres* ("Introduction," 1815). In these works he used his chemistry to provide a materialistic definition of life and to explain its maintenance and appearance (both through reproduction and through spontaneous generation), and the way in which living organisms gradually evolved, including the emergence of the higher mental faculties.

Lamarck's antagonism to Lavoisier and the new chemistry took several forms. On the simplest level, he felt that his ideas were being ignored because of limited personal interests of those in favor of the new chemistry. By the time of the *Hydrogéologie* (1802), he could accurately accuse them of taking control of the French educational institutions and most periodicals with the aim of promoting their system at the expense of any opposing ones.[25] Because of his personal experience, Lamarck was very sensitive about his chemical views.[26] Lamarck also tried to criticize the new chemistry on the basis of a peculiar kind of eighteenth-century positivism. He charged that the new chemists were inventing systems and hypotheses based only on a few new experiments. Lamarck claimed that, in contrast, he would consider all known facts before putting forth any theories. This consideration leads to the basic reason for his hostility. Lamarck thought of nature as an organic whole in which everything was closely interrelated; such a view was an outgrowth of his natural history orientation. He frequently stated that while the facts were important, the most essential thing was the whole.[27] Methodology was thus intimately related to a conception of nature. Lamarck's chemistry was an integral part of this total view. Any attack on the chemistry was therefore a danger to his whole system. Although he offered to abandon his theories if they definitely could be proven false, he

24. The first memoir appeared in the *Journal de physique, de chimie, et d'histoire naturelle,* 48 (1799), 345–361. The second, "Mémoire sur la matière du son," appeared in the same journal, vol. 49 (1799), 397–412.

25. *Hydrogeology,* trans. Albert V. Carozzi (Urbana, Ill.: University of Illinois Press, 1964), pp. 120–121:

26. For examples, see "Avertissement," *Recherches,* I, vii–xvi, and *Mémoires,* pp. 408–410, and *Hydrogeology,* p. 91.

27. "Avertissement" and "Discours Préliminaire," *Recherches,* I, viii–xii and 7–8; *Hydrogeology,* pp. 78–80. See also Gillispie, "Formation," pp. 331, 334–335. Crosland, p. 441, noted this orientation in Lamarck and associated it with the natural history tradition; he traces it back to Buffon's idea of nature.

could not accept the new chemistry without rejecting his conception of nature. In this respect, Lamarck is very close to a whole group of people at the end of the eighteenth century; many of the other representatives of "Jacobin science" shared this view.

Lamarck also defended his theories on the grounds of utility for the new citizens of France, a criterion which was of increasing concern in the late eighteenth century.[28] He frequently pointed out the usefulness of his chemical theories for medicine and agriculture. Finally, in Enlightenment faith, he believed that truth would ultimately triumph and reason prevail and his theories would finally receive their due recognition.

Let us now turn to an analysis of Lamarck's chemistry. Because his views remained the same, with minor exceptions, throughout his works, our discussion will proceed in a topical rather than chronological way.[29] Lamarck was quite aware that his chemical views were eclectic, but he claimed that their value derived from a new understanding of phenomena and a resulting higher synthesis of previously separate theories.[30] His definition of matter in general was basically Newtonian. Lamarck went on to observe that matter was not homogeneous, since compounds existed in nature.[31] There must be several different simple kinds of matter which combine to form the substances we observe. These simple types of matter are elements. While Lamarck said that man can never be sure that he knows the "true elements of bodies," provisionally he should accept as elements "every substance that he finds inalterable, that his faculties never allow him to decompose, and that are never found in greater simplicity in any natural phenomena."[32] Despite the "positivistic" phrasing of this definition and its resemblance to Lavoisier's, Lamarck concluded that the substances which met these qualifications were the traditional four elements: earth, air, water, and fire. He attacked those new chemists who maintained that water was itself a compound, and attempted to discredit their experimental results. He had similar criticisms of the new theories maintaining that air was a mixture of gases and not an element and that fire was a kind of movement.[33] Light for Lamarck, as for Buffon, occupied a peculiar position in this system. It was

28. See especially his "Au peuple Français," *Recherches,* I, v–vi, and *Mémoires,* pp. 408–410. The stress on utility was closely linked to the idea of progress in Lamarck's writing. Crosland, pp. 435–437, discusses the growing interest in the utility of chemistry in the eighteenth century.

29. Lamarck did develop some of his points in the course of his publications, but an analysis of these developments would be too detailed for this paper and is not relevant to our major concerns here.

30. "Avertissement," *Recherches,* pp. x–xii.

31. A particularly clear treatment of this and the following points may be found in the *Recherches,* I, 315–318; see also I, 19–63, and II, 390 ff. In the *Mémoires,* see the first memoir and pp. 368–370.

32. *Recherches,* I, 315–316; there is a similar statement in *Mémoires,* p. 54.

33. See, for example, *Recherches,* I, 8 and 316 (note).

not exactly a fifth element since it was not a constituent of any compound and acted primarily as an instrument. Light was found everywhere; it penetrated all bodies and, depending on the way it interacted with them, produced different colors (Lamarck's theory of colors, probably derived from Rouelle, was very different from that of Newton). Light could also interact with fire in a very strange way, so as to replenish the amount of active fire in the world. Light from the sun and other stars performed this necessary function.[34]

Lamarck followed Rouelle's Stahlianism to maintain that the nature of a given compound resided uniquely in the essential molecules of that compound, as the nature of the four types of simple matter resided uniquely in their "integral molecules."[35] The latter, in different combinations, formed the essential molecules of the various compounds. Destruction of these essential molecules destroyed the nature of the substance they represented. Differences between compounds depended both on the number and proportion of the elements and on the degree of intimacy in the union of the elements. Sensible masses were produced by aggregations of essential molecules of one or several types. Aggregation was produced by Newtonian attraction; the different sizes and shapes of the molecules determined how solid a body would be. This view was very similar to Buffon's. Chemical change did not involve these aggregations but only the essential molecules.

Lamarck adapted Rouelle's element-instrument theory to his own purposes. He maintained that each of the four elements had a natural state in which it was free and demonstrated its real properties, which were usually masked in any combined form.[36] In every compound, however, at least one of the elements was not in its natural state but rather modified away from it in one of several possible states of "condensation." Each degree of modification away from the natural state increased the amount of strain or tension on the element to return to its natural level. The image one might use to understand Lamarck's conception is that of a spring; each amount the spring is stretched from its "natural state" increases the strain on the spring to return to that natural state. Lamarck then reasoned that matter could not

34. *Recherches,* I, 317–318, and *Mémoires,* pp. 183–188.

35. Rappaport's "Rouelle and Stahl" provides a good account of Rouelle's views (pp. 73–83) and the ways in which they differed from the theories of Stahl and Boerhaave (pp. 83–95). In the last part of the article (pp. 96–102), she shows that Rouelle's modified Stahlianism dominated French chemistry from the 1760's to the 1780's, when it began to be replaced by the new chemistry. See also Hélène Metzger, *Newton, Stahl, Boerhaave, et la doctrine chimique* (Paris: Félix Alcan, 1930), Part II, chaps. 5 and 6. The discussion which follows is based on *Recherches,* II, parts 2 and 5 and pp. 394–395, and *Mémoires,* first memoir and pp. 368–371.

36. *Recherches,* II, part 2 and pp. 390–391, and *Mémoires,* fourth memoir and pp. 369–374.

modify itself, since the modifications were away from its natural state. In other words, matter alone could not form any combinations in nature; the affinity theory of chemical combinations accepted by most of his contemporaries was therefore absurd. There must be an external cause which modified matter. When this cause was removed, the elements tended ("avoir une tendance") to return to their natural state: "The principles that any sort of cause has forced to submit to the state of combination, necessarily have in themselves a real tendency to disengage from this state. This tendency has an intensity or an energy relative either to the very nature of each combined principle or to the state of its combination in the compound which contains it."[37] The only possible form of chemical reaction therefore was decomposition. The compounds which decomposed most easily Lamarck termed imperfect. The imperfect compounds were those which contained the elements most easily disengaged: water was most easily disengaged, then the elastic principles air and fire; earth was the most difficult.

The different stages of decomposition of compounds produced all the inorganic substances known in the world; this strange theory seems to have been one of Lamarck's original conceptions. To illustrate his point, Lamarck suggested that if man had complete knowledge and control of chemical analysis, "it is clear that, in producing all the degrees of possible alterations upon a compound in which the four elements (I suppose always that there are only four) abound up to a certain point, by the smallest changes in the proportions and in the number of the principles, especially if he could operate in a way so as to exhaust all cases, with this single compound, he would obtain successively all the combinations which nature, in the minerals, and chemistry, in the products of its operations, can furnish examples of."[38] This view is an expression of the Great Chain of Being, particularly the principle of continuity as applied to the mineral kingdom. In another place Lamarck explicitly stated that there were no species among minerals, only a chain of individuals in which differences arose from imperceptible shadings.[39] It is interesting that at this point (prior to 1800) he would not extend this idea to the level for which it had been developed in

37. *Mémoires,* p. 375. Lamarck was not alone in his attacks on affinity theory. Metzger (p. 89) claims that affinity theory was violently attacked at its height of popularity by Rouelle and his students Monnet, T. Baron, and G. F. Venel. Rappaport, "Rouelle and Stahl," p. 83, n. 34, says "Rouelle's refusal to speculate about the causes of affinity has given him an undeserved reputation as a repudiator of the whole idea of affinity." She goes on to point out that Rouelle has been confused with his pupils, Venel and Monnet, who were against affinity theory. Crosland, p. 405, adds Demachy, another student of Rouelle.

38. *Recherches,* II, 9. Metzger can only characterize Lamarck's decomposition theory as "so bizarre" (p. 98). Crosland cites (p. 402) this theory as an extreme example of the natural history orientation in chemistry; for Lamarck "historically, if not methodically, chemistry was a branch of natural history."

39. *Recherches,* pp. 8–10, and *Mémoires,* pp. 104–105 and 372.

natural history (by Buffon and others), to living organisms. Lamarck's elaborations of the principle of continuity in the mineral kingdom were in direct opposition to Lavoisier's conception of specific elements. Lamarck's firm adherence to a view of the "whole" (tout) in the mineral kingdom may have paved the way for his later extension of this idea to the rest of nature.

In another sense, however, Lamarck already had a view of nature as a whole in process.[40] One side of nature was constituted by the compounds of the mineral kingdom; this was the decaying side of nature. What was the constructive or synthetic side of nature; what was the external cause which modified the elements away from their natural state? Having rejected affinity theory, Lamarck asserted that living organisms alone could produce chemical combinations or modifications in opposition to natural tendencies. Vegetables only could combine directly the four free elements by their "vital" action to form compounds. This view did not reflect the contemporary work on photosynthesis, but rather the earlier ideas of Baumé, whom Lamarck cited with great respect.[41] Using the combinations produced by vegetables, animals could then elaborate them into more complex substances. This ability of animals to form more complex compounds than plants was due to their more intricate physiological structure.

Here then we have a cycle involving all of nature. Animals and plants represented the constructive part of the cycle and minerals the destructive; biology (Lamarck's later term) and physiology were the branches of knowledge which treated the synthetic aspect of nature, while chemistry and geology were the disciplines which dealt with the analytic facet of nature. To complete the cycle, Lamarck envisioned a transfer from the living realm to the dead through two processes: elimination of waste products during life and decomposition of an organism following death. The cause of death was also an integral part of Lamarck's total system. Essentially, living organisms silted-up gradually until they died. There was an accumulation of the earthy principle which was the hardest to get out of combination.[42] This process was also related to the health of the organism.

One large question remains. What was it about living organisms that allowed them to produce chemical syntheses? Lamarck spoke of "vital" principles or organic action which directed the assimilation of substances in plants and animals. This process was called life: "Life, in the bodies which are endowed with it, consists not only of the exercise of organic movements essential to the conservation of each individual, but also in the faculty that

40. *Recherches,* II, part 4; *Mémoires,* seventh memoir and pp. 374–376 and 386–392.

41. *Recherches,* II, 306–307. Lamarck first cited Baumé in his article "Classes," *Dictionnaire de botanique* (Paris: Panckouke, 1786), II, 36. On Baumé's views, see Metzger, p. 98, and Partington, III, 90–95.

42. *Recherches,* II, part 4, art. 1 and p. 398; and *Mémoires,* pp. 386–388.

the organs of these beings have of being able to execute their functions."[43] Animals, in general, could produce more complex substances than plants because they have differentiated systems of organs. Each type of organ, in the course of exercising its functions, produced different substances. Another factor which entered in was the strength or weakness of the organ. It would be very easy to move from this particular idea to Lamarck's later theory of the use and disuse of organs. Other factors which affected the process were the health of the individual and the climate: "High temperatures favor and accelerate organic activity, whereas in low temperatures the organs are capable only of slow, weak actions."[44] Here we have the suggestion of the powerful influence of environment on organisms.

As for the faculty which directed the functioning of the organs, Lamarck was not explicit until later in his biological works. There, the so-called "vital" principles were revealed to be the imponderable fluids, especially those associated with fire, such as electricity and caloric.[45] We shall see from the following discussion of his ideas about fire that aspects of its relationship with life processes were suggested in his chemical works.

The most important of the four elements in Lamarck's chemistry was fire. His ideas about it were a synthesis of many contemporary views.[46] Fire existed in both natural and modified forms of which there were two: fire in a state of expansion (as instrument) and fixed fire (as element). The natural state of fire, or "etherial fire" (feu éthéré), was that to which the various modifications of fire had a tendency to return. It was a free fluid existing everywhere. Lamarck suggested that it was the material cause of sound; in addition, by some unknown modifications it became magnetic fluid and electrical fluid. Lamarck seems to have derived this relationship between the matter of fire and that of electricity from Franklin, whom he cited with great admiration on this point.[47] It is interesting that this electrical fluid was one of Lamarck's so-called vital fluids following Alessandro Volta's work.[48]

Fire in its natural state could be acted upon by light from the sun and raised to the first of the modified states, that of expansion or the "caloric

43. *Mémoires,* pp. 386–387.

44. *Hydrogeology,* p. 90.

45. Lamarck first relates the action of these fluids to the evolution of organisms in his 1802 *Recherches sur l'organisation des corps vivans* (Paris: Maillard).

46. Discussion which follows is based on *Recherches,* I, part 1 and pp. 318–342, and II, 391–394; and *Mémoires,* sixth memoir and pp. 377–385. See especially the table following p. 226; it is reprinted from *Réfutation*, p. 36. See also *Hydrogeology*, ch. 4, especially pp. 83–84, 88–89, 119–120; and "Mémoire sur la matière du feu, considérée comme instrument chimique dans les analyses," *Journal de physique,* 48 (1799), 345–361. Specific references will be indicated in other footnotes. Gillispie, "Formation," pp. 333–334, has a brief summary of Lamarck's ideas on fire.

47. *Recherches,* I, 138, 198–199, 203.

48. See his interesting reference to Volta in *Hydrogeology,* p. 80.

fire" (feu calorique).[49] This process maintained the heat of the earth. Caloric fire, considered alone, produced the different colors of the spectrum.[50] It also could act as an instrument in two main ways. First, as an atmosphere around isolated molecules of water, it produced vapor and around molecules of metallic or other heavy matter it led to volatilization. Second, it could act as an instrument by penetrating the interstices between the molecules of a body (without combining with those molecules) to produce a number of very important effects: heating, dilation, luminosity, liquefaction, vaporization ("réduit en gaz"), and burning. Also between the molecules, but combining with them, caloric fire caused calcination and produced salts and metals.

The third state of fire, and the second modified form, Lamarck termed the fixed state ("feu fixé"). In this state of modification, fire was a constituent principle of compounds. The presence of fixed fire in bodies determined their colors when acted upon by light and fire in a state of expansion.[51] Fixed fire was produced by vegetables with the aid of light.[52] Here Lamarck seemes to have a faint idea of photosynthesis, although we have previously seen that he had not related this action to the general role of plants in nature. Fixed fire was responsible for the maintenance of animal heat; animal heat was "produced by the continual release of fixed fire which passes ceaselessly into the blood by way of the foods which animals use."[53] During chewing and digestion, that portion of fixed fire which was most weakly bound in the food passed directly to the blood. The rest got into the blood by way of the chyle "in which fixed fire is one of the constitutive principles."[54] In health there was a constant body temperature due to the balance between the assimilation of fixed fire and the general force of decomposition. Here we have other instances of the essential role of Lamarck's chemistry in his total view of nature.

There were a number of things that Lamarck had to explain if he held that fixed fire was a constituent principle of substances. He was forced to state that fixed fire existed in two forms, "acidific" and "carbonic." In its

49. In the *Recherches,* the term "fire in the state of expansion" was used originally. At the time of publication (1794), Lamarck inserted "caloric fire" in parentheses, and this term was used in preference in his subsequent works, probably in an attempt to demonstrate that his theories could incorporate some of Lavoisier's work.

50. See, for example, *Mémoires,* pp. 236–237. Section 3 of the *Recherches* (II) was devoted to an exposition of his theory of colors. The attack on Newton is similar to Rouelle's. Lamarck developed his color theory at the end of the third memoir and in a supplement to it where he presented a graduated series of natural colors; his language suggests a parallel with the Great Chain of Being.

51. See note 40 above and *Recherches,* I, 333–334, Corollaries 19 and 20, and II, 397.

52. *Recherches,* I, 333, art. 19; *Mémoires,* pp. 292–298; and *Hydrogeology,* p. 87.

53. *Recherches,* I, 236.

54. Ibid.; see also *Mémoires,* pp. 309–310.

"acidific" form it was found in imperfect compounds which were more or less soluble in water and incombustible. Thus, for example, among gases it was found in mephitic, nitrous, sulfurous, muriatic, acetic, and floric gases. It was present in such liquids as acids and floric acids. It was present in such liquids as acids and alkalis. Among solids it could be found in what he called the "concrete salts," the calx of chalk, and metallic calxes. Fixed fire in its second, or carbonic, form was a constituent of perfect compounds which were insoluble in water and, most important, combustible. Lamarck stated that this form (fixed carbonic fire) was the same as phlogiston and carbon, the names differing only according to the conditions under which it was observed. Fixed carbonic fire was a constituent of gases such as inflammable air (hydrogen) and azotic air (nitrogen); of fluids such as petroleum and the "fatty oils"; and of solids such as sulfur, carbon, and metals. It was this last part of Lamarck's theory which dealt with fixed fire that required special defense since it was in direct opposition to Lavoisier and the new chemistry.

Lamarck criticized Priestley's defense of the phlogiston theory and said that one must add his own idea about caloric fire and phlogiston as temporary states only.[55] His explanations of combustion were similar to previous phlogiston theories except that the fixed carbonic fire was modified into either the acidific fixed fire or the caloric fire. Although he believed it was absurd to think that the air rather than the "combustible" burned, he nevertheless recognized the need for air for combustion; he gave air a physical role as an instrument in combustion.[56] He then had the problem of explaining the increase in weight which accompanied calcination. This he did in a very unusual way, by first citing sections of Lavoisier's *Opuscules* which stated that metallic calxes are heavier than the metals from which they were formed and that these calxes contained a large amount of fixed air. Having invoked the authority of Lavoisier, he then proceeded to explain these facts in a very odd way to "save the appearances": "When one calcinates a metal, a part of the fixed fire of the metal is separated and emitted or dissipated by the column of ascending air. But at the moment of recooling . . . the nondissipated portions of fire in expansion combine with the surrounding air and form a gas which the residue of the metal absorbs in greater quantity than the fixed fire the metal lost in calcination."[57]

Lamarck's monistic conception of nature, to which his chemistry was the key, can perhaps best be understood by taking an example and following it

55. *Hydrogeology*, pp. 119–120; much of the above discussion is based on *Mémoires*, sixth memoir.

56. *Recherches*, I, 150–160 and 330, Corollary 17; see also *Mémoires*, pp. 192–209.

57. *Recherches*, I, 300–301; see also p. 298, where he cites Hales, Priestley, Black, D. Macbride, and Lavoisier in connection with the idea of air as a constituent element. In addition, see p. 294, reference to Buffon, and p. 312, reference to D'Arcet's experiments on the diamond; and *Mémoires*, pp. 209–224.

through the whole cycle. Carbon (fixed carbonic fire) is a good example because it was involved in a great number of processes.[58] Carbon was formed directly by plants from caloric fire (which was produced by the action of light on fire in its natural state) and small amounts of water and air with the aid of light. In this synthetic process, plants also formed other food and substances needed for growth and maintenance. Carbon was essential to plants because it served as the basis of their solid parts and also was necessary in their fluids. Vegetable synthesis of carbon was the only way nature employed to compensate for the continuous loss of this substance. Some of the carbon produced by plants was transported to the bottom of the earth's external crust where it became the supply for all the earth's volcanoes; this was an interesting linkage of Lamarck's chemical and geological theories. Much of the carbon returned to the caloric state during the combustion of vegetable products. Another portion of it was transformed by fermentation (which for Lamarck was a form of decomposition) into several types of gases or fixed acidific fire. Some of the carbon from plants was elaborated by the animals eating them to form their constituent substances. Carbon in grass, for example, was eaten by a horse and modified and elaborated into its various tissues and parts through the action of the animal's different organs. We have indicated previously factors affecting the production of different substances in animals. In the animal, some of the carbon not employed for the above processes was used to maintain a given body temperature. Carbon-containing substances were released from the organism through excretion and the death and decay of the animal. The natural tendency of degradation then came into full operation, and as these products decomposed gradually, new substances resulted. Among these were metals which were formed when the released carbon became intimately combined with an earthy base under favorable circumstances and perhaps over a long period of time; these two factors were also very important in Lamarck's evolutionary theories. Lamarck modified his previous statement about the ability of chemical combinations to form without the aid of living organisms: "When the constitutive principles of a compound are separated by the action of a substance causing it to change, in short, by any medium, they react on one another. They develop new bonds because they are still in a favorable condition for such a process, but the new compounds formed by them progressively display a lesser concentration or a lesser modification of their constitutive elements."[59]

The earthy bases with which carbon combined to form metals in this process were of two types: chalk from animals, which in combination with

58. Discussion which follows is based on *Recherches,* parts 4, 5; *Mémoires,* sixth and seventh memoirs; and *Hydrogeology,* Chapter 4.

59. *Hydrogeology,* p. 95; on the formation of the mineral kingdom, see *Mémoires,* pp. 316–367, especially table following p. 348.

carbon produced barite, and clay from plants, which in combination produced magnesium. Further degradation and modification by union with other substances led to the production of various acids, gases, alkalis, calxes, alcohols, and aromas. From these in turn were produced more or less inflammable and saline substances with varying amounts of carbon.[60] When a metal was calcinated, in addition to decomposition to a calx in which the carbon was in the acidic fixed state,[61] some of the carbon was further decomposed into caloric fire. Finally, caloric fire, given the opportunity, returned to its natural state of the "etherial fire" and the cycle of composition and decomposition was completed.

From this example we see how much Lamarck's chemistry was an integral part of his whole view of nature. It is not only necessary for an understanding of the total view but also for his theories regarding living organisms and geology. In the latter case, all the substances studied were some form of degradation of animal or plant products. The different geological formations resulted from depositions of these substances by various processes, as sedimentation from water.

In Lamarck's later evolutionary theory, his chemistry was used to explain the origin of life, its maintenance, and the evolution of living organisms. These matters were first taken up in detail in 1802 in the *Recherches sur l'organisation des corps vivans.* For Lamarck, life originated from a soup of the elements in which small mucilaginous and gelatinous bodies formed; these gradually became celllike structures and then the simplest animals and plants. This process of spontaneous generation, which Lamarck held was still going on, was caused by the action of heat, light, electricity, and humidity. Life was thus a physical phenomenon dependent on imponderable fluids. We have seen how life, once established, was maintained through synthesis and assimilation. Lamarck's theory of evolution rested on the mechanism of the movement and concentration of certain ponderable and imponderable fluids in the organism. The imponderable fluids were also found in the environment and could exert a decisive influence on an organism. In his attempt to account for the emergence of the higher mental faculties, Lamarck provided a physical basis of feeling through a nervous fluid which was a form of the electrical fluid. As fire was the key to Lamarck's chemistry, so also was it central to his biological theories.

There are a number of possible influences of Lamarck's chemistry on the development of his evolutionary ideas. The chemical system was established in all its major points before Lamarck put forth his evolutionary theories. Gillispie overstated his case when he suggested that the transference or

60. See Lamarck's attack on the new chemistry on this point, *Hydrogeology,* p. 118; also tables in *Mémoires* following p. 226 and p. 348.

61. In Lamarck's system it was possible to explain the "reactivation" or reduction of a calx by the addition of more fixed carbonic fire, *Hydrogeology,* p. 119.

generalization from the chemical to the biological realm in connection with the primacy of fire, the continuity among minerals, and the antagonism between living organisms and the physical environment was the decisive factor in the formation of Lamarck's evolutionary thinking.[62] One must also take into account Lamarck's work in such other fields as botany, geology, invertebrate zoology, and paleontology. There were however, some general influences of Lamarck's chemistry on his later thinking. In his early work in chemistry, he developed a monistic conception of nature which stayed with him throughout his life. Lamarck thought of his various studies as part of a coherent system whose pieces he would gradually fill in. In addition to the breadth of his focus, Lamarck retained a consistent methodology; he was always concerned with the general concepts at the expense of the specifics, which he felt were not as important. There is one further possible general influence of the chemistry on the formation of his biological theories. Lamarck first presented his evolutionary ideas in the form of a descending order from the most complex animal to the simplest (1800).[63] The process described in 1800 is analogous to his view of "degradation" in the mineral kingdom. In 1802, Lamarck inverted the order for animals and plants so that it would correspond with the order nature had followed in time. He then needed a mechanism to propel change up the evolutionary ladder, and he introduced what was to be the major feature of his evolutionary theory, a natural tendency toward increased complexity. This natural tendency was the reverse of the one operating in the mineral kingdom, where he had first thought in terms of dynamic processes. Here the chemical theories may have been the inspiration for the biological ones.

It has been suggested that one reason for the hostility to Lamarck's evolutionary theories was a rejection of the chemistry on which they were based.[64] Lamarck's chemistry was generally ignored during his lifetime, but there was a very interesting attack on it after his death in the "Eloge" by Georges Cuvier in 1832.[65] Cuvier is known as the great opponent of Lamarck's evolutionary theories and his uniformitarianism. It is therefore somewhat surprising to find the longest attack in the "Eloge" directed against Lamarck's chemical theories. In addition to specific opposition to the latter, Cuvier associated the chemistry with Lamarck's evolutionary ideas and geology. Burkhardt has attempted to show how Lamarck's growing sensitivity from the reception accorded his chemical theories was ex-

62. Gillispie, "Formation," pp. 328–334, and "Lamarck," pp. 272–275.

63. Gillispie, "Formation," p. 325, points out the form of Lamarck's first presentation of evolution and stresses the influence of Condillac. He does not suggest any relationship between this idea and Lamarck's chemistry.

64. Ibid., pp. 323–324.

65. Georges Cuvier, "Eloge de M. de Lamarck, lu à l'Académie des sciences, le 26 novembre 1832," *Mémoires de l'Académie royale des sciences de l'Institut de France*, 13, 2 ser. (1831–1833), i–xxxi.

tended to his evolutionary theories.[66] Lamarck became convinced that the French scientific community was deliberately ignoring everything he said and would not accept his evolutionary conceptions no matter how much evidence he provided to support them. Lamarck's reaction, according to Burkhardt, was to present his evolution without the evidence he could have provided so easily. He was therefore partly responsible for the hostile reception of his evolutionary ideas.[67]

Lamarck's attitudes toward the established scientific community could be described as "anti-corporate," but his hostility to Lavoisier and the new chemistry ran much deeper and must be traced to certain intellectual changes in mid–eighteenth-century France. At that time, Buffon, Diderot, and others separated natural history from French "positivistic" Newtonian science. They claimed a new methodology for their natural history which specifically rejected any use of mathematics and instead stressed observation, experiment, and a wider role for hypotheses. The resulting field of natural history, which included the chemistry of Rouelle, has sometimes been described as "anti-Newtonian" or "Jacobin" science. Lamarck and Lavoisier grew up in this tradition, but Lavoisier had training in quantitative techniques. Lamarck, in contrast, absorbed the ideas about the unity of nature and the new methodology so thoroughly that they dominated all his work. Lavoisier was ready to appreciate the work of the English pneumatic chemists and to study gases, especially oxygen, and their reactions in precise quantitative ways. In Lamarck's mind, the progress of science required the rejection of the physico-mathematical tradition, whereas for Lavoisier progress lay in following that tradition along new paths.

66. Burkhardt, pp. 280–284.
67. Ibid., pp. 287–291.

II

SCIENCE, SOCIETIES, AND THE ENLIGHTENMENT

LAPLACE AND THE VANISHING ROLE OF GOD IN THE PHYSICAL UNIVERSE

ROGER HAHN

THERE is a widely told anecdote about God in the physical universe which often serves as a symbol for his disappearance. Napoleon, in one of his more tranquil moods, is purported to have asked his friend the astronomer Pierre-Simon Laplace about his recent book on the system of the world: "Newton spoke of God in his book. I have perused yours, but failed to find His name even once. How come?" Laplace retorted with a feigned reverence and a slight tone of arrogance: "Sire, I have no need of that hypothesis."[1]

Historians are justifiably wary of such anecdotes, often discarding them as rhetorical inventions dreamed up to illustrate a point or to defame Laplace. Nevertheless this legend, while not authentic in every detail, is probably a fair representation of a datable historic event. William Herschel recorded in his diary a visit to the French First Consul at Malmaison on 8 August 1802 which was witnessed by the chemist and then-Minister of Interior Jean Antoine Chaptal, the physicist Count Rumford and Laplace. Bonaparte asked Herschel "a few questions relating to astronomy and the construction of the heavens, to which I made such answers as seemed to give him great satisfaction. He also addressed himself to M. La Place on the same subject and held a considerable argument with him in which he differed from that eminent mathematician. The difference was occasioned by an exclamation of the First Consul's, who asked in a tone of exclamation or

A version of this paper was read at the Fourth International Congress on the Enlightenment, New Haven, 14 July 1975.

1. Augustus de Morgan, *A Budget of Paradoxes* (London, 1872), pp. 249–250 (originally appeared in *The Athenaeum*, no. 1921, 20 August 1864, pp. 246–247); and in Hervé Faye, *Sur l'origine du monde: Théories cosmogoniques des anciens et des modernes* (Paris, 1884), pp. 110–111. Its meaning is discussed without resolution by George Sarton "Laplace's Religion" and Jean Pelseneer "La religion de Laplace" in *Isis,* 33 (1941), 309–312, and 36 (1946), 158–160.

admiration (when we were speaking of the extent of the sidereal heavens) 'and who is the author of all this?' M. de La Place wished to shew that a chain of natural causes would account for the construction and preservation of the wonderful system; this the First Consul rather opposed."[2]

The event deserves to be taken seriously, not merely because it happened, but for its symbolic meaning as well. At the turn of the eighteenth century, Laplace was considered the most accomplished physical scientist of his time, having written extensively on the system of the world, on the mathematics of probability, and contributing to physics and chemistry in smaller but no less respectable ways.[3] He received the supreme accolade from his peers in 1827 when eulogists dubbed him the "Newton of France," one hundred years to the month after Sir Isaac's Westminster funeral.

No one disputed the display of his scientific prowess, but there were some who could not bear the juxtaposition of an atheistic, Jacobin-associated French Laplace with the pious Newton.[4] Already in 1804, John Robison, the influential professor of natural philosophy at the University of Edinburgh, had attacked Laplace for having openly parodied the Newtonian tradition. Robison pointed out that "M. de la Place carefully abstains, through the whole of this performance [the *Exposition du système du monde,* 1796], from all reference to a Contriver, Creator, or Governor of the universe, particularly in the present reflections, *which are so pointedly contrasted* with the concluding reflections of the great Newton."[5] Robison's observation was accurate. We know Laplace carefully expurgated from his writings all "irrelevant" references to the very Deity, who, for Newton, had been a direct object of concern in the famous General Scholium of the *Principia.* In the century that separated Laplace from Newton, God had vanished from the physical universe, and Laplace could well do without "that hypothesis."

How shall we assess and explain this change? To begin with, we must propose an interpretation for Laplace's reticence to discuss God and decision to exclude Him from the *Exposition du système du monde* and the *Mécanique céleste.* The task is delicate, for Laplace was secretive and more guarded in his comportment than Newton. The French Newton never published an antireligious General Scholium.[6] Indeed there is reliable contem-

2. John L. E. Dreyer, ed., *The Scientific Papers of Sir William Herschel,* 2 vols. (London, 1912), I, lxii.

3. Robert Fox, "The Rise and Fall of Laplacian Physics," *Historical Studies in the Physical Sciences*, 4 (1972), 89–136; and Henry Guerlac, "Chemistry as a Branch of Physics: Laplace's Collaboration with Lavoisier," ibid., 7 (1975), 193–276.

4. Jack B. Morrell, "Professors Robison and Playfair, and the *Theophobia Gallica:* Natural Philosophy, Religion and Politics in Edinburgh, 1789–1815," *Notes and Records of the Royal Society,* 26 (1971), 43–63.

5. John Robison, *Elements of Mechanical Philosophy* (Edinburgh, 1804), p. 685.

6. Laplace's most explicit reaction is in Book V, Chapter 6, of the *Exposition du Systéme du Monde, Oeuvres complètes* (Paris, 1878–1912), VI, 479–480. A manuscript he never intended

porary testimony that he deliberately opted for prudence when making public pronouncements.[7] He even instructed his younger colleague François Arago to refrain from using the Bonaparte anecdote when writing a biographical article.[8] Nonetheless, by piecing together private and unpublished correspondence, by reading between the lines, and above all by setting Laplace in the intellectual context of his times we may hope to reconstruct his position and thereby begin to grasp the historic development which he epitomizes. For while he may have avoided public statements about the relationship of religion to science, Laplace was no less aware of the metaphysical implications of his activities than all thinking men.

One should not be surprised at Laplace's inner feelings. Like so many other Enlightenment scientists, he had been trained in theological disputation as a youth, graduating from the University of Caen with an M.A. in the Faculty of Arts in 1769 on his way to ordination.[9] Jean d'Alembert, who was Laplace's most prominent and effective scientific sponsor, referred to him as an abbé as late as August of that year.[10] But soon thereafter, Laplace abandoned the priesthood for a career in mathematics and astronomy, retaining a private interest in metaphysical questions on the nature of human will and on causality, but maintaining a disdainful and critical attitude toward the hundreds of hack theologians and evangelical publicists he encountered throughout his life. Laplace followed the Voltairean line, believing that religion—specifically Christianity—was useful for maintaining the social fabric of society, but not necessary for the elite.[11] He considered his aged colleague the astronomer Joseph Lalande a fool, not so much for holding an atheistic position derived from reading the baron de Holbach, Claude Adrien Helvétius, and Charles François Dupuis' *Origines de tous les cultes,* but for proclaiming it publicly in the revised *Dictionnaire des athées,* in newspapers and even in free public lectures under the Pont-Neuf. Religion, like science, was a matter reserved for rational discussion only among experts. The exalted *demi-savant* was intolerable in either realm.

---

for publication, and which was most likely inspired by reading Newton, is discussed in Roger Hahn, "Laplace's Religious Views," *Archives internationales d'histoire des sciences,* 8 (1955), 38–40.

7. Augustin Cournot, *Souvenirs* (Paris, 1913), pp. 85–86; and letter of Laplace's son Emile to Jean Baptiste Dumas dated 8 February 1869 in Archives de l'Académie des Sciences, Paris, dossier Laplace.

8. Faye, *Sur l'origine du monde,* p. 111, n.1.

9. Archives Départementales du Calvados, Caen, MS D 1047, fol. 43r; and Léon Puiseux and E. Charles, *Notices sur Malherbe, La Place, Varignon, Rouelle, Vauquelin, Descotils, Fresnel et Dumont-D'Urville* (Caen, 1847), p. 36.

10. Letter of d'Alembert to Le Canu dated 25 August 1769 in Institut National de Recherche et de Documentation Pédagogique, Paris.

11. Pierre Sylvain Maréchal, *Dictionnaire des athées anciens et modernes* (Paris, an VIII), pp. 231–232; and in a copy of a letter written by Baron Jean Frédéric Théodore Maurice in Archives de l'Académie des Sciences, Paris, dossier Laplace.

Moreover the timing of Lalande's campaign ran counter to Bonaparte's rapprochement with Pius VII that eventuated in the 1801 Concordat. By using his authority as a scientist to speak on religion, Lalande was proving himself to be an annoyance to the ruler who had reasons to support both science and organized religion in France.

Laplace did not share the political responsibilities of the First Consul and could afford a different outlook. He had asserted that the progressive discovery of nature by scientific means was the supreme example of civilized man's intellectual capabilities. In a way reminiscent of Bernard de Fontenelle, d'Alembert, the Marquis de Condorcet, and Adam Smith, Laplace used the history of astronomy to illustrate his conviction that advancement in science came in part by the gradual abandonment of superstitious beliefs about the universe in favor of accurate and rationally discovered laws of the system of the world.[12] Science progressed through disagreements between experts who nonetheless agreed on a rational and experimental method of inquiry. By contrast, religious opinion was mired in hopeless scholastic debate which he had once rehearsed as a student of theology, but that held no promise of resolution in the future. Religious arguments could easily be equated with the inconclusive debates between Gottfried Wilhelm von Leibniz and Samuel Clarke or the interminable metaphysical arguments over the nature of force between advocates and opponents of *vis viva.* Laplace wholeheartedly endorsed d'Alembert's dismissal of such activities as fruitless, considering traditional philosophy a waste of time for those seeking to advance our understanding of nature. He represented and spearheaded that group of antimetaphysical scientists whose philosophical position Henri Gouhier has labeled prepositivist.[13] For Laplace the business of science is the advancement of well-founded, positive knowledge. The historical record suggested that metaphysical concerns often misled well-meaning natural philosophers, thereby hindering their progress. As a professional scientist he felt justified in deliberately excluding theology and philosophy from science whenever he could. In his own professional writings, there are no more references to the Deity than to the philosophic premises of René Descartes, Baruch Spinoza, Leibniz, Nicolas Malebranche, or David Hume, which one might have expected in the early days of the Enlightenment.

This deliberate refusal to engage in abstruse issues is both an example and a consequence of the growing autonomy of science from other realms of human concern. It was made possible by the confidence French scientists

12. Book V of the *Exposition du système du monde* is devoted to a short history of astronomy. It was printed separately as the *Précis de l'histoire de l'astronomie* (Paris, 1821).

13. His position is best displayed in the "Mémoire sur la chaleur" written with Lavoisier in 1783 in which he refuses to take sides on the differing theories about the nature of heat. See also Georg Misch, "Zur Entstehung des Französischen Postivismus," *Archiv für Geschichte der Philosophie,* 14 (1901), 1–39, and Henri Gouhier, *La Jeunesse d'Auguste Comte* (Paris, 1933), I, 123 *et seq.*

had developed during the late Enlightenment in their ability to establish laws of nature by their special methods of inquiry. But this justifiable pride of accomplishment was also tempered by a sense of the limits of the scientific approach. In the *Exposition du système du monde,* every time Laplace approaches an ontological issue, he reminds his reader that certain kinds of knowledge are inaccessible to science.[14] Human understanding is such that we will never know what the nature of force really is; nor what, if anything, stands behind the universal laws of gravitation; nor what explains molecular attraction. Following the tradition of skepticism, symbolized in France by Michel de Montaigne and reaffirmed by Fontenelle, he urges his fellow-man to understand his human condition and to resign himself to a state of perpetual uncertainty.

For Laplace and his peers, the most annoying aspect of Newton's assertion in the General Scholium and some queries of the *Opticks* was the claim that one may and ought to learn about religious matters from a scientific examination of nature. Following a remarkable series of assertions about the Lord, Newton had said: "And thus much concerning God, to discourse of whom, especially from the appearance of things belongs to experimental philosophy."[15] How could the sublime discoverer of the universal law of gravitation be so utterly reckless in his assertions, Laplaceans wondered? Perhaps, as Jean Baptiste Biot reported, Newton had suffered a mental setback after the exhausting effort required to produce the *Principia.*[16] Laplace was fond of pointing out that these probably mistaken and certainly misplaced notions appeared only in the second edition of the *Principia* in 1713, not in the original one.[17]

One of the meanings of the famous Laplacean phrase that God was a superfluous hypothesis may be understood against this Newtonian backdrop. The *Principia* was originally written without the benefit of a General Scholium. It took genius and heroic efforts to produce this grand synthesis, but Newton's reflections on God were a mere appendage, an afterthought. The discovery of the laws of the universe, which for Laplace is

14. Laplace, *Oeuvres,* VI, 152, 343, 392, 449, and 470.

15. Newton, *Mathematical Principles of Natural Philosophy and His System of the World,* trans. Andrew Motte, ed. Florian Cajori (Berkeley, 1966), p. 546. For a discussion of the British Newtonian tradition, see Hélène Metzger, *Attraction universelle et religion naturelle chez quelques commentateurs anglais de Newton* (Paris, 1938); F. E. L. Priestley, "Newton and the Romantic Concept of Nature," *University of Toronto Quarterly,* 17, no. 4 (July 1948), 323–336; J. E. McGuire, "Force, Active Principles, and Newton's Invisible Realm," *Ambix,* 15 (1968), 154–208; and Margaret C. Jacob, *The Newtonians and the English Revolution* (Ithaca, N.Y., 1976).

16. Jean Baptiste Biot, *Mélanges scientifiques et littéraires* (Paris, 1858), I, 264–289. I. Bernard Cohen refutes this contention in "Isaac Newton's Principia, the Scriptures, and the Divine Providence," *Philosophy, Science, and Method,* ed. Sidney Morgenbesser, Patrick Suppes, and Morton White (New York, 1969), p. 523.

17. Laplace, *Ouvres,* VI, 479, n. 1.

the proper business of the professional scientist, could be and indeed was made without invoking the attributes of God. He was in practice a separable entity, just as religion was separable from natural philosophy.

Behind these professional attitudes based upon a concept of the historic process required for the advancement of knowledge stood an even more fundamental notion about the epistemological status of modern science. How did the scientist of the late Enlightenment perceive his newly restricted realm of activity? For Laplace, the answer was framed in terms of a philosophic issue that was much debated in his youth. Science was still considered, as it always had been, an activity whose goal is the grasping of certain knowledge. But since essences are unknowable, knowledge must be based upon the accumulation of ascertainable facts and generalization from them to laws of nature. On this central issue, Laplace followed the implications of Newton's mathematical and empirical philosophy, rejecting the older Thomistic tradition of deducing nature from the essence of things. In mid-century, Enlightenment scientists such as d'Alembert and Condorcet had attempted to reconcile these traditions by postulating two kinds of laws in the universe, one essential and necessary, the other contingent and phenomenological.[18] The implication was that no matter how similar the laws appeared to be, they were separated by an ontological chasm that was unbridgeable. Starting with this difficulty, Laplace inverted the conundrum by a brilliant new approach that absorbed necessity into contingency.[19] In the process of transforming this philosophic issue, he was also brought face to face with the role and function of God. In spite of later attempts to abstain from discussing the subject, his view has remained in print for posterity to scrutinize.

Initially, d'Alembert had raised the issue rather sharply in the second edition of the *Traité de dynamique* (1758): "It is not a matter of deciding if the Author of nature could have provided it with laws other than the ones we observe; as soon as one allows an intelligent being capable of acting on matter, it is evident that this being can at any instant move or stop it at will, . . . The question raised thus comes down to knowing if the laws of equilibrium and motion we observe in nature are different from those which matter, left to itself, would have followed."[20] The job of the investigator then is to compare what ought to be with what is, or, to put it in its historical context, to decide if his empirical findings match what is logically

18. Giorgio Tonelli, "La nécessité des lois de la nature au XVIIIè siècle et chez Kant en 1762," *Revue d'histoire des sciences,* 12 (1959), 225–241; Roger Hahn, *Laplace as a Newtonian Scientist* (Los Angeles, 1967), pp. 14–18; and Keith M. Baker, *Condorcet: From Natural Philosophy to Social Mathematics* (Chicago, 1975), pp. 95–109.

19. Roger Hahn, "Determinism and Probability in Laplace's Philosophy," *Actes du XIIIè Congrès international d'histoire des sciences, 1971* (Moscow, 1974), I, 170–175.

20. Jean le Rond d'Alembert, *Traité de dynamique,* nouvelle édition (Paris, 1758), pp. xxiv–xxv.

derivable from essences. D'Alembert, following a tradition handed down by Descartes, associates necessity with God's rational creation, and contingency with his will, the latter being ascertainable through a study of phenomena. He asserts that the natural philosopher can in fact learn a good deal about God:

> If they are different from one another, he will conclude that the laws of Statics and Mechanics derived from experience are contingent truths since they follow the particular and express will of the supreme being; [or] if on the contrary the laws derived from experience correspond to those determined by reason alone, he will conclude that the observed laws are necessary truths—not in the sense that the Creator might not have been able to establish altogether different laws, but in the sense that he did not deem it worthwhile to establish laws other than those resulting from the existence of matter itself.[21]

Several years later, in a publication Laplace knew quite well, Condorcet elaborated this position even more clearly: "Bodies in motion seem to be subject to two essentially different types of laws. One is the necessary consequence of the concept we have of matter, the other seems to be the product [*l'effet*] of the free will of an intelligent Being who has willed the World to be as it is rather than any other way. The whole and consequence of necessary laws constitute mechanics. We call the whole and consequence of the other laws the 'System of the World,' which we could ascertain only if we knew those [the laws] of all phenomena."[22] He continues, as did d'Alembert, by comparing the two sets of laws, and draws a remarkable inference: "Perhaps these laws differ among themselves only because, given the present relationship between things and us, we require more or less shrewdness [*sagacité*] to know them; such that . . . one could conceive [of the universe] at any instant to be the consequence of what had to befall to matter once it had been set in a specific order and left to its own devices. In such a case, an Intelligence knowing the condition [*l'état*] of all phenomena at a given instant, the laws to which matter is subjected, and their consequences at the end of any given time, would have a perfect knowledge of the 'System of the World.'"

It was left to Laplace to transform this speculation, articulated to allow a view through the intricate web of inconsistent traditions, into the famous statement of materialistic determinism which was later often associated with his atheistic position.

The language Laplace used was obviously borrowed from Condorcet, though now the words were reshuffled to express a new idea:

> The present state of the system of Nature is evidently a consequence of what it was in the preceding instant, and if we conceive of an Intelligence who, for a

21. D'Alembert, pp. xxvi–xxvii.
22. Marie Jean Antoine Nicolas Caritat de Condorcet, *Lettre sur le systême du monde et sur le calcul integral* (Paris, 1768), pp. 4–5.

given instant, apprehends [*embrasse*] all the relations of beings in this Universe, it [the Intelligence] will be able to determine for any time of the past or in the future the respective positions, motions, and generally the disposition [*affections*] of all these beings.

Physical-Astronomy, that subject of all our understanding most worthy of the human spirit, offers a notion, albeit imperfect, of what such an Intelligence would be.[23]

There, in a nutshell, is a statement of Laplace's conception of the Supreme Being, now omniscient, rather than omnipotent as Newton had presented him.

If Laplace continued to hold a view of God in the image of such an Intelligence, are we fully justified in speaking of God's "vanishing role" in the physical universe? For Laplace personally, there are reasons to doubt that he freed himself completely from thinking about God when writing on fundamental aspects of science. But if we take a larger historical perspective, we may confidently answer in the affirmative, because of the implications stemming from the triumph of omniscience over omnipotence. For once scientists abandoned the interventionist God and replaced him with eternal laws, the idea of a universe without God became logically conceivable and followed naturally. Once created, the world could run by itself and as Jacques André Naigeon put it, "God is like an excess wheel [*une roue de luxe*]."[24]

This is no place to review the century-long debates between the necessitarians and voluntarists which continued to engage theologians throughout the Enlightenment.[25] It was one of the central issues that separated Leibniz from Newton's mouthpiece, Samuel Clarke, and it remained unsettled despite their epistolary exchanges.[26] Newton's God demonstrated His will, among other ways, by intervening in the solar system setting it straight so that it could not run down.[27] Leibniz' God created a cosmic clock which did not require His intervention. But God was necessary not merely once for the universe's creation, but constantly to maintain its operations following the three fundamental principles He infused into nature: sufficient reason,

23. Laplace, *Oeuvres,* VIII, 144. For details, see Roger Hahn, "Laplace's First Formulation of Scientific Determinism in 1773," *Actes du XIè Congrès international d'histoire des sciences, 1965* (Warsaw, 1967), II, 167–171.

24. Jacques André Naigeon, *Adresse à l'Assemblée Nationale sur la liberté des opinions* (Paris, 1790), p. 31, and in *Encyclopédie méthodique, Philosophie* (Paris, an II), III, 374–384.

25. Léon Brunschvicg, *L'expérience humaine et la causalité physique* (Paris, 1922); Reijer Hooykaas, *Religion and the Rise of Modern Science* (Edinburgh, 1973); and Margaret J. Osler, "Nature, God, and the Mechanical Philosophy," a paper presented at the December 1976 meeting of the History of Science Society.

26. F. E. L. Priestley, "The Clarke-Leibniz Controversy," *The Methodological Heritage of Newton,* ed. Robert E. Butts and John W. Davis (Toronto, 1970), pp. 34–56.

27. David Kubrin, "Newton and the Cyclical Cosmos: Providence and the Mechanical Philosophy," *Journal of the History of Ideas,* 28 (1967), 325–346.

plenitude, and preestablished harmony. Both Clarke and Leibniz, with good reasons, claimed a share of religious orthodoxy.

What Laplace and his contemporaries did was to eliminate the need for an interventionist God by showing the solar system to be stable just as Leibniz craved to do; and they turned their backs on the ontological concerns of the Leibnizians by following Newtonian empiricism. The former story needs no telling, for it is a standard part of the annals of astronomical research. It was one of Joseph Lagrange and Laplace's great achievements to have demonstrated (at least to their satisfaction) the stability of the solar system under the universal law of gravitation. The latter story remains to be understood. Let me suggest its elements.

Instead of concentrating on the rationalism of the universe illustrated by the phenomena of the system of the world, Laplace assumed the existence of invariable laws and set as his primary task their discovery on the basis of evidence. He invented (or rediscovered) Bayesian inductive probability, and by a strict application of its mathematical principles identified secondary causes through the study of their effects, disentangling real from spurious ones.[28] His immense success at this game of celestial mechanics—shared by Lagrange and lesser mathematical astronomers—gave him confidence that he was approaching the kind of knowledge he had postulated for the omniscient Supreme Intelligence. Theoretically, with sufficient evidence and the persistent use of rigorous analytic techniques, Laplace and his successors could come as close as they pleased to determining the laws of nature without recourse to an ontological metaphysics. One need no longer worry about the issues raised by d'Alembert and Condorcet. The example of celestial mechanics further provided psychological reassurance that mankind, using the sophisticated tools developed by scientists, and given sufficient time and data, could eventually grasp the laws of the universe with greater certainty than philosophers or theologians could ever hope to do. The methods of science, though never able to provide absolute certainty, seemed to promise a higher probability than those employed by system builders or devout believers.

The consequences of this position were quite far-reaching. Science was now in command of a powerful new approach to nature. Despite incomplete evidence, there was now a probabilistic calculus permitting the scientist to induce probable causes from known consequences. One could for example obtain the probability that a particular law governed a set of phenomena

28. Charles C. Gillispie, *Les fondements intellectuels de l'introduction des probabilités en physique* (Paris, 1963), and "Probability and Politics: Laplace, Condorcet, and Turgot," *Proceedings of the American Philosophical Society,* 116 (1972), 1–20; Hahn, "Determinism and Probability,"; Baker, pp. 165–171; and L. E. Maistrov, *Probability Theory: A Historical Sketch,* trans. Samuel Kotz (New York, 1974), pp. 87–100.

even though one might never possess all the evidence to make the judgement absolute. Hence it was no longer worthwhile for scientists to concern themselves about ontological questions. Laplace was in a way echoing the Malebranchian tradition of considering scientific laws as descriptive relations rather than as statements about the essence of nature.[29] But unlike Malebranche, he saw no chasm between the two types of laws.

A second and even more startling consequence followed. Since the scientist was unable to arrive at essences, and since secondary causes were in theory easily discoverable from empirical evidence, nothing was gained by searching for first causes. Laplace often noted that assertions about first causes were a convenient rhetorical technique for hiding man's ignorance. He reasoned that if the search for first causes was destined to failure, there was no need to look for such hypotheses to grasp the workings of nature! One should willingly abandon the illusion of the certainty of first causes for the reality of highly probable secondary ones.

The most fruitful illustration of Laplace's approach is evident in his search for a unitary cause of the origins of the solar system. The details of his hypothesis are less important than the procedure he followed to claim its plausibility. Far from asserting that God created the solar system (and was therefore the First Cause), Laplace merely offered a likely hypothesis for its beginnings, initially as a mere footnote.[30] By contrast to Newton who used the same regularities of the behavior of planets as evidence for design, Laplace employed the data, treated in a Bayesian fashion, to assert the high probability that a single physical cause was responsible for the origins of the solar system. He deliberately did not refer to the Supreme Being, but offered the nebular hypothesis as a plausible set of initial conditions from which the configuration of the solar system might be derived.

If we refer to Herschel's description of the confrontation between Bonaparte and Laplace, we will now grasp its overall significance better. Whereas the First Consul wanted to know *who* created the solar system, Laplace "wished to shew that a chain of natural causes would account for the construction and preservation of the wonderful system." Philosophers, theologians, and even political leaders might still be concerned with Who was the Author of this world. That question could not be answered by

29. Thomas L. Hankins, "The Influence of Malebranche on the Science of Mechanics during the Eighteenth Century," *Journal of the History of Ideas,* 28 (1967), 193–210.

30. Laplace, *Exposition du système du monde* (Paris, an IV), pp. 301 *et seq.* The important changes in subsequent editions are discussed by Remigius Stölzle, "Die Entwicklungsgeschichte der Nebularhypothese von Laplace," *Abhandlungen aus dem Gebiete der Philosophie und ihrer Geschichte. Eine Festgabe zum 70. Geburtstag Georg Freiherrn von Hertling* (Freiburg im Breisgau, 1913), pp. 349–369; and Jacques Merleau-Ponty, "Situation et rôle de l'hypothèse cosmogonique dans la pensée cosmologique de Laplace," *Revue d'histoire des sciences,* 29 (1976), 21–49, and 30 (1977), 71–72.

Laplace. But scientist could strive to explain *how* the solar system was likely to have been organized and how it functioned.[31]

One must agree with Robison's assessment that God's role as "Contriver, Creator or Governor of the Universe" was missing from Laplace's system of the world. But must one necessarily assume that this turned Laplace into an atheist? Nowhere in his writings, either public or private, does Laplace deny God's existence. He merely ignores it. For Laplace as an astronomer God has truly become a superfluous hypothesis.

31. Victor Monod, *Dieu dans l'univers* (Paris, 1933), pp. 190–209.

NEWTON'S INSIDE OUT! MAGIC, CLASS STRUGGLE, AND THE RISE OF MECHANISM IN THE WEST

DAVID KUBRIN

DESPITE the outpouring of Newtonian scholarship in recent decades, Isaac Newton, the culminating figure of the seventeenth century scientific revolution, remains an enigma, his legacy anything but clear. While William Blake saw Newton as one of the major figures responsible for the eighteenth century's enthronement of Reason, Lord Keynes insisted rather that Newton was one of the last of the *magi*.[1] Since Keynes's time rationalizers have tried to deemphasize the occult aspects of Newton's thought, but his magical and Hermetic interests, researches, and aspirations are, I think, beyond question.[2]

Still not adequately explained, however, is what role, if any, did Hermetic or magical ideas play in Newton's natural philosophy—in his theory of the cosmos, of matter, of causation, of space and time? Additionally, why has his interest in such matters been until recently a well-kept secret, known only, as it were, to the initiates? These inquiries form the heart of my essay.

I argue that despite the clouds of obscurity surrounding them, the various aspects of occult thought (Hermetic, magical, alchemical, and so on) were crucial to what is meant by Newton's natural philosophy; moreover, his deep interest in magic and its significance for his thought were hidden from

This work was done on the basis of support from the John Simon Guggenheim Foundation.

1. Lord Keynes, "Newton the Man," *Newton Tercentenary Celebrations* (Cambridge, Eng.: Royal Society, 1947), pp. 27–43.

2. See, for example, Marie Boas and A. Rupert Hall, "Newton's Chemical Experiments," *Archives internationale d'histoire des sciences, 11* (1958); also, Richard Westfall, "Isaac Newton: Religious Rationalist or Mystic?" *Review of Religion,* 22 (1958), 155–170; see, however, the latter's more recent assessment in "Newton and the Hermetic Tradition" in *Science, Medicine, and Society: Essays to Honor Walter Pagel,* ed. Allen G. Debus (New York, 1972), II, 183–198. More recently see J. E. McGuire, "Neoplatonism and Active Principles: Newton and the *Corpus Hermeticum,*" in Robert S. Westman and J. E. McGuire, *Hermeticism and the Scientific Revolution* (Los Angeles, 1977).

the public until very recently not merely nor even mainly because of Newton's eighteenth-century literary executors' trying to preserve his good name by selecting for publication only those manuscripts which did nothing to lessen his image as a pious member of the Church of England. Even more important was Newton's *own* censorship, his repression of his own insights, ideas, visions, and grand plan of the cosmos. The reason for this repression, I shall argue, was largely social, and stemmed from the fact that Newton realized the dangerous social, political, economic, and religious implications that would be associated with him should he dare reveal his true thoughts.

I

Isaac Newton was born in 1642 at Woolsthorpe in Lincolnshire, as England's civil war was breaking out. His father having died before he was born, Newton was forced to help his mother run the small manor house and farm she owned. Even then, in school and on the manor, young Newton showed a fascination with things mechanical, building toy clocks, sundials, and treadmills, for example. His enormous temper and rages were also evident at an early age. So too was his dreaminess, his tendency to become totally oblivious to his surroundings when thinking.

At the end of his schooling in nearby Grantham with the local apothecary, from whom it may be presumed Newton had an early introduction to chemical phenomena (and possibly chemical philosophy), he proceeded to Trinity College at Cambridge University in 1661. We know that soon after arriving he sought out the Cambridge Platonist Henry More, also from Grantham, for discussions on theological matters. More, who had played a crucial role in popularizing Descartes' philosophy in England, was a decisive influence on Newton; his writings about the immateriality of the soul, the existence of immaterial forces, the immanence of God, the nature of space, and the insufficiency of inertia to explain all the behavior of matter, were read and absorbed by the young Newton.[3] Newton's early college career was relatively obscure. In his last two years at Cambridge, however, he began studying with the natural philosopher and mathematician Isaac Barrow, under whose influence his talent and genius began to emerge.[4]

By the mid 1660's when Newton graduated from Cambridge, a complex code of behavior had emerged to govern the still young study of natural philosophy in England. At the time of the 1660 Restoration, the settlement of the civil war and the end of the Interregnum, there had been a powerful move by the propertied and educated classes to prevent a recurrence of the frightening class strife that surfaced during the war, by wiping out what was

3. See the references in Newton, *Question[e]s quaedam Philosophiae*, Cambridge University Library MS Add. 3996 fols. 85–135.

4. Frank Manuel, *A Portrait of Isaac Newton* (Cambridge, Mass., 1968).

seen as the intellectual source for that strife—"enthusiasm." Not only was the Clarendon Code used to suppress the various non-Anglican sects (many of which had emerged as revolutionary forces during the civil war), driving many of them out of existence, and denying non-Anglicans access to the churches, schools, and governing bodies in England, but an extensive, if less visible, campaign against enthusiasm was undertaken, a campaign which was just as important in the establishment and perpetuation of law and order at the Restoration. Leading the drive against enthusiasm was a strange alliance of prominent members of the Church of England and the scientific community, seeking to ensure the triumph of "moderate" over "enthusiastic" ideas.[5]

I want to return to this campaign later, but first I want to discuss in some detail just who the parties were and what questions were involved in the English civil war and the Commonwealth period that followed the beheading of the Stuart King Charles I until the Restoration. Though the major conflict was between the Royalists and the Parliamentarians, there was a more fundamental fight going on which pitted the people of property, on both sides, against an angry mass of the "masterless" people in England, as Christopher Hill has called them.[6] Hill's brilliant book, *The World Turned Upside Down,* clearly shows how fundamentally important for the whole civil war this latter conflict between the possessors and the dispossessed in seventeenth-century England was, especially since it was necessary for Parliament to get allies for the fight against Charles I from the masses of people. The contradiction between the propertied classes and the various beggars, vagabonds, squatters, immigrants, criminals, peddlars, hired laborers, and many small craftspeople sometimes was literally explosive, as when Cromwell's New Model Army in 1645 and 1649 had to fight armed units of angry "masterless" people who were trying to raise—by words and by arms alike—different questions from those debated publicly by the Parliamentary spokespeople.[7] For example, Gerrard Winstanley, the Digger spokesman, complained about the limits put on political debates in Parliament: "While this kingly power reigned in one man called Charles, all sorts of people complained of oppression. . . . Thereupon you that were the gentry, when you were assembled in Parliament, you called upon the poor common people to come and help you. . . . That top bough is lopped off the tree of

5. See M. C. Jacob, "The Church and the Formulation of the Newtonian World-view," *Journal of European Studies,* 1 (1971), 128–148; David Kubrin, *How Sir Isaac Newton Helped Restore Law 'n' Order to the West* (privately printed, 1972), hereafter referred to as *HSINHRL'n'OTTW*—a copy has been placed in the Library of Congress; P. M. Rattansi, "Paracelsus and the Puritan Revolution," *Ambix,* 11 (1963), 24–32. An important contemporary attack on "enthusiastic" ideas was Henry More, *Enthusiasmus Triumphatus* (London, 1662; 1st ed., 1656; reprinted ed. Los Angeles, 1966).

6. Hill, *The World Turned Upside Down. Radical Ideas during the English Revolution* (New York, 1972), pp. 12, 32.

7. Ibid., p. 20.

tyranny, and the kingly power in that one particular is cast out. But alas, oppression is a great tree still, and keeps off the sun of freedom from the poor commons still."[8] Even when the fight between the moderates and the more radical, dispossessed classes was not so outwardly visible, when there were no armed confrontations, it still seethed beneath the apparently calm surface.

As victims of the changing economic and social conditions in seventeenth-century England, the masterless people not surprisingly insisted on discussing questions intimately connected with their deteriorating situation. The rapid development of the English wool industry and the huge market for land which developed from Henry VIII's confiscation and selling of the monastic holdings brought about a rapid transformation of agriculture in England in the sixteenth and seventeenth centuries. Vast amounts of new land for the profitable sheepfarming were also made out of formerly arable land, newly drained fens, or recently deforested areas. This process forcibly removed countless peasants from areas on which they had depended for food by hunting or fishing in the forests and fens or by raising vegetables and livestock in the common land now being enclosed by the large landowners.[9] Most of those displaced became either landless farm laborers, forced for the first time to live off wages they could sometimes earn by working for the gentry, or beggars and vagabonds, wandering around until perhaps finally settling in one of the towns. The towns in turn became engorged with the human waste of this economic development. When many of these people were unable to find employment in the towns and cities, they justifiably became angry; and at a time when *some* of the fundamental institutions and values of the society were being challenged, they demanded that *all* of them be questioned.

Some of these displaced people, especially in southwest England and East Anglia, eventually became workers in the rapidly expanding wool industry. Though in the seventeenth century England was not a capitalist society, certain advanced sectors of its economy were organized along capitalist lines, and of none was this more true than the cloth industry. Under the direction of the clothiers, wool was produced by an increasingly defined division of labor—for example, spinners, weavers, dyers, fullers, shearers, and others. Already by the first part of the sixteenth century it was common in many places for all these different kinds of cloth workers to work in one building.

Mining was another rapidly changing industry organized along capitalist lines. Increased trade necessitated greater amounts of precious metals, increased warfare, greater amounts of iron and lead. In England, however,

8. Quoted in Ibid., p. 107.

9. Ibid., p. 42; Karl Marx, *Capital* (London, 1970), vol. I, part 8, ch. 27–30; A. L. Morton, *A Peoples' History of England* (London, 1968), pp. 166–70.

coalmining especially developed rapidly in the sixteenth and seventeenth centuries, the tonnage of coal mined increasing roughly tenfold from 1590 to 1680. Not only were more mines opened, but each mine exploited more seams and went deeper in its search for coal,[10] many of the mines financing their expansion by selling stock on the market, consolidating the role of investment capital in this critical field. Other consequences of the deeper mines were poisonous and explosive gases replacing the air in the shafts, flooding of many of the shafts, and cave-ins becoming increasingly common. Here too a more complex division of labor was necessary, as well as the articulation of an emerging hierarchy.

The rise of the cloth industry, coalmining, and large-scale farming (sheep and arable as well) are important for us in two regards. First, they indicate the development of sectors of the economy along capitalist lines in the sixteenth and seventeenth centuries, and help us understand some of the ideas and demands which emerged in the course of the civil war among workers in these industries and others. Second, in all three of these areas, it was necessary to establish a whole new relationship between the productive forces and nature, a relationship which would eventually come to dominate capitalist relations of production in general. That is to say, for the first time in England the earth was seen primarily as a source of profits by an increasingly powerful sector of the economy. The shift from an economy largely of subsistence and balance to one of consciously pursued growth and profits was brutal—no less so to the earth than to the many people forced off their land and into beggary or wage-labor.

It is my conjecture that this relationship with the earth—not wholly new in European history but for the first time a major driving force in England's historical development—underlay the transformations of the ways European natural philosophers began to conceive of the nature of space, time, and matter. It was at this time that scientific discourse was taken over by the new mechanical philosophy, which saw nature as composed only of totally inert, dead particles. When the basic constituents of the cosmos—the atoms and molecules—were seen as already dead and purposeless, it was, I think, harder to argue against the owners of capital who were driving both their workers and the earth itself to the breaking point—literally, for in many of the mines, cave-ins were frequent.

II

I will come back to the question of the nature of matter after outlining, as described by Christopher Hill, the ideas and ideologies of the various

10. J. U. Nef, *The Rise of the British Coal Industry* (London, 1932), vol. II, part 4, ch. 3–4.

masterless people set in motion by the process of economic development and by the civil war.[11]

The forces of the left were largely organized into Independent sectarian religious groupings (Familists, Muggletonians, Behemists, Fifth Monarchy Men, Levellers, Diggers, Anabaptists, Ranters, Seekers, Friends, among others) whose power and positions shifted in rapid kaleidoscopic succession during the chaotic period of the 1640's. In their world of intense political and moral fervor, people easily diametrically changed allegiances and views. For example, militant Puritans pursued their quest for righteousness so deeply that they ended up questioning the very existence of God, many wondering whether there existed any force in the universe save nature. The sects demanded religious toleration, and the abolition of tithing as well as an end to the state church. They thought that ministers, like their congregation in general, should earn their way through work; the regular clergy, at the time, was held in fairly general contempt by masses of people. Many people saw the Bible as an allegory, and, like the Seekers, began to doubt all religious dogmas, ceremonies, and churches.[12]

Women played an important role in the left groupings, many serving as ministers or spokespersons for their congregations. They led an attack on the scarcity of educational opportunities for women and on the monogamous family. Then, as now, the raising of questions having to do with the social role of women also brought about a reexamination of sexuality and morality in general. One woman sectarian claimed that adultery (as well as murder, and theft) was not a sin.[13] "Whoring" and drunkenness were defended, especially by the Seekers, some of whom called those acts sacraments. Nakedness was not uncommon, even during church services. The Ranters were said to hold Dionysian orgies.[14]

Some of the radicals claimed sin was not merely an illusion, but was

11. I will have to simplify greatly—rather than giving a detailed historical analysis and description of the different groupings which emerged in the years of civil war and Commonwealth, I can merely give a composite picture of the kinds of ideas and demands found among these people. Since my purpose is mainly to emphasize the significant *class* questions that were raised by the civil war, like Hill I will consider their voices in isolation from their many contemporaries who were raising altogether different demands. It is not my intention to imply that these were the only questions brought to the fore by the civil war.

12. Hill, *World Turned Upside Down,* pp. 153–154, 138–140, 148, 168, 183, 78–81, 23–26, 82–83, 22, 154.

13. Ibid., pp. 250–251, 254, 110; K. V. Thomas, "Women and the Civil War Sects," *Past and Present,* no. 13 (1958), 42–62.

14. Hill, *World Turned Upside Down,* pp. 60, 151, 160, 172, 256, 161, 255. Many pre-Christian groups in England and Europe had practiced sex-magic, in which sexual acts were seen as sacramental. While suppressed by Christianity, this tradition did not altogether die out, and, as we see here, seems to have flowered in times of social upheaval. For an illuminating discussion of this tradition, see Arthur Evans, *Witchcraft and the Gay Counterculture* (Boston, 1978).

caused by class society, invented by the ruling class to keep people in submission and to justify their own private ownership of property. Winstanley called for an end to class rule, private property—the basis of Adam and Eve's Fall—and wage labor, which was the cause of sin. Lawrence Clarkson, the Ranter, saw the nobility and the gentry as the cause of oppression. Sectarians in 1650 demanded an end to enclosures—the process by which country land held in common and used by the laboring and poor people for their subsistence for hundreds of years was being fenced in and expropriated by the gentry and aristocracy for their private use—land reform, and collective farms. Abiezer Coppe, the Ranter, demanded a redistribution of property so that all had an equality of goods and lands. George Foster, besides calling for an expropriation of the property of the rich, predicted an international revolution which would lead to the destruction of the Pope and the beginnings of a classless society everywhere.[15]

Not only were the propertied classes the enemy, all hierarchy in society was illegitimate. Winstanley saw God as everywhere immanent, in nature as well as in people. The Fifth Monarchy Men thought that the execution of Charles I in 1649 had prepared the way for Jesus Christ to appear, to reign over the Kingdom of Heaven in England: several times the Fifth Monarchists took up arms to remove ungodly magistrates they saw as obstructing this process.[16]

What is most important for us here is the view of nature held by the left movements. There was a veritable explosion of works on the occult during the years of social upheaval, no doubt partly because of the end of censorship.[17] The works of many continental *magi* and alchemists such as Jacob Boehme, Agrippa, and Paracelsus, were translated into English, as were some of the Rosicrucian manifestos. Even Robert Fludd, the English Rosicrucian, was able to get his works published in England for the first time. The decade from 1650 to 1660, in fact, saw more Paracelsian, alchemical, and astrological books translated into English than in the whole preceding century.[18]

More than a temporal link between the general interest in these kinds of works and the ideas of the sectarians existed. The sectarians' ideas about nature stemmed in part from the view they had of God. To many of the left

15. Hill, *World Turned Upside Down,* 134–135, 166, 131–132, 116, 106, 172, 96, 100–101, 103–105, 170, 180.

16. Ibid., pp. 22ff, 111–113, 143, 77–78, 102; David Kubrin, "Providence and the Mechanical Philosophy: The Creation and Dissolution of the World in Newtonian Thought. A Study of the Relations of Science and Religion in Seventeenth-Century England," (Ph.D. diss., Cornell University, 1968), chap. 4.

17. Allen G. Debus, *The Chemical Dream of the Renaissance* (Cambridge, 1968), p. 26; Hill, *World Turned Upside Down,* pp. 14, 71–72; Rattansi, "Paracelsus and the Puritan Revolution."

18. Debus, *Chemical Dream,* p. 26.

sectarians, God was not an outside agent—(s)he was a force inside all people, or at least in those who believed. Not only was God within people, so too was (s)he a force found in other creatures, and indeed in all objects too—chairs, rocks, and matter in general according to Clarkson. Matter as such is good. To the Ranters and Muggletonians, and George Fox in the 1640's, the only God was nature. Winstanley said that the only preachers needed were the various objects of creation. The Ranters asserted the unity of all created things.[19]

The widespread pantheism—the identification of God with nature—found among the left sectarians seemed to make many of them sympathetic to various forms of occult and magical thought. Among the classes from which the sectarians came a tendency toward occult ideas had long existed. Squatters in the forested or pastoral regions had long been sympathetic to witchcraft, for example, and women healers or "cunning men" who healed were commonly found in the villages; magic and alchemy were often associated with popular healing.[20] Many sectarians were enthusiastic readers of German mystical theology, or of the ideas of Jacob Boehme, an itinerant shoemaker-preacher-alchemist.[21] Members of the Familists, as well as many of the followers of Boehme, thought that alchemy was an outward symbol of the internal regeneration of truly pious mystics as they learned more and more of the secrets. One alchemist thought that the widespread production of gold, by destroying the basis of exchange, would destroy the foundations of the antichristian Beast of Bablyon.[22]

The influential astrologer William Lilly perhaps had Familist leanings. Familists were often linked with alchemy and astrology, according to Keith Thomas' work on seventeenth-century magic, especially in the work of John Everard (1575–1650), the translator of the works of Hermes Trismegistus, the mythical source of the magical corpus. Lawrence Clarkson, the Ranter leader, was a practicing astrologer.[23] Nicholas Culpeper, who fought in the New Model Army, was a follower of the alchemy of Paracelsus and a practicing astrologer, and was denounced as a Seeker.[24] Gerrard Winstanley described the true spiritual path in terms very close to alchemical: "To know the secrets of nature is to know the works of God . . . and indeed if you would know spiritual things, it is to know how the spirit or power of wisdom and life, causing motion or growth, dwells within and governs both the several bodies of the stars and planets in the heavens above; and the

19. Hill, *World Turned Upside Down,* pp. 160, 165, 172, 139–140, 143, 165, 112, 119–120.
20. Ibid., p. 70.
21. Ibid., pp. 118, 141, 302; also, Debus, *Chemical Dream.*
22. Hill, *World Turned Upside Down,* p. 233.
23. Ibid., pp. 72–73, 149, 232–233, 240.
24. Ibid., p. 240; F. N. L. Poynter, "Nicholas Culpeper and His Books," *Journal of the History of Medicine,* 17 (1962), pp. 159–160.

several bodies of the earth below, as grass, plants, fishes, beasts, birds and mankind."[25] William Dell and John Webster, during the civil war both severe critics of the monopoly over education held by the propertied classes, were both also enthusiasts of magic.[26]

I do *not* want to claim or even imply, however, that all (or necessarily even most) of the people who practiced or sympathized with the various occult fields were on the left *or* belonged to the dispossessed classes in England at the time. Unquestionably there were representatives of the propertied classes and people with conservative politics with similar interests. What I *am* claiming, however—and I think it needs emphasis—is that magical teachings were *widely viewed* by contemporaries as both *radical in politics* and *lower class* in following. In 1664, John Heydon the Rosicrucian wrote that the English civil war "admitted stocking-weavers, shoemakers, millers, masons, carpenters, bricklayers, gunsmiths, porters, butlers etc. to write and teach astrology and physic."[27] In the reaction against the previous explosion of magical interest, which began to set in in the mid 1650's and continued through the Restoration, critics such as John Wilkins and Samuel Parker saw a strong tie between the various occult ideas they attacked and levelling, Familism, and the enthusiastic political sectarians in general.[28]

III

Why did such a link exist between the popular forces which emerged during the civil war and turmoil of the 1640's and 1650's and the ideas of magic and Hermeticism? For what reason did the radicals, in the midst of important debates regarding collective farming or the abolition of private property, have an interest in magic?[29] Though the subject cannot be fully discussed here, I can tentatively suggest three links, one organizational, another ideological or philosophical, and a third psychological.

First, both the political radicals and the proponents of magic were united in their desire to destroy the state church. For the radicals, the state church was a fundamental barrier preventing any real political, economic, or religious democracy from developing. The church hierarchy both reinforced the existing political and property relations and extended their hegemony to all spheres of life. The sectarian groupings tried to build democratic alterna-

25. Quoted by Hill, *World Turned Upside Down,* pp. 114, 240–241.
26. Ibid., pp. 233, 240, 242, 244.
27. Quoted in Ibid., p. 240.
28. Rattansi, "Paracelsus and the Puritan Revolution"; I have dealt with this subject in greater detail in *HSINHRL'n'OTTW,* pp. 17–22; see also Hill, *World Turned Upside Down,* pp. 237, 244.
29. Winthrop S. Hudson, "Mystical Religion in the Puritan Commonwealth," *Journal of Religion,* 28 (1948), 51–56.

tives in their churches.[30] Similarly, the religious mysticism of the Hermeticists was based on the notion that any individual—woman or man, poor or rich—was able to attain spiritual enlightenment; a rigid, national hierarchy contradicted this aim.

Second, and more germane to the questions of natural philosoply being discussed here, the political radicals and the Hermeticists had a similar analysis of the world and method of logic. That analysis and method, in brief, was dialectics. Recent research has shown that the philosophy and method of the Hermeticists and other occult schools of the civil war era were based on a dialectical approach to reality.[31] That is to say, entities and relations were not viewed statically, but as in a perpetual process of development. This perpetual development was seen to occur primarily because every entity and relationship, in itself, was inconsistent, contained within itself its own opposite. Thus, whereas traditional Aristotelian logic claims that something must be *either* "A" *or* "not-A," dialectical logic holds that something can, as a result of its internal inconsistency, be *both* "A" *and* "not-A," both bad and good, hot and cold, wet and dry, or destructive and constructive. Because of this, opposites are not seen as static, as always existing in opposition, for at times cold and hot, or constructive and destructive, pass over into one another in what later philosophers were to call a "unity of opposites." For example, in 1652 the alchemist Thomas Vaughan wrote of the first matter of the Creation that it was "a miraculous *substance* . . . of which you may affirme *contraries* without *Inconvenience.* It is very *weake,* and yet most *strong,* it is excessively *soft,* and yet there is nothing so *hard.* It is *one* and *all: spirit* and *body; fixt* and *volatile, Male* and *Female: visible* and *invisible.* It is *fire* and *burns* not: it is *water* and *wets* not, it is *earth* that runs, and *Aire* that stands still."[32]

One crucial consequence of the internal contradiction is its resultant tension, which is the fundamental reason why all things and situations change. Thus, the fundamental cause for change is *internal* to an entity or situation, and not some agent acting from *outside*; things change *because of themselves*. Political, social, and religious radicals might well have been attracted to dialectical logic because of this emphasis on the impermanence of all things and the notion that the impetus for change comes from within.

30. Hill, *World Turned Upside Down,* p. 240.

31. Kubrin, *HSINHRL'n'OTTW;* A. L. Morton, *The Everlasting Gospel: A Study in the Sources of William Blake* (London, 1958), chap. 2; Joseph Needham and Wang Ling, *Science and Civilization in China,* vol. 2, *History of Scientific Thought* (Cambridge, 1956), chap. 10 and especially pp. 75–77; Hill, *World Turned Upside Down,* p. 287; Carolyn Merchant, *The Death of Nature: Women, Ecology, and the Scientific Revolution* (New York, 1980). I have greatly benefitted from talking about this subject, and about her researches for an earlier period than mine, with Ms. Merchant.

32. [Vaugha]N, [Thoma]S, *Aula Lucis, or the House of Light . . .* (London, 1652), p. 14.

In times of social turmoil, I would suggest, those favoring the status quo argue, wishfully, that things are permanently fixed, seeking to prove it on metaphysical, scientific, and logical grounds. On the other hand, those who are trying to overturn the existing social order eagerly seek out a different metaphysics, science, and logic, which demonstrate the world's mutability, which dialectical logic does quite well. In both instances, the frame of nature is not only a suitable, it is a preferred, referent.

The importance of all this should not be overlooked. For if this model is correct, political radicals two centuries before Karl Marx were using a logic based on dialectics—though, to be sure, it was more often than not a dialectical idealism rather than a dialectical materialism.[33] G. W. F. Hegel, Marx's source for his dialectical logic, seems to have understood this early history, for in his history of philosophy, philosophers who are today considered of primary importance, such as Thomas Hobbes or George Berkeley, are discussed in only a few pages (four and five respectively), while on the other hand Hermetic thinkers such as Giordano Bruno and Jacob Boehme, largely ignored by scholars today, receive lengthy discussions (eighteen and twenty-nine pages, respectively). Hegel, interestingly enough, was highly critical of the magical philosophers, however, because of their "enthusiasm."[34]

Third, it would seem that this phenomenon of "enthusiasm" has its foundation in the conception of the world held by Hermeticists. What is "enthusiasm"? Why was it so feared by the propertied classes? Perhaps most alarmingly, enthusiasm seemed to mean a call to, or at least a justification for, rebellion. Religiously, it was the free spirit let loose, without authority, or without any kind of authority save what the enthusiast was able to find in his or her own spirit; it was often a speaking in tongues. It meant affirming the testimony of the senses "in excess"—whatever that meant in a mid-seventeenth-century context. At a time of popular uprisings it meant also land seizures, strikes, armed confrontations, and insubordination in the armed forces. I would suggest that the central phenomenon of *activism* of all kinds among the enthusiasts found support in the Hermetic concept of the inher-

33. According to the English Marxist Joseph Needham, an important parallel to this existed in China, in the ideology and political practice of the Taoists, who for centuries led antifeudal rebellions and peasant resistance, and who also had a dialectical logic and science. In passing Needham also comments on the English phenomena my essay explores. (Needham and Wang Ling, *Science and Civilization,* chap. 10 and especially pp. 89 ff. and 97–98.)

34. Hegel, G. W. F., *Hegel's Lectures on the History of Philosophy,* trans. E. S. Haldane and Frances H. Simson (London, 1963), pp. 115–116, 122. It is not clear whether Marx realized this early history of dialectical thinking, though his doctoral dissertation in many respects was an attempt to trace the concept of dialectical change in ancient Greek atomism. Mao-Tse-Tung in his essay "On Contradictions" does mention that dialectical thought has roots in popular thought reaching back hundreds of years. (Mao Tse-Tung, "On Contradiction," *Selected Readings from the Works of Mao Tse-Tung* [Peking, 1967], p. 74.)

ent activity in all matter. That is, a conception of the world's being inherently active, full of gods,[35] and constantly changing helped develop people's self-confidence, and perhaps encouraged them (as did, indeed, Protestantism in general, but not always as "extremely" as enthusiasm did) to step forth to act, to transform the world, rather than allowing themselves to remain passive in the face of the great social transformations then sweeping England. Above all, enthusiasm was life-affirming. What a contrast with its successor, the mechanical philosophy!

IV

The reaction against "enthusiastic" science which took place in the mid 1650's and early 1660's easily fit in with the growing conservatism of much of English society, especially among members of the propertied classes, who were alarmed by disarray in society and the raising of far-reaching questions about society, religion, class, and property. By 1660, this conservatism had grown strong enough so that with the death of Oliver Cromwell, and the failure of anyone else to replace him, a newly called Parliament invited Charles II back to England to reestablish the monarchy.

The Restoration Settlement of 1660, however, was not simply a turning back of the clock. Many of the crucial institutions—the monarchy, the House of Lords, the Church of England—were reestablished, but on wholly new grounds. In short, the civil war had accomplished substantial changes to England, especially to its social structure. The power of the monarchy, and of the remnants of the feudal aristocracy, was severely limited. Power was transferred, in effect, to Parliament, a Parliament to be dominated for a long time by the landed interests of the gentry. By the end of the seventeenth century and on into the eighteenth century their new position as the dominant social force in England became increasingly clear. Sharing that power with them, as junior partners, were the propertied classes of the city, representing the new role of merchants and mercantile capital; the fight between capital (by then, industrial and finance capital) and the landed forces for control would be deferred until later on in the nineteenth century.[36]

As early as 1654 many of the leading natural philosophers were discovering the intellectual basis for rejecting "enthusiastic" science, turning to the new mechanical philosophy of René Descartes, Pierre Gassendi, and (to a much lesser extent in England) Thomas Hobbes. From the mid-1650's on, we see a series of conversions to the new philosophy of "matter and motion," very often by natural philosophers who had earlier been sympathetic

35. The root for "enthusiasm," *enthousiasmos,* means "full of gods." (Robert M. Pirsig, *Zen and the Art of Motorcycle Maintenance* [New York, 1974], p. 296.)

36. Morton, *Peoples' History of England,* chaps. 12–13.

to alchemical and Hermetic notions.[37] According to the new cosmology of the mechanical philosophy, matter itself (sometimes in motion, sometimes at rest) existing in empty space (to the atomists) or in a *plenum* (to Descartes) is all there is, all that underlies the whole of the sensate world of phenomena. Changes in the phenomenal world all arise out of the "matter and motion" of the underlying molecular or atomic world, each of the atomic or molecular particles in itself having *only* size, shape, and its state of motion—all quantitative entities—as its attributes. The world, in essence, is colorless, tasteless, soundless, devoid of thought or life. It is essentially dead, a machine, if you will (by whom or by what it is run, to this day, remaining pretty much a mystery).

Along with this mechanical philosophy as a new metaphysical basis for moderation, a new methodological approach, Baconianism, began to emerge in England as the official guide to research. Some spokespeople for the new science, such as Thomas Sprat, thought the new method would ensure that the study of natural philosophy would reinforce a respect for law and order and an unwillingness to tamper hastily with the basic institutions of society.[38]

It is in this light that we should view the extensive attempts at language reform spearheaded by the newly formed Royal Society. The reform tried to reduce language to a one-to-one correspondence with "real" objects—"so many things in the same number of words" was how Sprat put it—so as to make virtually impossible ambiguity and nuance. But the target of this campaign was, once more, "enthusisam," which was seen to be encouraged by the use of figurative or visionary language, and the passions in general. This conception of language would be pathetic in its naïvette, were it not for its quasi-totalitarian implications—Bishop Samuel Parker in 1670 actually proposed that Parliament outlaw the "use of fulsome and lushious Metaphors" by preachers—and the very real influence it had in drying up the wellsprings of poetic imagination after the Restoration.[39]

37. Robert H. Kargon, *Atomism in England from Hariot to Newton* (New York, 1966); Nina Rattner Gelbart, "The Intellectual Development of Walter Charleton," *Ambix,* 18 (1971), 149–168; Rattansi, "Paracelsus and the Puritan Revolution."

38. I have examined this theme in more detail in *HSINHRL'n'OTTW,* pp. 17–33, 36–39. See Thomas Sprat, *History of the Royal-Society of London for the Improving of Natural Knowledge* (London, 1667; St. Louis, 1958), ed. Jackson I. Cope and Harold W. Jones, p. 57. Sprat went so far as to claim that the widespread teaching of the scientific method would allow social debate to occur while diminishing the "danger of a *Civil War*"—this seven years after the Restoration of the Stuart monarchy had brought to a close the period of chaos and increasing demands by the unpropertied classes which had begun with the civil war.

39. Richard Foster Jones, *The Seventeenth Century. Studies in the History of English Thought and Literature from Bacon to Pope* (Stanford, 1951), especially the essays "Science and English Prose Style in the Third Quarter of the Seventeenth Century," "The Attack on Pulpit Eloquence in the Restoration: an Episode in the Development of the Neo-Classical Standard for Prose," and "Science and Language in England of the Mid-Seventeenth Century." The quotations are from pp. 108n and 118.

This general movement toward moderation in science, and its concrete expression in the mechanical philosophy and Baconianism, were no doubt evident to the young Newton in his studies at Cambridge, as a friend and avid reader of Henry More, who, in the 1650's and early 1660's at least, had pointed out the usefulness of Descartes' mechanical philosophy for supporting moderate religion.[40] But while Newton learned that Descartes' metaphysics was useful, his a priori methodology was all too similar to the methods of the dogmatists and fanatics themselves; accordingly Bacon's empiricism was to be preferred.

It is certainly true, then, that Newton was highly influenced by both the mechanical philosophy, especially Descartes' idea of the inertia of matter, and the methodology of Francis Bacon. That is not at issue here. What I want to emphasize, however, is that he was *also* heavily involved with ideas from an altogether different and ultimately *inconsistent* source—the alchemical and magical notions of Hermetic philosophy. Yet despite the fact that Hermetic ideas were a crucial source for his theories, Newton himself hid from the public any indication of their interest to him. Indeed, it seems Newton offered his ideas primarily in their garb of the inert bodies of the mechanical philosophy and of Baconian methodology, both, at best, partial truths (and therefore also partial lies) about his real thoughts. One consequence was the resultant confusing legacy Newton left his disciples and posterity. He so censored himself, I think, because he was aware of the onus that would fall on him were he openly to reveal the true nature of his ideas.

Newton's paranoiac character has been beautifully portrayed by Frank Manuel in his biography.[41] Suspicious beyond measure of both his associates and his rivals, Newton was always on the verge of rage, always on the defensive, always fearful of attack. Reacting to criticism of his views, in fact, several times he tried to withdraw from all future philosophical work. His paranoia intensified, I think, what already would have been as evident to Newton as it was to any student at Cambridge just after the Restoration, namely, that Hermetic knowledge was widely viewed by his contemporaries as an inducement to enthusiasm, social chaos, and the rising up of the lower orders, and that extreme caution must be exercised with such ideas.

Despite this danger, there is no doubt that Newton continued to toy with

40. On the adaptation of Cartesian ideas for English purposes in the seventeenth century, see especially Sterling P. Lamprecht, "The Role of Descartes in Seventeenth-Century England," *Studies in the History of Ideas* (New York, 1935), III, 181–243; Marjorie Nicolson, "The Early Stage of Cartesianism in England," *Studies in Philology*, 26 (1929), 356–374; and *HSIN-HRL'n'OTTW*, pp. 34 ff. See also B. Hessen, "The Social and Economic Roots of Newton's *Principia*," in *Science at the Cross-Roads* (London, 1932), pp. 27–42.

41. Frank Manuel, *Portrait of Newton;* also, see a series of penetrating articles by Richard Westfall: "Short-Writing and the State of Newton's Conscience, 1662 (I)," *Notes and Records of the Royal Society*, 18 (1963), 10–16; "Newton's Reply to Hooke and the Theory of Colors," *Isis*, 54 (1963), 82–97; and "Newton Defends His First Publication: The Newton-Lucas Correspondence," *Isis*, 56 (1966), 299–314.

Hermetic ideas, sought the answers to his most profound questions in their recesses, read manuscript after manuscript, treatise after treatise, from Hermetic authors and spent many years pursuing the keys to the Hermetic treasures. As a thorough and persevering student of the cosmos, Newton would have had trouble not delving somewhat into Hermetic sources, so widespread were Hermetic ideas at the time and so connected were they with some of the phenomena he was studying.[42] Even so, Newton was remarkable for his persistence and devotion in the study of them. Yet he steadfastly put his published work into mechanical, not Hermetic, terms. The combination of an atmosphere of philosophical moderation at Cambridge, indeed in all of England, aimed at enthusiasm and meant, ultimately, to quell the tempers and conflicts of the civil war and Commonwealth period, *and* Newton's own paranoia had brought about a self-censorship which was as effective as if some outside authority, church or state, had suppressed his ideas. In fact because it was a self—hence less visible—censorship, it was in many respects all the more effective, Newton thus presenting his ideas to the world in mechanical terms, for the most part, keeping to himself and a few select disciples his inner, Hermetic, visions.

V

If this is so, if Newton purposely obscured his Hermetic interests and leanings, where can we see these visions of his? First (though to be sure, others besides the *magi* wrote on Biblical prophecies), we can see Newton's interest in Hermetic studies in his work on Revelations and the Book of Daniel, both of which were important to him because they revealed God's plan for earth; Newton, who considered the successes of his science to have come about because he was a prophet in the tradition of Moses, felt knowledge of that plan to be crucial.[43]

Second, we see Newton's Hermetic interests in his published work on the ancient temple of Solomon and the size of the ancient cubit.[44] A long Hermetic tradition which emerged as the basis of much of eighteenth-century Freemasonry had taught that the ancient magical knowledge of the cosmos was hidden by the *magi* in symbolic mathematical relations built

42. As one outstanding example, I will mention the tradition that some of the Pythagoreans, a school of mystical philosophers (see below), had suggested a heliocentric model of the heavens with some similarity to the one put forward centuries later by Copernicus and adopted by Newton.

43. P. M. Rattansi, "Newton's Alchemical Studies," in *Science, Medicine, and Society: Essays to Honor Walter Pagel,* ed. Debus, II, 171, 174; J. E. McGuire and P. M. Rattansi, "Newton and the Pipes of Pan," *Notes and Records of the Royal Society,* 21 (1966), 126; Newton, *Observations upon the Prophecies of Daniel and the Apocalypse of St. John* (London, 1733); also, Newton, *Chronology of Ancient Kingdoms Amended . . .* (London, 1728).

44. Newton, *Chronology of Ancient Kingdoms Amended* pp. 332–346.

into the very structure of ancient holy buildings such as Solomon's Temple. To unlock these secrets of the cosmos, it was necessary above all to find out the exact size of the basic units used in primitive times to measure volume, length, time, and weight. In his work on Solomon's Temple, Newton tried to reconstruct both the plan of the temple and the size of the basic units used.

Third, there is Newton's related interest in the Great Pyramid of Cheops. Students of the mysteries also knew that the pyramid embodies fundamental cosmic proportions in its structure. In the late 1630's an English mathematician, John Greaves, went to Egypt to see if his measurements taken on location could support this tradition; Greaves's *Pyramidographia* reported his favorable findings. Newton read the work, and wrote a treatise himself on the basic measure, the cubit, used by the Egyptians as the fundamental constant in the whole pyramid.[45] A recent work by Peter Tompkins and Livio Stecchini has shown how measurements of the pyramid, in fact, reveal a fundamental unit incorporating a value for the mathematical ratio *phi,* defined by a numerical series supposedly first investigated by the twelfth-century mathematician, Leonardo Fibonacci. According to Tompkins and Stecchini, the Egyptians saw *phi* as fundamentally important, a symbol of the *logos,* the male sperm, and the creative force. An easy relationship allowed *pi* to be obtained from *phi.* In addition, the pyramid, based on this unit, was built as a model of the Northern Hemisphere on a scale of 1:43,200 or one to the number of seconds in twelve hours.[46] As such the Great Pyramid was used in geodetic work and for mapping. The perimeter of its base, for example, is the length of one half of a minute of latitude at the equator, correct to five significant figures, according to Tompkins and Stecchini. Stecchini claims that the Egyptians were greatly interested in the flattening of the earth, as was Newton, and that the Great Pyramid was used

45. Greaves, *Pyramidographia, or a Description of the Pyramids of Aegypt* (London, 1646); Newton, *A Dissertation upon the Sacred Cubit of the Jews and the Cubits of Several Nations; in which, from the Dimensions of the Greatest Egyptian Pyramid, as taken by Mr. John Greaves, the Antient Cubit of Memphis is Determined* in Greaves, *Miscellaneous Works of..*, (London, 1737), vol. 2. Peter Tompkins, *Secrets of the Great Pyramid,* with an appendix by Livio Catullo Stecchini (New York, 1971), chaps. 15–16, pp. 30–31, 201–203, 206–207, 210–212. In the 1690's, Newton wrote concerning the ancients that they "would have [this one God] dwelt in all bodies whatsoever as in his own temple, and thence they shaped ancient temples in the manner of the heavens," just as the "entire universe was rightly designed a Temple of God." (Quoted in J. E. McGuire, "Force, Active Principles, and Newton's Invisible Realm," *Ambix,* 15 [1968], 200.)

46. Besides the example of the Great Pyramid, the figure 43,200—or factors of it—is found in several widely separated ancient contexts. Heraclitus gives 10,800 (1/4 × 43,200) as the duration of the Aiōn and the Babylonian Great Year is said to last 432,000 years; the *Rigveda* has 432,000 syllables; the Indian fire altar is made of 10,800 (1/4 × 43,200) bricks; Angkor Wat has 108 (1/400 × 43,200) stone figures along each of the five roads leading to each of its gates; and in the *Grimnismal* 432,000 *einherier,* warriors who fell in combat on Earth and who now reside in Valhalla, emerge from the 540 gates of Walhalla to do battle. (Giorgio de Santillana and Hertha von Dechend, *Hamlet's Mill: An Essay on Myth and the Frame of Time* [Boston, 1969], p. 162.)

to determine the different lengths of a degree of latitude. A value for the earth's circumference thus computed was accurate, by today's standards, to one part in four hundred.

Some historians of science today might be skeptical of such claims, but they should realize that Newton was not. In fact he took them very seriously. His works on the units used in the pyramid and his later statements that the Egyptian priests knew many of the same secrets of the universe that he had revealed in his *Principia* and *Opticks* indicate just how much credit, after having investigated various of its sources, Newton gave to this tradition.[47]

Fourth, we see Newton's Hermetic interests revealed in his well-known interest in alchemical studies: his tireless reading of alchemical treatises and his endless copying of their passages, with long commentaries, his many hours spent on alchemical experiments, often not sleeping or eating for days on end while he tended his fires. Because this part of his work on the occult is somewhat better known, I will not discuss it at any length here.[48]

Fifth, most historians of science see Newton's interests—as well as those of his contemporaries—in the theories and work of the Pythagoreans, the ancient Greek mathematical and religious cult, in terms only of the profound mathematical knowledge the Pythagoreans established. But, given all the other indications of Newton's Hermetic interests, there is absolutely no reason to ignore the Hermetic roots of the Pythagorean cult, or the influence these roots probably had on the many natural philosophers—in particular Newton—who saw themselves as part of the Pythagorean tradition. Legend, as Newton recorded in his notes, teaches that Pythagoras studied for years under Egyptian priests to learn the essence of his own later teachings on mathematical and philosophic subjects.[49]

All these interests may *seem* peripheral to Newton's science. But the sixth aspect of his Hermetic interest, and obviously the most important, was the *central theme* of the whole of Newton's natural philosophy, the theories of universal attractive and repulsive forces existing between all bodies in the cosmos and tying them all together. Not only was Newton in debt to Hermetic notions in his persistent use of these forces, but in his long search for an explanation of these mysterious forces, as he pondered the strange nature of matter, the ways in which it underwent transformation, the various means by which activity is initiated in the world, he came to see a

47. McGuire and Rattansi, "Newton and the Pipes of Pan," pp. 109, 136–138.

48. See Rattansi's "Newton's Alchemical Studies" for a good discussion of the literature and an assessment of the role of alchemy in Newton's work. I have also benefitted from private correspondence in 1970 with Rattansi on Newton's alchemy. I have not yet had a chance to see the recent work on Newton's alchemy by Betty Jo Teeter Dobbs, *The Foundations of Newton's Alchemy, or The Hunting of the Green Lyon* (Cambridge, 1975).

49. Newton, "[Notes] Out of Cudworth," William A. Clarke Memorial Library MS.

creative, active principle behind all such activity which had much in common with the *archeus,* or principle of activity of the alchemists.[50] This active principle, whose elucidation Newton labored on and discussed at some length in his famour Thirty-first Query (to the *Opticks*) and the General Scholium (to the *Principia*), the two most extended public discussions of his underlying world-view, was a source of divine energy in the world.

In fact, in the final analysis, to Newton the world was more to be understood in terms of this divine energy than it was to be in terms of the material atoms on which, on the surface, he based his analysis. That is, Newton came to emphasize that the atoms themselves could be seen as arrays of miniscule points held apart merely by repulsive forces. As he continued to pursue this line of reasoning, the amount of actual matter he thought was contained in all of the miniscule points in the whole of the cosmos, compared with the vast amount of force-filled space, approached zero. So little real solid matter existed, in fact, that the world could readily be seen mainly in terms of immateriality.[51]

Despite this basic strain of immaterialism inherent in his thought, Newton partially misled his readers by presenting his natural philosophy in terms of the matter and motion tenets of the mechanical philosophy, basing his *Principia,* for example, on a set of laws deriving only from the inert qualities of brute matter. Yet, as Newton emphasized repeatedly, his notion of causality, of the initiation of change in the apparently brute matter filling the world, saw the ultimate causes as akin to the effects of a willful mind—God's, a person's, or the "willful" mind of the active principles—on the material world. That is to say, Newton saw the ultimate source of change in the universe to lie in *the mystery by which mind could control matter,* a central concern of occult thought down through the ages.

Considering the prevalence of change and activity in the cosmos—digestion, fermentation, putrefaction, growth, heat, gravity, efflorescence, and so on[52]—Newton once asserted, apparently too hastily, "We cannot say that all Nature is not alive." His hesitancy in claiming such a heretical notion is conveyed in the double negative, and by the fact that this sentence was dropped when he published the Queries. In another unpublished draft he wrote, "since all matter duly formed is attended with signes of life . . . those laws of motion arising from life or will may be of universal extent."[53]

50. Rattansi, "Newton's Alchemical Studies," pp. 176–178.

51. Arnold Thackray, "Matter in a Nut-Shell: Newton's Opticks and Eighteenth-Century Chemistry," *Ambix,* 15 (1968), 29–53; McGuire, "Force, Active Principles, and Newton's Invisible Realm," p. 157; Newton, *Opticks or A Treatise of the Reflections, Refractions, Inflections & Colours of Light* (New York, 1952), p. 394.

52. All of these were examples which Newton repeatedly cited, some as early as his student notebooks in the early 1660's.

53. Newton, Cambridge University Library MS 3970, fols. 619r–620v; Margaret Candee Jacob, "John Toland and the Newtonian Ideology," *Journal of the Warburg and Courtauld Institutes,* 22 (1969), 323.

Though he later apparently reconsidered this notion of nature being alive, Newton's image of nature was certainly compatible with the idea of the cosmos as being some kind of animal, perhaps more so, in fact, than it was with the machine universe which, historians tell us, was Newton's legacy. Nature was a "perpetual circulatory worker," Newton said in an early statement of his philosophy, and it was a conception he retained throughout his career. Nature delights in transformations, Newton insisted, and everywhere one saw her protean quality. Individual objects changed and the cosmos as a whole underwent transformations, Newton pointed out. Nature as a whole as well as its parts seemed inherently dynamic to him.[54]

VI

I do not want to suggest, however, that we should see Newton in a static way, merely as operating at two levels, the deeper, Hermetic, vision which was closer to his true feelings about nature, and his open, mechanical, vision, which he used as a front, so to speak.[55] Rather, important changes to his philosophy of nature, which reflect on the tensions between the two approaches to nature, were taking place in the two decades following the publication of the *Principia*. What is especially important is the way Newton began to qualify his notion of a dynamic and transmuting nature in later years in two important respects, indicating, I think, Newton's growing awareness of just how dangerous were the social implications of the Hermetic aspects of his philosophy.

As mentioned above, in the 1660's, after the Restoration, influential clergy, doctors, lawyers, and natural philosophers had emphasized the necessity for a sober natural philosophy, based ultimately on the hard inert particles of the mechanical philosophy, to cut the ground out from under the enthusiasm of the sects. This campaign seemed to work, for one saw and heard little of the sects in most of the 1660's and early 1670's. However, the rising tensions over whether the Catholic James II would succeed Charles II and the attempts by Lord Shaftesbury and others to prevent him gave rise to a new oppositional movement, which in the 1670's and 1680's brought out from hiding the old Leveller and republican sentiments. The aristocratic leaders of the new movement tried to keep it moderate in its ultimate aims, but (just as in the 1640's) in order to obtain power they had not only to accept the support, but also to tolerate some of the radical ideology, of the lower middle classes, who had never really lost their allegiance to republican and levelling politics.

54. See Kubrin, "Newton and the Cyclical Cosmos: Providence and the Mechanical Philosophy," *Journal of the History of Ideas,* 28 (1967), 325–346; Newton, *Opticks,* p. 374; Kubrin, *HSINHRL'n'OTTW,* pp. 61f.

55. Compare Westfall, "Newton and the Hermetic Tradition," pp. 190–191, 195.

As the late 1640's and 1650's when the radical political and social movement had been parallelled by the rise of Hermeticism, so too the new radical social movement of the 1670's and 1680's saw a revival of Hermetic ideas. One of the most important exponents of the new Hermeticism was John Toland, an Irishman educated in Edinburgh in the mid-1690's by the Newtonian David Gregory and associated with the political and social movement of Shaftesbury and the others.[56] In the 1690's, after leaving Edinburgh, Toland associated with deists, atheists, freethinkers, skeptics, and others whose notions were scandalizing the orthodoxy and magistrates. There was little question but that the rise of this group of unorthodox thinkers was part of a larger social movement which ultimately would try to question not just religion but the running of civil and class society as well, posing a concrete danger to the English crown.

In an attempt to construct a philosophical system that would justify the social, religious, and political activities of the freethinkers and rebels, Toland, like his civil war predecessors, turned to Hermeticism. Toland's major source for his ideas was the Renaissance *magus* Giordano Bruno, though he rephrased some of Bruno's mystical notions in a more rational language.[57] Toland, however, was also quick to see and to point out in his books the magical implications of some of Newton's ideas.[58] Seeing how Newtonianism was being used to justify the moderate religious and political settlement in England after the Glorious Revolution of 1688, Toland was especially cynical about the public claims by Newton and his disciples (such as Gregory) that Newton's natural philosophy was based on the hypothesis that the basic corpuscles of matter were inert.[59] Toland emphasized that it made more sense to see that the Newtonian attractive and repulsive forces implied that matter everywhere—far from being inert—was sensate, filled with self-activity, and found everywhere in constant motion. Nature was not passive, but active, infused, as it were, with God. God had instilled an inherent energy and harmony in the cosmos, as gravitational attraction so well showed.

Toland thus attacked Newtonianism using Newton's own principles. To make matters worse, his attack was widely understood to be aimed ultimately at justifying the politics of the republicans and the religion, or lack of it, of the freethinkers. Toland also turned to nature to demonstrate society's need for a revolution. For the universe was composed of things such that "all these depend in a Link on one another so their Matter (to speak in the usual Language) is mutually resolv'd into each other; for Earth and Water,

56. Jacob, "John Toland and the Newtonian Ideology," pp. 310–314.
57. Ibid., pp. 309, 312–313.
58. Ibid., pp. 320–321.
59. Ibid., pp. 313–314, 331. I have greatly benefitted from conversations with M. C. Jacob and J. R. Jacob about the material in this paragraph.

and Air, and Fire, are not only closely blended and united, but likewise interchangeably transform'd in a perpetual Revolution; Earth becoming Water, Water Air, Air Aether, and so back again in Mixtures without End or Number."[60]

All the more alarming that the ideas which Toland insisted on reading into Newton's philosophy were not, at least on one level, very much different from Newton's own privately held thoughts! To be sure, Newton was no republican, let alone a freethinker. Yet he did deny the trinitarian beliefs of the Church of England, a belief which cost his disciple and successor in the Lucasian Professorship of Mathematics at Cambridge, William Whiston, his position in 1707. And when Newton was not careful to censor his ideas, they did, in fact, seem to be similar to some of those of Toland. Like Toland, Newton was deeply influenced by Hermetic notions. Like Toland, Newton saw the cosmos as infused everywhere with God. In his *Opticks* he had actually written that the world *was* the sensorium, or organ of sense, of the all pervasive and immanent God. But, as it was going to press, Newton evidently changed his mind and had printed in its place that the world was *like* God's sensorium.[61] As mentioned above, Newton similarly had written in a draft for the Thirty-first Query that all nature could be seen as alive, though last-minute second thoughts led him to drop this idea too from the text.

Newton, therefore, at least at one level of his consciousness, agreed with Toland that nature was alive, sensate, infused with self-energy (identified with God), and, in fact, pervaded by the literally omnipresent God. But with the association between the concept of nature of Toland and the radical political and social ideology of the republicans and freethinkers, it was at all costs imperative for Newton to avoid being linked with Toland's ideas. It was this fear, I think, that accounts for Newton's marked reluctance to publish his real thoughts. His specific personality and his paranoiac fears of attacks on his ideas were clearly responsible for Newton's sensitivity, but the specific social and political environment in which he lived—with its tremendous anxiety over radical political and religious movements—certainly provided the context that enabled this sensitivity to lead Newton to hide some of his most important ideas.

Newton reacted very clearly to the threat he saw from the possible association between his theories and those of Toland. In a series of lectures for the influential Boyle sermons given and published in 1704, his disciple and public spokesperson, Samuel Clarke, singled out Toland's notions along with those of Hobbes and Spinoza as particularly dangerous to religion. Two years later when Newton published the Latin edition of his *Opticks*,

60. Quoted in Ibid., p. 319.

61. I. Bernard Cohen and Alexander Koyré, "The Case of the Missing *Tanquam:* Leibniz, Newton, and Clarke," *Isis,* 52 (1961), 555–567; Newton, *Opticks,* pp. 370, 403.

which was translated by Clarke, he added a number of Queries to it, including what eventually became the famour Thirty-first Query.[62] Here he wrote at length concerning his metaphysics in a careful attempt to answer his critics and to distinguish his principles from those of Toland (and others, such as Hobbes and Spinoza), in the end not *quite* succeeding in hiding hints of his own pantheism, despite his hasty last-minute revisions. In addition, we can find other, more fundamental, changes introduced into Newton's natural philosophy in the 1690's and early 1700's.

Newton had always insisted on the protean quality of nature, its endless transformations testifying to the hegemony of change. Yet, while in his earlier writings, such as, for example, the first edition of the *Principia* in 1687, Newton had included among the kinds of transformations to be found in nature the idea of the transmutation of one element to another, of one kind of body to another kind, in the 1690's he rejected this notion. Working on a revision of the *Principia* for a stillborn new edition, Newton came to emphasize that while nature still "delighted" in transformations, certain ones were not allowed. Instead, all bodies now had certain "essential" qualities such as extension, inertia, and impenetrability, which could never be removed, even in theory. That is to say, Newton developed a theory of matter, a notion of the primary qualities of bodies that formed the psychological theory of phenomena at the basis of the mechanical philosophy, like that of his friend John Locke, whose *Essay on Human Understanding* had been published in 1689.[63]

According to Newton's new theory, only the secondary qualities of a body, its color, smell, taste, and so forth, were now subject to transformations. Newton began to make this distinction, according to J. E. McGuire, in order to be able to intimate the possibility of gravitational attraction being an essential aspect of bodies. Newton at the time was having enormous difficulties justifying his use of such a concept as universal attraction, his contemporaries seeing in it the occult or magical roots that in fact it partly rested on and that Newton was now trying so hard to deny.

In making his new distinction between the primary and the secondary qualities of bodies, however, Newton had crossed a crucial watershed, for

62. The whole of this article, as well as my familiarity with some of the manuscript sources, has greatly benefitted from a seminar on the changes to Newton's *Opticks* given by Professor Henry Guerlac at Cornell University in 1962. See Jacob, "John Toland and the Newtonian Ideology," pp. 321ff.; Newton, *Opticks*, pp. 375ff. I do not mean to suggest, as one critic has thought, that the political and social basis for the opposition to magic existed only in England, or that Newton's continental critics (Huygens, Leibniz, and so on) were irrelevant to Newton's attempts to cover his magical tracks, only that in England that basis was more clearly articulated and understood. Continental critiques of magic also had their political and social basis. On this, see Carolyn Merchant, *The Death of Nature*.

63. C. B. McPherson, *The Political Theory of Possessive Individualism: Hobbes to Locke* (Oxford, 1962); this paragraph and the next rely upon J. E. McGuire, "The Origin of Newton's Doctrine of Essential Qualities," *Centaurus*, 12 (1968), 233–260.

the notion of certain essential unchanging qualities of bodies erected a formidable barrier to conceiving of nature as being in all respects transformable. Now, indeed, certain aspects of nature, certain aspects of her building blocks, were to be considered permanent. And for the first time to Newton a theoretical barrier to considering certain kinds of alchemical transformations of matter existed. Matter was, in effect, stabilized, made more permanent. With this change, Newton entered more fully into the mechanical camp.

The meaning of this change may become clear after I discuss a second, related change to Newton's conception of the world. Throughout the 1680's and 1690's, various English natural philosophers, including several who were undoubtedly speaking with Newton's concurrence if not his outright encouragement, were trying to construct theories of cosmogony—that is, theories of how the world was created and developed in time.[64] Many such theories were emerging, and the public interest in them in England at the time was keen. When in the early 1690's these cosmogonical theories became linked to freethinking and republicanism, however, the religious hierarchy mounted a public attack on the very idea of cosmogony. By showing how the earth's geological forms had changed in time, these theories might further undermine the idea that nature was everywhere stable; some of the High Church spokesmen seemed to worry that people might not be able to expect a stability for the state and church that went beyond the stability of nature itself. Newton had not only inspired several cosmogonic writings (those of his disciples Edmond Halley, William Whiston, and David Gregory, and to a lesser extent that of Thomas Burnet), but as I have shown elsewhere, a rather detailed cosmogonic framework was slowly built by Newton himself, and gradually inserted by bits and pieces in the pages of the later editions of the *Principia* and the *Opticks*.[65] Nevertheless, in 1706 in his first work published after the public campaign against cosmogonic theories had been mounted, Newton publicly came out in his Latin *Opticks* with a condemnation of speculation about how the world had developed in time. It is sufficient to know merely that God had created the world out of a chaos, Newton said. To try to explain either how the creation occurred or how the world developed subsequently was unscientific.[66]

That is to say, in the 1690's Newton decreed that there were limits to the transformability of the natural world, that not only did it not include the

64. See Kubrin, "Providence and the Mechanical Philosophy" (Ph.D. diss., Cornell University); and *HSINHRL'n'OTTW*, pp. 46–84.

65. Kubrin, "Newton and the Cyclical Cosmos"; Kubrin, "John Keill," *Dictionary of Scientific Biography;* some of the ideas in this paragraph came out of discussions with M. C. Jacob, though I take full responsibility for them.

66. Newton, *Opticks*, p. 402. As I have emphasized in "Newton and the Cyclical Cosmos," this was Newton's public position only. Privately, and by means of hints in his published works, Newton continued to speculate about cosmogonical change.

fundamental, essential qualities of the primordial bits of matter of the cosmos, their impenetrability, extension, or inertia; but furthermore, in 1706, a decade and a half later, his ban on the transformability of nature was extended to include the cosmos itself. Newton asserted now that the world was created and would eventually be destroyed, but that it was not the business of natural philosophers to ask questions about either of these epochal events or of the development that lay between them.

Newton came in later times, then, to emphasize the fundamental stability, rather than the transformability, of nature (though the latter was still an important, albeit heavily qualified, aspect of his philosophy) or at least of the two key aspects of nature—the basic particles of the cosmos and the cosmos itself. His change in emphasis, I think, was basically for much the same reasons why, despite his deep indebtedness to Hermetic visions of the cosmos, Newton chose publicly to emphasize the mechanical and Baconian aspects of his natural philosophy. Newton was deeply anxious lest his natural philosophy be identified with the outlook of any of the radical sectarians. And if this was an issue in the 1660's. when he was putting together the broad foundations of his natural philosophy, it was even more so later on, in the 1690's and early 1700's when Newton was being called on to defend his notions against many critics, both Continental and English. The uncomfortable similarity between some of his notions and those associated with radical religious and political ideologies made it necessary for Newton to construct a body of philosophy that was deceptively one-sided and which conveyed only a part of his real vision.

VII

I hope I have demonstrated that Hermetic ideas played a central role in the development of Newton's natural philosophy, and that he realized that should this role ever be publicly recognized, his philosophy would be subject to even stronger attacks than was already the case. He was therefore led not only to mask some of his central ideas, but in later years, in fact, to abandon some of the fundamental concepts upon which his natural philosophy was based.

That Newton felt compelled to hide so much of what he had to say is of course a great loss to science and culture. It has made our own attempts to understand his ambiguous legacy all that more confused, to both our losses—though we, unfortunately, are the ones who have to survive somehow the present wasteland which in large measure is the result of that confusion.

But I would not want to end the discussion with this point alone. I think that a similar kind of distortion of ideas quite likely has occurred numerous other times within the field of natural philosophy, and it is important to understand what it represents. In the case of Newton I have tried to indicate the degree to which his published views were molded by social, political, and economic considerations, hoping thereby to show the extent to which science (and culture in general, I think) exists within the context of a class society and the extent to which it enters, willingly or not, into the struggles between these classes. Because of censorship or the normal passivity of people in times of calm in our society, that struggle, as well as the role science plays in it, is often difficult to see. In times of civil war, or even the threat of one, however, the priorities and mechanisms of a society are made clearer to us. I hope I have been able to show a case study of that in the English civil war and the scientific revolution of the seventeenth century.[67]

Finally, it is important to realize that the mechanical philosophy, which a one-sided interpretation of Newton's principles of nature points to, became enshrined after him as *the* metaphysical underpinning of all rational thought. Mechanism, as a metaphysics and an epistemology, not only spread from physics to chemistry and biology, but also to physiology, philosophy, psychology, religion, poetry, ethics, political theory and art. The eighteenth-century Enlightenment spread mechanical thinking throughout Western Europe and to the Americas, Russia, and even parts of Asia. To this day, in fact, in universities and in educated circles, mechanical thinking dominates most academic subjects—with the notable exception, ironically, of physics, where it began—and many nonacademic disciplines as well.

Nor do I think its longevity has been accidental. Mechanical thinking has lasted so long because it has served as the foundation of what has been called "bourgeois ideology"—that is, the set of assumptions, categories, methodologies, rules, and ideas about reality (the world, people, how things

67. Scholars have unfortunately largely ignored the subject of the vast potential of science—in its role as definer of "reality" and judge of what can and cannot be possible "causes" of change in the world—to reinforce certain social forms over others. Compare Marx and Engels, who claimed: "The ideas of the ruling class are in every epoch the ruling ideas: i.e., the class, which is the ruling material force of society, is at the same time its ruling intellectual force. The class which has the means of material production at its disposal, has control at the same time over the means of mental production, so that thereby, generally speaking, the ideas of those who lack the means of mental production are subject to it. . . . In so far, therefore, as [the individuals in the ruling class] rule as a class and determine the extent and compass of an epoch, it is self-evident that they do this in their whole range, hence among other things rule also as thinkers, as producers of ideas, and regulate the production and distribution of the ideas of their age: thus their ideas are the ruling ideas of the epoch." Karl Marx and Frederick Engels, *The German Ideology* (New York: International Publishers, 1947), p. 39. Unfortunately, with rare exceptions, the attempt of scholars since Marx's time to debate this thesis has more often than not been marked, on the one hand, by a crude mechanical reductionism which has ended up by inverting the central Marxist concept of *consciousness*—thereby transforming it into an object and negating its historical and political significance to Marx; or, on the other hand, by a rather petty, nit-picking, and close-minded scholarship which begs the central questions and seeks enlightenment in mere pedantry—becoming the narrowest kind of empiricism, one whose sole purpose sometimes seems to be to demonstrate that there are no patterns, *none whatsoever,* in the course of human affairs.

change) which serves as an intellectual justification and explanation for the rule of bourgeois property relations over the whole of capitalist societies. The mechanical foundation of bourgeois ideology had the historical development that it did and was ultimately as successful as it was as a result of a fundamental ideological struggle—a class struggle—between it and an alternative vision of reality about the basic elements making up the world and of the causes that allowed them to change from one thing to another. As I have indicated, this alternative vision was based on dialectical, rather than mechanical, logic. The victory of one vision rather than the other in important respects reflects the relative strengths in the seventeenth and succeeding centuries of the classes that supported the two visions. I would like to think, however, that as those classes change their respective strengths—as has been happening during the twentieth century and especially in the last two decades on an international scale—a whole new chapter in this story will unfold. But I will leave that to future historians, writing in happier times.

CLANDESTINE CULTURE IN THE EARLY ENLIGHTENMENT

MARGARET C. JACOB

LIBERTINES and freethinkers of the early eighteenth century necessarily practiced secrecy as a way of life. Clandestine behavior, found in greater or lesser degrees in every European country including The Netherlands, infused and protected the culture of the early Enlightenment. We now know about some aspects of this clandestine culture, in particular the work of Ira Wade and others has charted the French circulation of subversive manuscripts.[1] Likewise the biographies of individual freethinkers, whose writings oftentimes flourished in that circuit, have recently become available in monographs and articles, including works on John Toland, Anthony Collins, Anthony Ashley Cooper, third earl of Shaftesbury, and Tyssot de Patot, as well as the most famous and important freethinker of this period, Pierre Bayle.[2] From this collective data a definition emerges for that elusive term,

This research was made possible by a Summer Grant in 1975 from the National Endowment for the Humanities. Various scholars and friends have assisted or advised in the preparation of this essay. My thanks go to the librarians of the University Library, Leiden, in particular C. Berkvens-Stevelinck, to J. G. A. Pocock, J. R. Jacob, Frances Yates, Catherine Barel, I. H. van Eeghen, and Dorothy Schlegel, and to M. Fidanza for typing and research assistance.

1. Ira O. Wade, *The Clandestine Organization and Diffusion of Philosophic Ideas in France from 1700 to 1750* (Princeton, N.J.: Princeton University Press, 1938); J. S. Spink, "La diffusion des ideés matérialistes et anti-religeuses au début du XVIIIe siècle: Le 'Theophrastus redivivus,'" *Revue d'histoire littéraire de la France,* 44 (1937), 248. Cf. F. H. Heinemann, "Prolegomena to a Toland Bibliography," *Notes and Queries,* 185 (1943), 184 *et seq.;* P. Marchand, *Dictionnaire historique,* I (The Hague, 1758), 318–325, article on *De tribus impostoribus.*

2. Dorothy Schlegel, *Shaftesbury and the French Deists* (Chapel Hill, N.C.: University of North Carolina Press, 1956); James O'Higgins, S. J., *Anthony Collins. The Man and His Works* (The Hague: Martinus Nijhoff, 1970); D. R. McKee, *Simon Tyssot de Patot and the Seventeenth-Century Background of Critical Deism,* Johns Hopkins Studies in Romance Languages, 40 (Baltimore, 1941); Aubrey Rosenberg, *Tyssot de Patot and His Work 1655–1738* (The Hague: Martinus Nijhoff, 1963–64); Margaret Candee Jacob, "John Toland and the Newtonian Ideology," *Journal of the Warburg and Courtauld Institutes,* 32 (1969), 307–331.

freethinker: it simply meant someone whose approach to learning and culture rejected the prevailing doctrinal orthodoxies of churches and states, who therefore distrusted traditional authorities, either political or ecclesiastical, and who embarked inevitably upon an intellectual odyssey that led generally to the skepticism of Bayle, the Platonized deism of Shaftesbury, or the pantheistic materialism of Toland. But these early progenitors of the Enlightenment should not be seen as isolated individuals; they sought and found converts and most important they created a social world, conscious of its goals and international in its character.

Whatever path the freethinker took he trod it cautiously and the discoveries made along the way were shared primarily with trusted friends and associates. Only with great caution, yet with considerable dedication, were they revealed in print to that always desired larger audience. Secrecy pervaded an international culture that adopted forms of expression still found among dissenting subcultures to this day: the clandestine yet repeatedly copied manuscript, the art of circumlocution (used with such finesse by Bayle in his *Dictionnaire,* 1697), and the formation of coteries or circles of initiates who met in a specially designated salon, tavern, or coffee house.

Art, as well as literature and social life, was shaped by the necessity for secrecy. In certain early eighteenth-century engravings we can discern the use of a special iconography, an ensemble of symbols, by which freethinkers could puzzle their enemies while communicating their beliefs and ideals to sympathizers. One particular symbolic ensemble turns up in engravings intended for the frontispieces and title pages of books written or edited by a single coterie of early Enlightenment freethinkers at work largely in The Netherlands. They consistently used representations of the pagan gods Minerva and Mercury, and the meaning of that symbolism will require explication.

This newly discovered coterie, active in the early years of the Enlightenment, provides the focal point for this essay. Although similar groups probably existed in London and Paris, this group of Protestant refugees and émigrés living in The Netherlands from roughly 1710 until their deaths, generally in the 1750's, has surfaced through the discovery of manuscripts located at the British Library and at the University Library in Leiden. As

For a critique of O'Higgins see David Berman, "Anthony Collins and the Question of Atheism in the Early Part of the Eighteenth Century," *Proceedings of the Royal Irish Academy,* 75, no. 5, 85–102. Cf. J. H. Broome, "An Agent in Anglo-French Relationships, Pierre Desmaizeaux, 1673–1745," (Ph.D. diss., University of London, 1949), and his "Bayle's Biographer Pierre Desmaizeaux," *French Studies,* 9 (1955), 1 *et seq.;* "Anthony Collins et Pierre Desmaizeaux," *Revue de littérature comparée,* 30 (1956), 161, *et seq.;* P. Hemprich, *Le journal littéraire de la Haye, 1713–1737* (Berlin: Fritz Herrmann, 1915). And see chapter six in my *The Newtonians and the English Revolution, 1689–1720* (Ithaca, N.Y.: Cornell University Press, 1976); Paul Vernière, *Spinoza et la pensée française avant la révolution* (Paris: Presses Universitaires de France, 1954), pp. 355–385.

might be expected, most of these French Protestants began their intellectual odyssey devoted to the memory and critical spirit of Pierre Bayle. But they did not stop there, and it would be useless to try to explain their intellectual development simply by recounting their application of Bayle's attitudes toward ancient learning, religion, or the clergy.

New sources sometimes require a reassessment of traditional historiographical assumptions and interpretations, both for their discovery and for their explication. Here it is assumed that if the Enlightenment in its origins is to be seen as a living culture, and not exclusively as a series of great books penned by young and daring philosophers and in turn taken up by their followers, then we must penetrate its social world. If we are to arrive at a social history of Enlightenment ideas then the historian must undertake to search for the individuals and groups who wrote or circulated the clandestine manuscripts, who consistently published or edited books or collections intended to promote Enlightenment, and who tried in their daily lives to give expression to freespirited inquiry and heretical beliefs. Understandably not all of those activities were willingly made public and efforts to uncover them sometimes mean that the historian must find a clue, a piece of evidence mistakenly (or perhaps even purposefully) preserved.

Some years ago at the British Library while going through manuscript remains of the English freethinker, John Toland (1670–1722), I found an extraordinary document. It was a two-folio meeting record revealing the existence of an unknown coterie of writers, editors-booksellers, and an engraver who lived for the most part in The Hague and Amsterdam, but who were clearly in contact with the freethinking and international subculture to which Toland belonged throughout most of his life. Written in French, this manuscript is entitled "Extract of the registers of the general chapter of the Knights of Jubilation, held at Gaillardin, house of the order, on 24 November 1710." This is nothing less than the record of events at one meeting of a clandestine group who called themselves the Knights of Jubilation. They enjoyed epicurean pastimes and had formalized these happy occasions and their association by subscribing to some sort of constitution mentioned in the document. Most important for the historian, the document is signed by the Knights who add their individual titles within the organization: G. Fritsch, *Grand Maistre,* M. Böhm, *Echanson* (butler) *de l'ordre,* G. Gleditsch, *Thrésorier* (treasurer) *de l'ordre,* Ch. Levier, *Arlequin & Bouson* (joker and drinker) *de l'ordre,* Bernard Picart, *Enlumineur* (illuminator) *de l'ordre,* M. de Bey, and P. Marchand, *Secrétaire de l'ordre.*[3]

3. British Library (BL), MSS Add. 4295, fols. 18–19. Fritsch's name has a line through it, but this is of no significance since in the Marchand MSS there are letters signed by him with that title. Picart has a second, but illegible title, and M. de Bey's title is also illegible. I know of only one mention of a similar, indeed possibly the same, organization; M. Arthur Dinaux, *Les sociétés badines, bachiques, littéraires et chantantes* (Paris, 1867), I, 421–425.

All the signatories were Protestant, either refugees or immigrants,[4] and this meeting held in late 1710 took place in all probability at The Hague where Toland and many of the signatories resided in that year. In the course of their lives as publishers, editors-booksellers, and in the case of Bernard Picart as a prominent engraver, the Knights lived in various places in The Netherlands and they traveled extensively, in the Low Countries, Germany, and France. This much can be said about their activities because of the excellent work by Dutch scholars on the early history of printing in The Netherlands. More can now be added on the basis of the first extensive search ever undertaken through the enormous cache of manuscript bequeathed by Prosper Marchand (d. 1756) to the University Library, Leiden. Marchand was secretary to the Knights of Jubilation, and he was also the editor of Bayle's correspondence and the heir to his *Dictionnaire.* Indeed his own revised and highly individual version of that work was published posthumously.[5] Marchand's vast manuscript collection, when combined with a close reading of his published writings, provides a series of clues that ultimately reveal an entirely new source for the history of clandestine culture in the early Enlightenment.

The good detective should pursue each clue as it is discovered and thus I must return to the first piece of evidence that revealed the existence of the Knights of Jubilation. What was Toland doing with that document in his possession? Aside from the meeting in 1710 which he probably attended, did his contact with this circle continue throughout his life? Further documentary evidence for Toland's involvement with the Knights and their friends is also unpublished. Like Toland, Marchand, Picart, and their circle appear to have been on good terms with the court of Prince Eugene of Savoy, the great military leader of the allied and largely Protestant forces during the War of the Spanish Succession. His headquarters were at The Hague and among his literary and intellectual admirers were the Knights of

---

I first discussed this manuscript in "An Unpublished Record of a Masonic Lodge in England: 1710," *Zeitschrift für Religions-und Geistesgeschichte,* 22 (1970), 168–171. At that time I mistakenly assumed that because this manuscript was amid Toland's papers that the meeting took place in London. I also misread a badly written text, that is, "l'ordre" for "Londres." The full text will be published in the appendix of my forthcoming, *The Radical Enlightenment: Pantheists, Freemasons and Republicans* (London and Boston: Allen and Unwin, 1980).

4. Short biographies of some of the individuals discussed in this essay can be found in E. and E. Haag, *La France Protestante ou vies des Protestants français* (Paris, 1854–59), vols. 1–9; I. H. van Eeghen, *De Amsterdamse Boekhandel 1689–1725* (Amsterdam: Scheltema-Holkema, 1962–67), vols. 1–4; E. F. Kossmann, *De Boekhandel te 's-Gravenhage tot het eind van de 18e eeuw* ('s-Gravenhage, 1937). For Picart see J. Duportal, *Bernard Picart, 1673 a 1733, extrait de l'ouvrage, Les peintures français XVIIIe siècle* (Paris, 1928).

5. P. Marchand, *Dictionnaire historique* (La Haye, 1758–59), 2 vols.; Pierre Bayle, *Lettres choisies, avec des remarques* (Rotterdam: Fritsch et Böhm, 1714). Cf. C. Berkvens-Stevelinck, "Prosper Marchand, auteur et éditeur," *Quaerendo,* 3 (1975), 218–234; *Prosper Marchand et l'histoire du livre, Academisch Proefschrift . . . Universiteit van Amsterdam* (Bruges, 1978).

Jubilation who in their letters discussed dedicating books to him. This adoration reached the level of apotheosis in an elaborate engraving done by Bernard Picart to celebrate Eugene's military victories. The prince had a reputation for intellectual interests and his intimate friend, Baron Hohendorf, whose own intellectual reputation was decidedly libertine, assembled for him one of the finest libraries in Europe.[6]

The manuscript letters of John Toland contain one letter dated March 1712 to Baron Hohendorf discussing the Roman Emperor Anastasius who, according to Toland, "was hated by the clergy because he cultivated peace, tolerated all sects and despised the self-importance of the priests as he hated their ignorance." This letter indicates that Toland and Hohendorf had been discussing both religion and political matters, in particular the Continental war. In a postscript Toland adds this comment, "The formula, or philosophical liturgy has not yet been written out: I will set it out in my next [letter]."[7] At this stage in our investigation we are concerned only with Toland's relationship with the Knights of Jubilation and their friends. What has a philosophical liturgy to do with that question? Indeed, one might well ask, what is a philosophical liturgy? Without for the moment speculating upon the ceremonial use to which Toland and his friends in The Netherlands may have put his liturgy, Toland's published work reveals that this was a liturgy in praise of nature, not God, to be chanted by a private club as part of their meeting ritual. Such liturgies, as we know, were practised by Freemasons in the eighteenth century, and indeed I shall be suggesting later in this essay, that the Knights of Jubilation and their friends, including Toland, constituted themselves as some sort of private Masonic lodge. If that is the case, they did so years before organized speculative Freemasonry is supposed to have existed on the Continent, or even for that matter in England.

In addition to their mutual involvement in the intellectual life of Prince Eugene's court, there were other links between Toland and the Knights of Jubilation or more precisely, with an extended and larger society to which the Knights and other Protestant intellectuals belonged from at least 1713

6. Marchand MSS 2, University Library (UL), Leiden, Fritsch to Marchand, November 1711; Teylers Museum, Haarlem, Picart engravings, portfolio 188, "Explication du sujet simbolique . . . pour servir d'accompagnement aux Armes du Prince Eugene de Savoye." See also W. Suchier, *Prinz Eugen als bibliophile* (1928); and this curious work, Mr. B** de B**, *La religieuse malgré elle, histoire galante, morale et tragique. Dédié a Mon. le Prince Eugene de Savoye* (Amsterdam: Claude Jordan, 1740). For a discussion of works by Giordano Bruno in Eugene's library, see Zeno, *Lettere di Apostolo Zeno,* 2d ed. (Venice, 1785), IV, 236–237; III, 465–470. See also *Bibliotheca Hohendorfiana; ou catalogue de la bibliotheque de . . . Baron de H.* (The Hague, 1720).

7. BL, MSS Add. 4295, fols. 19v–20r; "Nondum est descripta Formula, sive Liturgia Philosophica, quam per proximos expediam Tabellarios." Marchand found the liturgy revealed by Toland in the *Pantheisticon* (1720) to be too extreme, or at least so he said to Thomas Johnson; see U.L., Leiden, Marchand MSS 62. Cf. Guiseppe Ricuperati, "Libertinismo e deismo a Vienna: Spinoza, Toland e il *Triregno,*" in *Rivista storica italiana,* 2 (1967), 628ff.

onwards. Marchand, the French deist Themiseul de Saint-Hyacinthe, the Dutch scientist Wilhelm Jacob van s'Gravesande, and the epicurean writer, A. H. Sallengre, belonged to a literary society which for many years published from The Hague an important journal, the *Journal littéraire.* The journal displays a militantly Protestant attitude toward the religious controversies of the day, and also attempts to give extensive coverage to the freethinking assault on Christianity. But the public stand of the editors is decidedly on the side of a liberal and tolerant orthodoxy. The journal also covered other intellectual matters extensively and, quite probably as a result of s'Gravesande's efforts, displayed a strong interest in science and natural philosophy. If the *Journal littéraire* were the historian's only evidence about Marchand and his circle, the freethinking posture and bacchanalian spirit displayed in their other publications, private unpublished letters, and meeting records would never be revealed. Although some champions of orthodoxy had suspicions about the treatment their books received from the editors of the *Journal littéraire,* basically it was intended as a respectable, comprehensive, and presumably profitable literary endeavor.[8]

One of the foreign correspondents in England for this journal was the French refugee, Pierre Desmaizeaux, whose unpublished correspondence contains a wealth of material on intellectual life in the period up to 1740. It also reveals that the English freethinker Anthony Collins, Desmaizeaux, and Sallengre, one of the editors of the *Journal littéraire,* were on good terms, and that Sallengre visited and discussed philosophy over sumptuous meals with Desmaizeaux and Toland at Collins' country estate in Essex.[9] At this stage in our research, this is all the evidence we have for personal contact, as distinct from intellectual influence or agreement, between Toland and the Knights of Jubilation and their friends. However it is enough to establish beyond all doubt that Toland, Marchand, Picart, Sallengre, Desmaizeaux, Collins, and others belonged to the same or overlapping circles, and probably read some of the same books and clandestine manuscripts.

Knowing that Toland and the Knights of Jubilation belonged to the same secret world, we now must attempt to deduce what the Knights of Jubilation intended to accomplish by their jovial camaraderie, which as the Marchand manuscripts reveal, continued for many years and expanded to include new members. Toland's manuscripts have already yielded valuable evidence, ma-

8. P. Hemprich, *Le journal littéraire,* pp. 96–104. UL, Leiden, Marchand MSS 1, Humphrey Ditton to the editors, 29 March 1714. In Prosper Marchand, *Dictionnaire historique ou mémoires critiques et littéraires . . .* (The Hague, 1758–59), II, 214–217, contains Marchand's account of that society. Cf. E. Hatin, *Les gazettes de Hollande et la presse clandestine aux XVIIe et XVIIIe siècles* (Paris, 1865), pp. 214–215.

9. J. H. Broome, "An Agent in Anglo-French Relationships, Pierre Desmaizeaux," pp. 159–165; cf. BL, MSS Add. 4282, fols. 224–226. There are letters from Thomas Johnson to Desmaizeaux in MSS Add. 4284. There are also letters from Collins to Levier and Marchand in the Marchand MSS.

terial which he and Desmaizeaux, his editor, obviously wanted to keep secret. Because Toland provided us with the most important clue about this coterie, it seems appropriate to turn to him for further assistance.

Toland was zealous for the cause of republicanism in Europe and one aspect of the struggle as he understood it entailed the destruction of orthodox and organized Christianity and its replacement by a new natural religion, the first principles of which Toland attempted to devise. Indeed, of all the freethinkers of his time Toland was probably the most original, and therefore, of course, the most outrageous. He began his intellectual career in The Netherlands where he studied in 1692–93, and there he frequented the salon of the English refugee and friend of John Locke, Benjamin Furly. Toland also associated with Jean Le Clerc and Philip van Limborch, the leading liberal theologians of the day, and he knew Pierre Bayle. He carried on a dialogue with Bayle on the subject of materialism, and Toland is the author of objections to Bayle's refutation of Dicearchus published in the *Dictionnaire.*[10] In short, Toland was well-established in the 1690's in liberal and freethinking circles in The Netherlands. As a result Furly sent letters of introduction for him to Locke, but Locke seems to have shied away from Toland although they had mutual friends at Oxford. It was at Oxford that Toland had begun to write his first infamous work, *Christianity Not Mysterious,* which he was only able to publish in 1696 as a result of the lapsing of the Licensing Act in the previous year. Locke knew Toland's book in its manuscript form and we now believe that he wrote *The Reasonableness of Christianity* (1695) in response to it.[11]

There is no necessity here to dwell on other facets of Toland's career, which I have already discussed elsewhere.[12] Rather we need to ask about what light Toland's ideas and activities shed on the clandestine culture of which he was a part, and more particularly on the Knights of Jubilation and their associates. Throughout his life Toland maintained that the true philosophy is Janus-like in its nature. One face, its public or exoteric face, could be revealed in print, and so Toland in his signed works attacked revealed religion and attempted to undermine the liberal and Newtonian intellectual establishment of his day. In public Toland preached a natural religion de-

10. C. Gerhardt, ed., *Die Philosophischen Schriften von Gottfried W. Leibniz* (Berlin, 1882), III, 68. This point is missed in A. Vartanian, *Diderot and Descartes: A Study of Scientific Materialism in the Enlightenment* (Princeton, N.J.: Princeton University Press, 1953), 124n, and 226; but the undoubted link between Cartesianism and materialism seems born out in Picart's development.

11. See *The Newtonians and the English Revolution, 1689–1720,* chap. 6; and John Biddle, "John Locke on Christianity: His Context and His Text," Ph.D. dissertation, Stanford University, 1972, p. 17.

12. Much of what I am saying here about Toland is documented in "John Toland and the Newtonian Ideology." My interest in Toland began in the spring of 1965 when Henry Guerlac suggested to me that he was a figure in need of careful study. That suggestion has proved invaluable.

void of superstition and revelation, yet deistic in its acceptance of a Supreme Being who created and was separate from nature. But Toland's private or esoteric philosophy, revealed in his anonymous books and unpublished manuscripts, was blatantly pantheistic and materialistic. Nature is the only force in the universe, and Toland was, to use a word he invented in 1705, a "pantheist," or perhaps more accurately, a pantheistic materialist.[13]

It should be asserted that to Toland, at least, these two philosophies were complimentary. The deistic message was intended for a larger, less disciplined and less-learned audience; the pantheistic one for an intellectual elite, a mandarin class. Yet I think that Toland imagined that the first, although an adequate and necessary alternative to orthodox Christianity, could spur the truly rational person on to the discovery of the second. The fact that Toland did publish the *Pantheisticon* (1720), albeit anonymously, indicates that he, if not his friends, wanted to achieve a social and religious order where the hylozoic and pagan wisdom of the ages would prevail. Paradoxically secrecy was an essential form of discipline and a means to the achievement of that universalist goal.

This dedication to the semiclandestine appears to go against the mainstream of English republicanism as advocated by Harrington, whose *Oceana* Toland edited in 1700. In the context of the 1650's when Harrington wrote, secrecy would have seemed like the traditional practices of royal absolutism. The desperate struggle of political and religious radicals for survival in the period after the Restoration must have made certain forms of clandestine behavior necessary and therefore intellectually acceptable. That Toland, an English republican, found no difficulty participating in the intellectual milieu of Continental Protestant refugees and European libertines, long schooled in the clandestine mode, may indicate the existence by the early eighteenth century of a set of circumstances and responses, as well as a common culture, shared by freethinkers on both sides of the Channel.

Toland drew his political and natural philsophical ideas from both English and continental sources. He assimilated the pagan naturalism of the late Renaissance, which he learned from the Italian writings of Giordano Bruno (d. 1600), and out of it he fashioned the foundations of a new religion, a rationalist and purely naturalist alternative to any of the prevailing forms of Christianity. For Toland, nature is God and, as he would put it, all things are One, and the One is All.

A new religion, if it is to pose a viable alternative to prevailing orthodoxy,

13. [John Toland], *Socinianism truly stated: being an example of fair dealing in theological controversies . . . recommended by a Pantheist to an Orthodox Friend* (London, 1705). Cf. Anthony Collins, *A Discourse of Freethinking* (London, 1713), pp. 47–48, 119, 150–151; and J. G. A. Pocock, *The Machiavellian Moment* (Princeton, N.J.: Princeton University Press, 1975), chap. 13.

must be practiced, and under the circumstances, practiced in secret. Before Toland returned to the continent in 1701, he belonged to a secret club of Whig republicans who met in London and whose leader, Sir Robert Clayton, has been acknowledged by the official historians of Freemasonry as having headed some sort of early Masonic lodge.[14] This lodge was early in the sense that it predates the founding of the Grand Lodge in 1717. Very little is known about Clayton's lodge except for the names of some of its members.

One of Clayton's associates, John Toland, attempted throughout his life, and indeed succeeded, in establishing a secret society, a "college" of pantheists who practised a secret liturgy and ritual, part of which was revealed by Toland in the *Pantheisticon.* This tract describes the beliefs and rituals of Toland's Socratic Brotherhood, which met in private, probably in London, intoned a liturgy to nature, and was mildly epicurean in its habits. I think that this society was Masonic, yet quite different in spirit and intention from the official Freemasonry of the Grand Lodge. Evidence that such organizations did exist can be found in other published sources as well as from items found in the Marchand manuscripts.[15] Just as Toland's essentially English group occupies no place in the official records of the Grand Lodge, so too the records for Dutch Freemasonry are incomplete and inchoate for the period prior to 1735.

Were the Knights of Jubilation also one of Toland's secret groups, although not necessarily one that accepted his extreme pantheism? Toland proselytized for his new religion during various trips to the continent, in 1701 and 1702 at the Hanoverian court, and in 1707 in Vienna. Then in 1708 he settled down at The Hague and in 1709 the English bookseller, Thomas Johnson, published his revealing Latin treatise entitled *Adeisidaemon.* Johnson became a close friend to the Knights of Jubilation, in particular to Marchand, who arrived in The Hague in 1709, and to the literary society spawned a few years later by Marchand and his friends.

14. A. S. Frere, *Grand Lodge, 1717–1967* (Oxford: Printed for the United Grand Lodge of England by Oxford University Press, 1967), p. 31; Lambeth Palace Library, London, MSS Add. 4295, fol. 28.

15. UL, Leiden, Marchand MSS 59, loose page from a book called *Destinée des Garçons* and inscription on back of page; Cf. Boniface Oinophilus, *Ebrietatis Encomium: or the Praise of Drunkenness . . . by the Example of Heathens, Turks, Infidels, Primitive Christians, . . . Free-Masons . . .* (London, 1723), a translation of a work by A. H. Sallengre. This could not have been sponsored by the respectable Grand Lodge. See also, Eugenius Philalethes, Jr. [Robert Samber], *Long Livers: A Curious History of Such Persons of both Sexes who have liv'd several Ages . . . Most humbly dedicated to the . . . Free-Masons* (London, 1722), and mentioned by the translator of Sallengre's work. Harry Carr, ed., *The Early French Exposures* (London, 1971), pp. 49–51. In this official reprint of an eighteenth-century "exposure" of Freemasonry, recognition is given to the existence of similar organizations, whose libertinism threatened the reputation of respectable Masons. Cf. J. M. Ragon, *Tuileur général de la franc-maçonnerie ou manuel d l'Initié . . .* (Paris, 1861), pp. 86–87, 98.

Long after Toland's death (in 1722) Johnson contemplated a new edition of the *Pantheisticon* and approached Marchand for information about it.

The manuscript notes Marchand supplied to Johnson on the *Pantheisticon* indicate that he believed that anonymous work to be by Toland—"it is a good example of his kind of style, and he oftentimes avowed it"—but then Marchand demures on the subject of Freemasonry: "Barely being familiar with the lodges of the Freemasons, I would not want to judge indiscretely, but it certainly does seem that their singular and bizarre ceremonies, both in spoken phrases and gestures, resemble those found in the *Pantheisticon* as well as in other bacchanalian liturgies, so much so that one might believe that they are one and the same thing under different names." Marchand had no particular reason to be frank to Johnson about the subject of Freemasonry; the publisher had never actually belonged to the Knights or their literary society. But in the early 1750's, Jean Rousset de Missy (1686–1762), who was one of the leaders of organized Dutch Freemasonry, wrote long letters on that subject to Marchand, openly praised "Pantheism," and asked him repeatedly to salute a mutual friend and grand master, "brother" Charles, to address him with "my fraternal compliments," and to embrace him "by the Masonic number," that is, three times.[16] Marchand could hardly have done so if he were not also himself a Freemason, not withstanding what he might have said to Johnson. It does now seem plausible, on the basis of this new evidence, that Toland and the Knights of Jubilation laid one of the pillars upon which European Freemasonry was later to be constructed.

Positions advocated by Toland in 1709 in *Adeisidaemon* do seem to have reflected beliefs held by freethinkers on both sides of the Channel, in particular by the coterie of followers of Eugene of Savoy. Briefly stated, their politics revolved around a projected victory for European Protestantism, which meant in their logic a major setback for monarchical absolutism and hence an indirect spur for the advancement of republican principles. For these republican freethinkers the battle had to be waged on two fronts: against Louis XIV and his armies, but also against the orthodoxies of the

16. UL, Leiden, Marchand MSS 62, unfoliated, labeled in Marchand's hand, "These for Mr. Johnson bookseller at ye Hague, Holland"; Marchand MSS 2, Rousset de Missy to Marchand, fols. 35–36; 46–47; fol. 59, Rousset on his wife, "elle vous salue et le fr. Charles"; fol. 60, "Je vous salue et le frère Charles"; fol. 61, "Je retourne les complimens au frère Charles et vous"; fol. 64, "Je m'étonné de l'approbation que vous donnez à l'objection tirée de notre obstination a révéler nos Mistéres. . . ." This letter concerns an apparent difference of opinion over the question of Masonic secrecy. Cf. fol. 78, 1 October, n.a. but certainly the 1750's: "mes complimens fraternels à Charles & croyez voy plusque je ne puis l'exprimer tout à vous. . . ." Marchand MSS 2, Fritsch to his friends, 17 Xbre 1711, refers to "frères Prosper Marchand et Gaspard Fritsch"; Marchand to Fritsch & Böhm, 3 November 1711, refers to "frère Jacobson" as well as to his manuscript notes on Spinoza's *Tractatus theologico-politicus* which Fritsch has directed him to make.

established churches, for they even more than secular rulers constituted for the freethinkers the bastions of a rigid and backward-looking social and political order.[17]

Adeisidaemon is the unsuperstitious man, and as Toland tells his English friend and fellow freethinker, Anthony Collins, to whom the work is written, he takes his inspiration from the ancient pagans, such as Livy, who were totally rational men and never believed for a second the various superstitions they recount in their histories. According to Toland they were far less superstitious than present-day Christians, but they were never irreligious. Livy and others knew an esoteric religion and practiced a cult of divinity, which is the true religion prescribed by God who is simply the Supreme Reason.[18] This true religion, Toland is quick to note, is not atheism which is dangerous to the state, although as Bayle had noted, less dangerous than superstition. Other ancient philosophers have known this true religion and Toland singles out the Chinese mandarins whom he praises because they believed in the eternality of the world and did not recognize any other divinity distinct from the material structure of the world. They also rejected all teachings concerning the future state of souls.[19] Toland seems particularly impressed with, even envious of, these mandarins for they possessed political power and administered the king's public affairs while at the same time practising their religion in secret.

The importance of *Adeisidaemon* lies in its advocacy of a secret pagan religion known only to initiates which is materialistic yet pantheistic in the sense that God is identified as Reason, a spiritual force in the universe, yet not distinct from it. There is no direct proof that in *Adeisidaemon* Toland is speaking for the Knights of Jubilation or for any other group. Yet clearly during his time at The Hague, Toland associated with an intellectual circle whose subsequent activities were thoroughly in keeping with the Janus-faced philosophy advocated by Toland. Indeed the figure of Janus was used over and over again by Bernard Picart in engravings intended to proclaim the philosophy of Enlightenment.

Throughout their lives the original members of the Knights of Jubilation, and their growing number of associates, dedicated themselves to the spread of Enlightenment culture. Prosper Marchand (1678–1756) and Bernard

17. The peculiar enthusiasm of republicans such as Toland for Eugene of Savoy is briefly mentioned in Franco Venturi, *Utopia and Reform in the Enlightenment* (Cambridge: Cambridge University Press, 1971), pp. 60–66; Marchand MSS 2, Fritsch to Marchand, Paris, 11 Xbre 1712, on procuring books for the prince and his librarian, Hohendorf.

18. *Adeisidaemon, sive Titus Livius . . . Annexae sunt ejusdem Origines Judaicae* (Hague-Comitis: Thomas Johnson, 1709), pp. 52–54. I want to thank Wilfrid Lockwood, University Library, Cambridge, for his translations. Toland's views on Livy were already well formulated by 1706. See Bodleian MSS, Oxford, Rawl. c. 146, fols. 47–48; and Giancarlo Carabelli, "Un inedito di John Toland: Il *Livius Vindicatus,* ovvero la prima edizione (mancata) dell' *Adeisidaemon* (1709)," *Rivista critica di storia della filosofia,* 3 (1976), 309–318.

19. *Adeisidaemon,* pp. 74–75.

Picart (1673–1733) became the best known members of this original circle. Marchand, a French Protestant, established a bookselling business in Paris during the 1690's. In 1709 he removed to The Hague, presumably for religious reasons. While in Paris he became associated with Bernard Picart from at least 1700[20] and both had businesses on Rue St. Jacques in the Latin Quarter. Picart, of course, is one of the finest engravers of the eighteenth century. He was trained by his father, Etienne, and by the late 1690's his unique talent had gained widespread recognition. He became an immensely prolific engraver and drawer, whose work displays not only extraordinary craftsmanship but also, and this is what is important from our perspective, distinct philosophical themes and interests. These interests are evident before his departure from Paris and indeed his departure was necessitated by those interests. Picart was born a Catholic, yet when he was an already established engraver with a lucrative business he converted to Protestantism. This at a time when Louis XIV's campaign against French Protestants was at its height.

Picart's intellectual odyssey only began with his conversion. His engravings and drawings depict philosophical themes commonly associated with the Enlightenment, and his use of an ensemble of symbols and pagan figures provides the iconographical key for understanding those themes. Minerva and Mercury appear time and time again; Mercury in flight showering the earth with books, Minerva triumphant, Minerva as patron of the new science, Minerva surrounded by engrossed *putti* who delve into half-opened books, gaze at partially covered globes, or play with the pots of the alchemist. Always in these philosophical engravings the tools of the mason and the engraver appear, the square and the compass. In one engraving Apollo is taming nature in the form of animals and at the same time building blocks are falling into place, the mason's work completed for him by the harmonious sounds of Apollo's lyre. A matching engraving shows the sun imposing a similar harmony on the earth and on nature.[21]

The basic question inevitably arises as to how closely we can relate these

20. In the Teylers Museum, Haarlem, portfolio 74, there is an "Ex libris" engraving done by Picart for Marchand and dated 1700.

21. For a list of his engravings see B. Picart, *Imposteurs innocents, ou recueil d'estampes* (Amsterdam, 1734). See also MSS lists in the Prints Room, the Rijksmuseum, Amsterdam. Engravings which I have found particularly revealing can be seen in the Teylers Museum, portfolio 237, "L'histoire composant le grand Dictionnaire historique"; portfolio 202, "Monument consacré a la Postériorité en mémoire de la folie incroyable de la XX année du XVIII siècle"; from the Metropolitan Museum of Art, New York, see "Erudit et Ditat," 1722; "Vivitur ingenio cetera mortis erunt," 1728; figure of Apollo and animals, nos. 16, 18, 1718, "Minerva Duce," 1722. From the New York Public Library, Prints Division, Bernard Picart, *Collection de vignettes, petites, estampes et portraits,* nos. 5, 8. Cf. Bernard Picart, *The Temple of the Muses; or, The Principal Histories of Fabulous Antiquity* (Amsterdam, 1733), this work belongs to a tradition of late Renaissance naturalism which glorifies the pagans. This English translation has a preface disclaiming any such intention; in the French edition of 1742 there is no disclaimer.

iconographical themes found in Picart's engravings to a conspicuously articulated philosophy held and practiced by the Knights of Jubilation. The Marchand manuscripts contain a collection of notes or minutes, dating from 1711 to 1717, made for "les membres de notre mémorable corps hermétique," as the society is described in one such memorandum. In these jottings the linkage between symbolism and philosophy is made clear and emphatic. One note appears to be a speech delivered, possibly by the Grand Master, to the brothers and it states: "My advice would be therefore, my very dear brothers, if you approve, that we [take] for the body of our motto [*notre devise*] a sapling, which Minerva plants, and Mercury waters to make it grow. A small shoot soon becomes a tree."[22] Later in the same speech the gathering is told that "what had made me choose this device is the connection it has not only with all the order in general, but also with each member in particular." In this context one member, brother Laurent, is described as "Mercury's bishop" and after polite references to the various abilities of all assembled, the speaker concludes:

> You are well aware, I know, that our order, although illustrious, is not at present the largest nor the most powerful in the world, but it could become so. This is why I compare it to a small shoot, which with time will grow into a large tree. It is planted by Minerva, goddess of wisdom and the Protectress of those who cultivate the sciences, to show that we have had solely the same aim in mind. . . . Riches help a lot in making states, commonwealths, and private persons important: thus our order will both appear to be and will be great when our treasury is full. But how is it to be filled? Brother Laurent has told us several times that it is by trade [*le Négoce*] that we can make our organization rich by trade; over which Mercury presides as God. I believed one couldn't do better than to introduce this into my device.[23]

The proposal was adopted and over and over again on the frontispieces of books published by one or another of the Knights, the Picart engravings of Mercury or Minerva appear. Even on freethinking treatises such as the 1734 Rouen edition of Tyssot de Patot's *Voyages et aventures de Jaques Massé,* a Picart engraving of Mercury entitled "Erudit et Ditat" appears.[24] These symbols apparently became recognized and used by the libertine and clandestine publishing world of the eighteenth century; the origin of that usage appears to lie with the Knights of Jubilation and their inventive and enterprising publishing organization. Picart, Marchand, and their friends combined their dedication to the philosophy of Enlightenment, expressed through carefully chosen symbols, with the calculated pursuit of material interests. This secret society of businessmen sponsored a literary journal and

22. Marchand MSS 1, *varia,* unnumbered, but fol. 3 if counted in sequence. I wish to thank Clarissa Campbell Orr for assistance with transcribing and translating these very difficult notes.

23. Ibid., fol. 5.

24. A. Rosenberg, *Tyssot de Patot,* pp. 94–96, 117.

numerous publishing ventures. In this fascinating instance libertine ideals and the necessities and rewards of business proved totally compatible.

All aspects of this cultural and mercantile endeavor, but in particular Picart's engravings, merit a separate study and it is only possible here to hint at their relationship with earlier intellectual traditions. So many of the symbols that appear in his work harken back to a tradition of Renaissance naturalism, associated most commonly with the school of Palladio, of whom Picart was an admirer.[25] One central aspect of that Hermetic tradition was the belief in a lost and esoteric wisdom, capable of being recovered and transmitted by initiates to other receptive thinkers.[26] The engravings of Picart reveal that he believed in such a wisdom, a special knowledge, an enlightenment, which in his time must be proclaimed not openly but opaquely by the use of complex symbols. What for the Renaissance hermetists such as Bruno had been a mystical and magical secret wisdom becomes in the hands of these early eighteenth-century freethinkers a philosophy of reason and enlightenment. There is still a magical and mysterious quality about this wisdom—the decorative *putti* are clearly intrigued by this magic—yet the truth is discerned not by *magi,* but by enlightened men, unafraid to use their reason and free to think, to discuss, to debate, and even to practice their beliefs in private. Minerva and Mercury are pagan symbols of this lost but now recovered or remembered wisdom which is not static but progressive.

Its progress depended heavily upon the printed word, and Picart, Marchand, and their associates in the Knights of Jubilation collaborated in printing, publishing, and editing ventures. When Marchand arrived in The Hague he quickly became involved with friends and associates of Pierre Bayle (d. 1706) and the task of putting out an edition of Bayle's correspondence fell to him. Böhm published and Marchand edited the 1720 edition of Bayle's *Dictionnaire historique,* and Marchand spent many years of his life revising Bayle's enterprise and eventually produced his own unique version of the dictionary which was published posthumously. The title page of the 1720 edition is decorated with a symbolic engraving by Bernard Picart.

In a highly original, if overly ambitious essay, Dorothy Schlegel argues persuasively that the use of iconography in certain eighteenth-century literary projects was intended as an elaborate device for proclaiming a secret wisdom of Enlightenment.[27] She argues that this symbolism is Hermetic and

25. The Teylers Museum, Haarlem, an engraving by Picart for an edition of works by Palladio.

26. See the work of Frances Yates, *Giordano Bruno and the Hermetic Tradition* (London: Routledge & Kegan Paul, 1964); *The Art of Memory* (London: Routledge & Kegan Paul, 1966).

27. Dorothy Schlegel, "Freemasonry and the *Encyclopédie* Reconsidered," *Studies on Voltaire and the Eighteenth Century,* 90 (1972), 1433–1460. Cf. Georges May, "Observations on

Masonic in its origins, and that its appearance on the frontispieces and title pages of three major works of the Enlightenment argues for the existence of a secret society of *philosophes,* which was Masonic, or more precisely, which subscribed to a type of Freemasonry that Schlegel calls Minerval Freemasonry. This iconographical evidence appears in Shaftesbury's *Characteristicks* (1714), the 1720 edition of Bayle's *Dictionnaire* published by Böhm, and the frontispiece for Diderot's *Encyclopédie.* Certainly these iconographic similarities require explanation.

It is most improbable that Diderot or his publishers belonged to some sort of secret and Masonic society when they were preparing the publication of the *Encyclopédie* (1751). It is more probable that in employing Masonic symbolism on the frontispiece they were seeking to sell books, possibly even appealing to the prosperous lodges to purchase copies. In so doing, however, they may have been acknowledging the force of an iconographical tradition that can in large measure be traced to Picart and his French and English associates. Indeed Picart was suggested to Shaftesbury as a replacement for his engraver. The 1720 edition of Bayle's *Dictionnaire* was the work of two of the Knights, and Schlegel presents quite convincing evidence, drawn from Shaftesbury's manuscripts, to prove that he intended the artistic symbols used on the treatise plates of the three-volume 1714 edition of *Characteristicks* to reveal, through the use of a secret symbolic code, a hidden message to be understood only by initiates.

What we have here is an intellectual tradition, not a historically continuous secret society, which was first given expression in pagan symbolism and which was strongly pantheistic and therefore materialistic in its orientation. Its source lies in this international republican coterie based in England and The Netherlands. Toland, Collins, the Knights, and Shaftesbury to a lesser degree, were at the matrix of that tradition. Certain common artistic symbols identified it and appear repeatedly in the literary works first grouped together by Schlegel: Janus-faced figures, decorative rose branches, fermenting pots, masonic tools, half-opened curtains, partially veiled faces or figures, rays of an illuminating sun, and of course and most important, the figure of Minerva. Many of these same symbols recur in other engravings by Picart, oftentimes intended to illustrate philosophical themes. A shared intellectual tradition, with a common symbolism that in the course of the eighteenth century became increasingly associated with Masonry, links Shaftesbury's *Characteristicks,* the 1720 edition of Bayle's *Dictionnaire,* and finally Diderot's *Encyclopédie.*

---

an Allegory: The Frontispiece of the *Encyclopédie,*" *Diderot Studies,* 16 (1973), 159 ff.; Felix Paknadel, "Shaftesbury's Illustrations of *Characteristics,*" *Journal of the Warburg and Courtauld Institutes,* 37 (1974), 290–312; Coste to Shaftesbury recommending Picart, October 1712, The Hague; PRO 30/24/45, fol. 581.

It should now be clear that the Knights of Jubilation and their associates occupied a central place in the clandestine culture that spawned this subversive and enlightened tradition. They were its earliest bookdealers, editors, artists, literary collaborators, and propagandists. From 1710 when Picart and Marchand set up business together in Hoogstraat at The Hague[28] and also made contact with Toland and later with Anthony Collins, they and their friends became the social nexus out of which that pantheistic tradition grew and flourished. While Marchand increasingly withdrew from business affairs and devoted himself to scholarship, other Knights of Jubilation pursued their international publishing interests. Gaspar Fritsch, the Grand Master, was a German Lutheran from Leipzig. Arriving in Amsterdam in 1706, he went into a publishing business with Michael Böhm three years later. That business spread to Rotterdam and Leipzig until in 1715 the partnership was dissolved. Fritsch went to The Hague while Böhm became associated with another Knight, Charles Levier.

As I have noted elsewhere, Levier had quite a career as a circulator of clandestine manuscripts. Indeed the infamous and explicitly pantheistic *Traité des trois imposteurs,* first published in a very rare edition by Böhm in 1721, can be traced to Levier and another associate of this coterie, Jean Rousset de Missy. Levier got his hands on spinozistic manuscripts in the library of the English republican, Benjamin Furly, in Rotterdam, and together he and Rousset created an impious text that labeled Jesus, Moses, and Mohammed as impostors. It is not possible here to explore that text; likewise we must wait until the end of this essay for a discussion of the contact between Levier and Voltaire, who in 1722 was attempting to find a publisher for his *La Ligue,* later called *La Henriade.*[29] Here it must be sufficient to say that the Marchand manuscripts at Leiden show Fritsch and his associates as aggressive publishers with distinct intellectual interests who circulated their books to a wide market that extended into France and Germany.[30]

As we look at the social history of the Enlightenment in one of its earliest and most radical manifestations, it seems essential to delineate, as closely as possible, the exact beliefs of this coterie and to indicate, whenever possible, its indebtedness to previous intellectual traditions. Toland's pantheistic ma-

28. I. H. van Eeghen, *De Amsterdamse Boekhandel,* III, 233. Picart was a widower at this time. Marchand was married. When Toland lived at The Hague he resided for a time on Honslaerdyke. The literary society met in a coffee house on Korte Voorhout.

29. J. S. Spink, "La diffusion des ideés," p. 254n; I. H. van Eeghen, *De Amsterdamse Boekhandel,* III, 130; J. Vercruysse, "Voltaire et la Hollande," *Studies on Voltaire and the Eighteenth Century,* 46 (1966), 29–31; cf. Margaret C. Jacob, "Newtonianism and the Origins of the Enlightenment: A Reassessment," *Eighteenth-Century Studies,* 11 (1977), 1–25.

30. For example, UL, Leiden, Marchand MSS 2, Fritsch to Marchand, 7 Août 1711, on selling the *Cymbalum mundi* in Rotterdam and Basle; 11 Xbre 1712, on sending copies of M. David's *Parodies bachiques.*

terialism seems representative enough of its beliefs; but can we be more precise? Certainly there is no evidence that the Knights ever chanted the ritual prescribed in the *Pantheisticon;* but they did sing ribald and political songs together, copies of which turn up amid Marchand's papers.[31]

Fortunately Marchand and Picart published both separately and together, and this literary evidence coupled with the artistic evidence already discussed should provide a more exact representation of the beliefs that guided this freethinking coterie. While still living in Paris, Picart made an engraving in 1707, at the request of a M. Brillon, entitled, "La Vérité recherchée par les philosophes." Intended as the frontispiece for Brillon's thesis in philosophy, this engraving makes a philosophical statement by the use of an ensemble of figures and symbols. At the center of the picture is Minerva, taken in this instance by contemporaries to symbolize science, and she is slaying the monster of ignorance with the assistance of the figure of Time. The figure of Truth, also one of her assistants, leads a parade of modern philosophers, with Descartes at the head, followed by Zeno, Socrates, Plato, and Aristotle, who are accompanied by Pythagoras, Epicurus, and Diogenes. Gracing the scene are various instruments or tools and *putti,* who are diligently inspecting a half-shrouded globe. This engraving exultantly proclaims the triumph of the new science which without any divine assistance has discovered truth. And lest the point be missed, the engraving is accompanied by a text that explains its meaning.[32]

A prominent Parisian professor (presumably at the Sorbonne, although this is not clear) thought he saw in the figure of ignorance with the ears of an ass, a representation of Aristotle and in the figure of Minerva, the queen of Sweden, Descartes' great friend and patron.[33] Picart and presumably Brillon found themselves in difficulties and the following year Picart did a retraction in the form of another engraving with the suitably pious title, "L'Accord de la Religion avec la Philosophie, ou de la Raison avec la Foi."[34]

The first engraving is important for two reasons: it shows the early use of symbols and figures which Picart would perfect and use on frontispieces and title pages of books with a philosophical message, and it also indicates his early commitment, that is, prior to his departure from Paris, to the new science and its power to enlighten. This 1707 engraving proclaims the pagan authors led by the moderns as guides who will show men through the use of

31. UL, Leiden, Marchand MSS 55, 59. These collections are terribly confused and possess no discernible order. It is not clear who has copied the songs and poems. The first collection is entitled "Vers et chansons sur la cour de France."

32. A copy of this engraving can be seen in the Picart collection at the Prints Room, Metropolitan Museum of Art, New York. It was probably made for Pierre Jacques Brillon who wrote, among other works, *Suite des caractères de Théophraste et des pensées de Mr. Pascal* (Amsterdam, 1701). He was some sort of Jansenist.

33. *Nouvelles de la république des lettres* (Amsterdam, August 1707), pp. 232–234. Published by J. Bernard, with contributions by P. Bayle et al.

34. A copy of this engraving is at the Teylers Museum, Haarlem, portfolio 131.

pure reason how to discover truth and to destroy ignorance. We do not know if Picart intended a direct attack on the theological faculty of the Sorbonne and its dedication to scholasticism, but clearly the engraving aroused instant suspicions. If the learned doctors had known what Picart had been reading in 1706, a year before he did that engraving for Brillon, they would have been even more disturbed.

In 1711, Prosper Marchand edited and published an extremely rare work, and in its time an extremely heretical one, entitled *Cymbalum mundi ou dialogues satyriques sur differens sujets* published originally in 1537 by the naturalist, Bonaventure Des Périers. The title page of Marchand's edition is adorned by a small engraving designed by Picart.[35] At the center is a large star surrounded by a constellation of seven smaller stars and the central star appears to symbolize the sun or light which is radiating onto the earth. Framing this central theme are two cornucopia yielding forth open books and tools, among them a ruler and compass, the tools of workmen such as masons and engravers. The words "Inter Omnes" appear at the top of the engraving, which if translated as "among all people" carries the clear implication that the book contains a universal message worthy of illuminating all who read it.

The volume is introduced by a letter written by Marchand to Picart in October 1706, and in it Marchand argues that despite its infamous reputation the *Cymbalum mundi* was neither impious nor atheistical. Marchand does admit that one of the author's secret intentions may have been "to turn into Ridicule whatever is believ'd in Religion,"[36] but in general Marchand simply claims that Des Périers' intention was "to laugh indifferently at all the World." He rightly places Des Périers in the naturalistic and satirical tradition of libertinism made famous by Rabelais, and indeed Des Périers apparently did know people in Rabelais' circle.[37] Marchand tells Picart that he agrees with Bayle: instead of trying to convert the ancient pagans we should laugh at them. Yet Marchand is not simply laughing at Des Périers' account of Mercury's difficulties during a trip to earth, he is also laughing with Des Périers' satire on the alchemists and with the spirit of the book in general. In this work Minerva is the only pagan deity with any sense; she sends a message to Mercury telling him to chastise earthly poets for squabbling among themselves and to get down to the serious business of writing about love.

To understand the significance that this book must have had for Mar-

35. Marchand's edition was published in Amsterdam. For a discussion of this text see Dorothea Neidhart, *Das "Cymbalum Mundi" des Bonaventure Des Périers* (Geneva and Paris: Ambilly-Annemasse, 1959), p. 48.

36. A convenient English translation of this letter and the *Cymbalum* appeared in London, 1712; *Cymbalum mundi. Or Satyrical Dialogues upon Several Subjects. To Which Is Prefix'd a Letter . . . by Prosper Marchand.*

37. See A. J. Krailsheimer, ed., *Three Sixteenth-Century Conteurs* (Oxford, 1966).

chand and his circle we must place it in the context of European Calvinism. John Calvin attacked Des Périers and his naturalistic philosophy as an impious atheism. Des Périers was regarded by contemporary Calvinists as a Protestant who had once seen the light but had abandoned Protestantism for the pagan naturalism of the late Renaissance which Calvin so bitterly despised. Des Périers' satire was interpreted by his contemporaries as a thinly-veiled assault on Protestantism, on Luther in particular, and also on Catholicism. Published only about a year after the first edition of Calvin's *Institutes* (1536), the *Cymbalum mundi,* by a man who had once been involved in Reformation circles at the court of Margaret of Navarre, established Des Périers as a notorious freethinker and atheist.[38] It is no wonder that despite their interest in this book when they still lived in Paris, Marchand and Picart waited to publish their edition in the more tolerant atmosphere of The Netherlands.

All the evidence we now possess about the Knights of Jubilation leads to the conclusion that publicly they were Protestants of sorts, committed to the cause of international Protestantism as were most French refugees. They were also followers of Bayle, to whom they often pay tribute in their writings. Privately, however, a richer and more complex picture unfolds. Although many of them retained church membership, the Knights at their secret meetings sang ribald and republican songs and wrote letters mocking the forms of organized religion. They also discussed philosophical subjects and they certainly planned literary and business projects. They corresponded with other freethinkers, for example, with Pierre Desmaizeaux and with liberal Huguenots such as David Durand in England who wrote a life of Lucillo Vanini.[39] Yet none of the major publications for which the Knights were responsible, such as the *Journal littéraire* (1713–1722), indicated the depth of their private unorthodoxy or the existence of their organization. Only in his last will and testament did Marchand permit himself to castigate orthodox religion, and he described its ceremonies "not only as vain and contemptible, but also as criminal and worthy of condemnation; they should be seen as the great abuse which they in fact are."[40]

But the abandonment of either Protestantism or Catholicism meant that something had to be put in its place. For Marchand, Picart, and their friends this meant a commitment to the spread of the new learning, specifically to

38. Ibid., p. 79. Cf. Lucien Febvre, *Origène et Despériers ou l'énigme du "Cymbalum mundi"* (Paris, 1942).

39. No Durand letters survive in the Marchand MSS, but he is frequently mentioned, e.g., Marchand MSS 2, C. Fritsch to C. Levier, 27 May 1711. David Durand, *La vie et les sentimens de Lucillo Vanini* (Rotterdam: Caspar Fritsch, 1717). (Fritsch's first name alternatively appears as Casparus or Gaspar.) This edition is adorned by a Picart engraving depicting Mercury spreading knowledge over the world and the motto "Terrarum ubique munera spargit." Cf. A. Barbier, *Notice sur la vie et les ouvrages de David Durand* (Paris, 1801).

40. E. F. Kossmann, *De Boekhandel,* 252.

the spread of Enlightenment publications. Hence their work on Bayle's *Dictionnaire* and their interest in books of every sort, an interest revealed in the hundreds of letters to be found in the Marchand manuscripts at Leiden. Marchand and his friends were not simply involved in the publishing trade as a business, although they were obviously skillful business men for they managed to survive in the very competitive world of Dutch publishing. For them the printed word provided a means to an end, and the goal, despite the necessity for secrecy, was universal enlightenment.

Marchand wrote one of the first histories of printing, the lengthy *L'histoire de l'origine et de premiers progrès de l'imprimerie* (The Hague, 1740). Once again on the engraved frontispiece of *L'histoire* we find Minerva and Mercury enthroned as the patrons of printers and their trade. In pagan naturalism can be found a universal religion of reason and brotherhood which the wisdom of the ancients vindicates and which holds the new learning, the new Enlightenment, represented by the figure of Minerva, as the only body of truth worth knowing or believing. This is a secret wisdom that only a chosen few are capable of discerning, but it is their obligation in turn to spread at least a portion of this learning through the art of printing and publishing.

Gradually the Knights of Jubilation chose new members and associates; the Marchand correspondence contains some details about this process of selection. The Knights constituted an original organization which spawned a literary society, the existence of which was known to contemporaries because it published the *Journal littéraire.* Yet the actual membership of this society and the range of its activities remained a well-kept secret, much to the annoyance of some contemporaries. The main participants in the editorial work of the journal were Marchand, Justus Van Effen, A. H. de Sallengre, Saint-Hyacinthe, a M. Alexandre, and s'Gravesande. Yet as the Marchand manuscripts indicate their literary society contained a far larger membership: Fritsch, Böhm, and Levier, along with Isaac M. Vaillant, a publisher at The Hague, F. le Bachellé in Utrecht who was a friend to Rapin de Thoyras, the Whig historian; a Mr. Boyd, and David Durand (1680–1763), the freethinking French Protestant minister, among others.[41]

Once again, as he had in the Knights of Jubilation's register, Caspar Fritsch signed himself "Le Grand Maistre" when writing to Alexandre, and Picart is mentioned in this correspondence, although it is unclear whether he belonged to the literary society. The Marchand correspondence contains hints that there may have been circles within this larger circle. For instance, Fritsch's letters to Marchand are often bawdy and irreverent, and contain

41. UL, Leiden, Marchand MSS 1, F. le Bachellé to Marchand, Fritsch and Böhm, Utrecht, 16 Septembre 1713; Marchand MSS 2, à Secrétaire de l'ordre, de Le Grand Maistre, à Bruxelles, 25 8bre 1712. The handwriting is that of Fritsch, as in Marchand MSS 2, Fritsch to Levier, Rotterdam, 7 Août 1714.

quotations jauntily pulled from the Latin Mass, while other letters to and from members are characterized by a certain distance and formality. Possibly the Knights and their friends shared secrets to which some members of the literary society were not privy. Secrecy can after all have its compensations, not least of which is the amusement and self-satisfaction derived from being more secretive than other practitioners of the art.

The literary society did produce one other joint literary effort generally attributed to Marchand, Sallengre, Alexandre, s'Gravesande, and most especially Saint-Hyacinthe. *Le chef d'oeuvre d'un inconnu* is an amorous and bawdy satire against the fanatic proponents of ancient learning.[42] This narrative poem or song recounts the amorous adventures of one Colin and in the process pokes fun at scholastic argumentation. It is a piece of badinage in keeping with the spirit of many of the letters found in the Marchand manuscripts. Before the text itself the authors publish various salutations sent to themselves, or actually in praise of one Dr. Matanasius, the supposed author of this masterpiece. These salutations are mostly anonymous, but in a copy of *Le chef d'oeuvre d'un inconnu* found in the Bibliothèque Nationale an equally anonymous hand has identified some of the authors of these encomia. They are: M. Alexandre, Thomas Johnson, publisher of *Adeisidaemon* and the *Journal littéraire,* and Justus Van Effen.[43] These predictable followers can be added to a list of well-wishers actually printed in the text which is headed by Henry St. John, Viscount Bolingbroke. From the point of view of political ideology, this was an improbable alliance not easily explained. It appears that a dedication to libertinism managed to transcend even the divisions provoked by the European wars.

As yet no other evidence exists to prove that Bolingbroke actually belonged to the literary society of Saint-Hyacinthe, Marchand, s'Gravesande, and others, but his entirely unnoticed association raises some interesting questions.[44] It implies that Bolingbroke was belatedly courting the Dutch followers of Eugene of Savoy whose cause he had done so much harm to at the bargaining table that concluded the Peace of Utrecht (1713). It is equally possible that this brief encounter with a republican coterie linked to radical and country Whigs in England may betray Bolingbroke's desperate attempt to forge a last minute alliance between disaffected Whigs and Tories against

42. Chrisostome Matanasius, *Le chef-d'oeuvre d'un inconnu. Poème, heureusement découvert & mises à jour avec remarques savantes & recherchées* (The Hague, 1714); published by "la Compagnie."

43. Bibliothèque Nationale, Paris, Rare Books, Res. 2. 2071, *Le chef-d'oeuvre* (The Hague, 1716). There are also hand written notes on Lucian's *History.* Perhaps this was the author's copy.

44. He is mentioned by Andre Lebois, ed., *Le chef d'oeuvre d'un inconnu, 1714, par Saint-Hyacinthe* (Avignon, 1965), p. 47. No mention is made in H. T. Dickinson, *Bolingbroke* (London: Constable, 1970), but see chap. 9; or in W. Sichel, *Bolingbroke and His Times* (New York, 1968), 2 vols., reprint of 1901–1902 edition.

a court-favored oligarchy. To these conjectures must be added others. If Bolingbroke knew this circle of freethinkers in The Netherlands prior to 1714 how closely was he associated with such groups of freethinkers in England, even perhaps before his exile? We know that he entered into this clandestine world after 1714 and his exile in France, and if Bolingbroke is associated with Saint-Hyacinthe and Marchand, what in turn was their association with similar groups in Paris? The Marchand manuscripts indicate that members of his circle made frequent business trips there, and indeed all over northern Europe. By undertaking an investigation of these comings and goings it should be possible to piece together a more complete picture of clandestine culture in this period than we now possess.

Bolingbroke belongs to the second generation of the Enlightenment, among those philosophes who came to intellectual maturity in the 1720's. That process of maturation was accelerated for him when in late 1716 he was introduced to Parisian intellectual circles by the Duke of Berwick.[45] Of course Bolingbroke was in exile for his sudden Jacobitism and he circulated in France among other English exiles of similar persuasion. The most recent research on the origins of Freemasonry in France has revealed that the first lodges there were started by Jacobite exiles.[46] We do not know if Bolingbroke ever became involved in this Freemasonry of the 1720's, but his connection with a literary society in The Netherlands, composed largely of French refugees, headed by a Grand Master and governed by a constitution, raises some unsettling questions.

At some point an explanation has got to be found for the relationship between the Knights of Jubilation and official organized Freemasonry of the liberal and enlightened variety. If in 1710 this group of freethinkers in The Netherlands had in fact established a Masonic lodge, however inchoate its organization and ritual, then this fact, if it is accepted as such, plays havoc with the established history of Freemasonry in England, France and The Netherlands.[47] First of all, it would force a radical revision in chronology; no more could the establishment of the Grand Lodge of London in 1717 be taken as anything more than the culmination of a transformation that had been well underway from at least the 1690's. More important, however, for the history of Freemasonry and the Enlightenment, would be the knowledge that speculative Freemasonry on the continent began with a group of republican freethinkers, imbued with the pagan naturalism of the sixteenth century, and not with the respectably liberal, court-centered and Newtonian

45. Dickinson, p. 156.

46. Pierre Chevallier, *Histoire de la Franc-Maçonnerie française* (Paris, 1974), I, 3–24.

47. In The Netherlands, Freemasonry supposedly began only in the 1730's. See Abbé Perau, *Le secret des Franc-Maçons,* Nouvelle Edition (Paris, 1744), 120–125, "Extrait d'une Lettre de Hollande du 17 Mars 1737." This Dutch masonry did have a symbolic aspect and used the rather standard symbols of the sun, moon, and opened compass; its membership was also in contact with French masons in Paris.

Christianity of Jean-Theophile Desaguliers, the Duke of Montague, and the Reverend James Anderson, the original founders of the 1717 Grand Lodge.[48] It is absolutely essential that we know more about the freethinking and republican culture of the early eighteenth century before we can assess its impact upon the formulation of Enlightenment beliefs, values, and social organizations.

One vital link exists between the Knights of Jubilation and the very center of Enlightenment thought and culture. In 1722, Bolingbroke became for a time Voltaire's philosophical mentor. Late in that year Voltaire visited Bolingbroke at La Source and read to him portions of *La Henriade.* When Voltaire made that visit in November or December of 1722 he had just returned from a trip to The Hague. There he had resided on the same street where Marchand and Picart had their book business and residence, and during that visit Voltaire had met with Charles Levier, "joker and drinker" of the Knights of Jubilation. Levier was trying to arrange a subscription edition of *La Ligue,* as *La Henriade* was called at that time. Voltaire wanted Picart to do the engraving for it but he had taken on too much work and could not comply with the request.[49] From what little we know about Voltaire's contacts in The Netherlands, it must be concluded that if he was involved in any circle there, it must have been with the Knights of Jubilation.

In a curious and anonymous manuscript poem dated 1726 and deposited in the Bibliothèque Nationale, Voltaire is accused of having attended services at a synagogue in Amsterdam where he played at a "culte secret" and where rabbis practised a "pantomine indiscret." Legend has it that Voltaire attended services of the Amsterdam synagogue in 1722, but no other evidence exists for his having been allowed to attend a Jewish service at that time, whether reverently, or irreverently as one biographer has claimed.[50] Could it be that Voltaire attended a ritualistic meeting of the Knights of Jubilation or a meeting of the larger literary circle some of whose members had promoted the *Journal littéraire*? It is at least plausible that contemporaries knew of the existence of such secret groups with their own private rituals and in turn easily and maliciously confused them with Jewish groups or sects. We shall probably never know if Voltaire attended a meeting of this clandestine circle, but it is certain that by 1736 Voltaire was on close terms with Prosper Marchand, whom he had known from the 1720's and whom he trusted to receive and keep his private letters.[51]

By the 1730's, however, we have arrived at a different world. The assault

48. Other evidence for this hypothesis exists; see *The Newtonians and the English Revolution, 1689–1720.*

49. T. Besterman, ed., *Voltaire's Correspondence* (Geneva, 1953), I, 119, 123.

50. Vercruysse, "Voltaire et la Hollande," p. 30, quoted from Bibliothèque Nationale, MS fr. 12654, fol. 163. In the 1716 edition of *Le chef d'oeuvre,* greetings are sent supposedly by two rabbis, one a moderator of the Amsterdam synagogue.

51. Besterman, ed., *Voltaire's Correspondence,* no. 1170, Voltaire to Marquis d'Argens.

of the philosophes against established orthodoxies was well under way, and while secrecy was still necessary, enlightened men had established their channels of communication, Masonic lodges were common enough in England and France, and a new culture had established its right to public existence. But the ground had been laid for that victory by the freethinkers of the early eighteenth century, whose clandestine culture, so carefully guarded as to be still obscure, nurtured the Enlightenment in it earliest and most precarious years.

CABANIS AND THE REVOLUTION: THE THERAPY OF SOCIETY

MARTIN S. STAUM

ANALYSIS of the political career of the physician P.-J.-G. Cabanis (1757–1808) reveals no simple derivation of political action from his science of man because his practical commitments preceded fully elaborated theory. Though he was very much the philosopher-participant in several phases of the French Revolution, he never wrote a planned treatise applying psychophysical correlations to physical improvement of the human species. Yet Cabanis' observations and assumptions about human nature, ethics, and society produced structural similarities in the concepts of Physiological Ideology and in ideas relating to public assistance and to education. These parallels should help show the distinctiveness of Cabanis in the sometimes diverse philosophical family of the Idéologues.[1]

Cabanis' basic empirical generalization about human physiology was that physical sensitivity is the most significant determinant of all behavior, including mental ability and moral judgment. To improve intelligence and character, moralists and legislators would need to affect sensitivity. Natural diversity and variation in sensitivity were inevitable because of differences in

This paper is an expanded version of a communication to the Fourth International Congress on the Enlightenment in New Haven in July 1975. I wish to acknowledge the assistance of a Canada Council research grant in the summer of 1974. Substantial portions of this essay appeared in another form in Chapters V and X of *Cabanis: Enlightenment and Medical Philosophy in the French Revolution* by Martin S. Staum, published by Princeton University Press,

1. See Sergio Moravia, *Il tramonto dell'illuminismo: Filosofia e politica nella società francese (1770–1810)* (Bari, 1968), pp. 97–107, 323–327, 305–313, 352–369, for topics discussed here, as well as the same author's *Il pensiero degli Idèologues: Scienza e filosofia in Francia (1780–1815)* (Florence, 1974), pp. 1–288. See also my brief survey, "Cabanis and the Science of Man," *Journal of the History of the Behavioral Sciences,* 10 (1974), 135–143, and "Medical Components of Cabanis's Science of Man," *Studies in History of Biology,* ed. William Coleman and Camille Limoges 2 (1978), 1–31.

age, sex, original temperament, and history of disease. Thus there were limits to the science of man—no one could induce the temperament of a healthy, sanguine adolescent girl in the body of an aged, ailing, phlegmatic man. But the goal of the Physiological Ideologist was both to be aware of natural limitations and to produce, if possible, an "acquired temperament" with the more acute perception and more refined sensitivity conducive to good citizenship. Use of the variable agents of climate and regimen would overcome unhealthful organic imbalances or physiological dysfunction which might handicap intelligence or produce antisocial passions.[2]

The medical practitioner of the science of man would certainly respect the basic maxim of Hippocratic medicine—that nature heals itself. Cabanis' history of medicine even described the origins of the art as a successful imitation of natural cures. But physicians in society must use reason to supplement instinctive determinations (I, 40–41). As a late eighteenth-century disciple of Hippocrates, Cabanis recognized that sometimes nature's "misguided efforts must be stopped or channeled in another direction" (I, 21). With techniques of clinical observation and "analytic" method, the therapist must act to cure disease (I, 19–21; II, 149). Similarly, the moralist and legislator need to enlighten natural self-interest and cultivate natural sympathy to achieve a harmonious society.

Cabanis' eclectic version of the Enlightenment search for a natural morality remained optimistic regardless of the human attribute he emphasized. When he followed the economists' stress on self-interest, he confidently reconciled private happiness with the public good. When he followed the Scottish moralists and Jean-Jacques Rousseau in writing about natural sympathy, he made social harmony almost a natural phenomenon by relating it to basic physical forces, from attraction to physical sensitivity (I, 576, 578; II, 321–322). Indeed, like Condorcet, he envisaged a society in which unethical behavior would be considered foolish or mad (II, 513–516).

In political life, however, Cabanis could not trust to inevitable natural harmony in a society so corrupted by Old Regime habits and, later, by revolutionary turmoil. In the areas of assistance to the poor and education, Cabanis' fundamental interest was secularization—replacement of church functions by the state, so that no priests or theology interfered with learning, character formation, or ideas about individual dignity. To help change old habits while still believing in natural harmony, Cabanis had to balance his stalwart defense of individual natural rights with a vision of public utility which permitted some government intervention. The conventional image of the Idéologues as bourgeois liberals should be refined to account for the insistence of spokesmen such as Cabanis on state action, if only to achieve

2. *Oeuvres philosophiques de Cabanis,* 2 vols., ed. Claude Lehec and Jean Cazeneuve (Paris, 1956), I, 196, 356–358, 618 (this edition hereafter cited in text by volume and page numbers).

natural equilibrium. As the revolution progressed, Cabanis acquired more interventionist views in education to stem the tide of reaction while he became more cautious about government assistance to the poor.

Cabanis' family background and intellectual formation predisposed him to belief in gradual social and technical reform. His father, an upper-bourgeois agricultural entrepreneur from a judicial family of Bas-Limousin, helped found in 1759 the Brive Society of Agriculture, later under the patronage of the intendant A.-R.-J. Turgot.[3] Doubtless, Cabanis' early introduction to the Paris society of Denis Diderot, P.-H. Thiry d'Holbach, and Benjamin Franklin in the salon of Madame Helvétius focused his attention on the injustice of Old Regime privileges. But in the early phase of the Revolution, Cabanis' elegant surroundings and professional status as a physician did not inevitably lead him to defend the propertied elite. In the spring of 1790 he appeared before the Paris Commune in defense of peasant rioters in his native region and sided with prorevolutionary Brive officials who deplored "atrocious excesses committed against the inhabitants of the countryside by the seigneur and other so-called privileged persons."[4]

Indeed, concern with the endemic population of beggars, the food shortages, and poverty at the root of such disturbances became a major theme of Cabanis' first major publication, *Observations sur les hôpitaux* (written in 1789, published in 1790; I, 3–31). During his service with the Department of Paris Hospitals Committee in 1791–1792 and during the Directory, Cabanis refined his views on public assistance. *Quelques principes et quelques vues sur les secours publics* appeared in 1803 as a revised and expanded version of Hospital Committee reports.

The remarks on inequality which appear consistently in Cabanis' works on public assistance complement his general philosophical views about the necessity of some natural variation and the unhealthfulness of extremely unbalanced sensitivity. Already apparent in 1789, the philosophy of inequality hardened considerably in a book review of 1798 before its complete formulation in the essay of 1803. In the fully developed exposition, Cabanis argued, somewhat as Rousseau had, that society should encourage "natural," and eliminate "artificial," inequality. The free exercise of natural faculties—such as the acquisition of property by the labor of the strong, skilled, diligent, and prudent accounted for natural inequality. Indeed in

3. Archives départementales (A.D.) de la Corrèze, Tulle, paroisse de Cosnac B.M.S. 1753–1792; Louis de Nussac, "La 'venue' de Cabanis," *Bulletin de la société scientifique, historique, et archéologique de la Corrèze,* 44–45 (1923), 243–270; René Lafarge, "La société d'agriculture de Brive," *Bulletin . . . Corrèze,* 27 (1905), 395–455; Registres de la Société d'agriculture de Brive, Bibliothèque du Musée Ernest Rupin, Brive: Archives nationales (hereafter AN) H[1] 1503.

4. AD Corrèze 6F 239, letter to Serre, 17 April 1790; *Actes de la Commune . . . ,* ed. Sigismond Lacroix (Paris, 1894), IV, 301–302, 322, 334–341, 508; Victor de Seilhac, *Scènes et portraits de la Révolution en Bas-Limousin* (Paris, 1878), pp. 117–150.

some passages of 1800 and 1803 Cabanis reads like the ideal spokesman for the entrepreneurial capitalist. He argued that any useful industry which brings wealth to an individual guaranteed public prosperity and resulted in only beneficial and self-correcting inequalities (II, 11, 18–19, 29, 391). He believed that the poor could only gain by the accumulation of wealth in "great new commercial and industrial enterprises—which spread life and abundance around them." Indeed, the middle class active in commerce and industry had been responsible for the "more equitable" distribution of property in the last two to three centuries (II, 18n.; 481).

Yet such self-correcting inequality could not solve the immediate, urgent problems of the poor. Old Regime privilege, wrote Cabanis, has established a "shocking disproportion of fortunes." Moreover, "if legislators and governments had not favored with all their power the maldistribution of wealth, would the earth have ever been covered with this crowd of indigents, whose cries accuse both nature which brought them forth, and the powerful who despoiled them before their birth?" (I, 28, 6n.). Like the Physiocrats and Adam Smith, Cabanis saw the relics of feudalism as the principal cause of inequality—monopolies, industrial regulations, inequitable inheritance laws, obstacles to labor mobility, and, most of all, hereditary privilege. The large population of beggars was the obverse side of great concentration of wealth. Moreover, revolutionary efforts had aggravated the plight of the poor since the poor suffered most from commercial regulation, forced requisitions, unemployment in the luxury trades, and the greed of the newly rich, and from the "senseless demagogues frightening property by doctrines subversive of all order" (II, 5, 11–14, 18).

In this situation, Cabanis formulated the case for public assistance, though he gradually added important qualifications to prevent misguided efforts to eliminate natural inequality. Public assistance was, first of all, a natural development of individual human faculties. Relief of the "poor whose lives and sufferings are sacred" was the logical development of the sympathy of a sensitive being. Humanitarianism was implicit in human nature. In addition, aid as a "calculation of interest" of the rich and the legislator also followed logically from self-interest, since hordes of beggars would threaten property and security. This latter motive had been the traditional justification for the English Poor Law as well as an important aspect of the "great confinement" of the seventeenth-century general hospital movement in France. The new Enlightenment principle was the consideration of aid to the poor not as a question of Christian charity but of social justice.[5] For Cabanis, natural inequality could not possibly justify existing poverty. Consequently, the state was obligated to intervene, apart from the

5. Camille Bloch, *L'assistance et l'état en France à la veille de la Révolution* (Paris, 1908), pp. 47–48; the term "great confinement" was popularized by Michel Foucault.

moral obligation of the wealthy to relieve the evils caused by the manner they have acquired wealth. Cabanis agreed with Siéyès, who otherwise certainly opposed social equality, that an appropriate article in the Declaration of the Rights of Man would include the right to assistance for "anyone who cannot provide for his own needs"—a right any wealthy family would grant an aged servant (I, 4, 6; II, 3–6).[6]

The right to assistance was the culmination of the view of eighteenth-century philosophes and administrators that individual laziness and improvidence were not the sole causes of poverty. The royal government already gave short-term employment subsidies after bad harvests to maintain production and to avoid social disturbances. Both Mercantilists and Physiocrats recognized that unemployment and begging meant a waste of human resources. During the eighteenth century, penalties remained harsh for "vagabondage," defined in 1724 as habitual or armed begging, or begging in a group or on false pretenses, and redefined in 1764 as being unemployed for six months without references or a domicile. After 1767, beggars guilty merely of being without resources were removed from "general hospitals" and confined to grim, overcrowded "depots of mendicity." These institutions sometimes unfairly detained migrant workers, but even if limited to "beggars" or "brigands," they could not possibly absorb the large bands living on the margin of society. At the same time some intendants established "charitable workshops" for the temporarily unemployed. While critics have charged that the state was obsessed with productivity, the "depot" and the "workshop" both showed that work, rather than mere confinement, was now the official remedy for begging. Ministers such as Turgot and Jacques Necker clearly realized that the poor needed social assistance in economic crises.[7]

At least one depot superintendent, the canon-turned-journalist Leclerc de Montlinot, had sincerely humanitarian aims. A protégé of Necker, he gained a considerable reputation at the model depot of Soissons. In 1791–1792 he was Cabanis' colleague on the Paris Hospitals Committee, and later the Thermidorian Convention and the Directory appointed him a hospitals administrator. In an essay on begging, Montlinot stressed the inability of the deserving poor to provide for old age or widowhood, or to deal with disabilities. How could one hope to enforce the harsh laws against begging when poverty persisted and master artisans were not compelled to hire hands? While there were incorrigible vagrants who could be chastened only

6. Siéyès, *Reconnaissance et exposition raisonnée des droits de l'homme et du citoyen* (Paris, 1789), art. 27, p. 47.

7. On the dépôts see Christian Paultre, *De la répression de la mendicité* ... (Paris, 1906); Jean-Pierre Gutton, *L'etat et la mendicité* ... (Lyon, 1973); Thomas M. Adams, "Mendicity and Moral Alchemy: Work as Rehabilitation" *Studies on Voltaire and the Eighteenth Century,* 151 (1976), 47–76.

by transportation to a penal colony, the deserving poor needed jobs—and a comprehensive public assistance program.[8]

Such ideas became commonplace in the Committee of Mendicity of the Constituent Assembly and found their way even into the moderate constitution of 1791, which expressed the intention to organize a "general establishment" to "furnish work for the able-bodied poor who have been unable to procure it for themselves" as well as relief for the infirm.[9]

Cabanis recognized that the state had to act, but to act wisely. Men were equal in rights, not abilities, hence any attempt to eliminate all inequality would be disastrous. He warned that Spartan communal ownership or a Roman-style "agrarian law" (a code word for the ideas of François-Noël Babeuf) would be "iniquitous and contrary to the purpose of society, which is the free exercise of the faculties of each, and the peaceful enjoyment of the goods those faculties procure" (I, 29; II, 17). Like Pierre-Samuel Du Pont de Nemours, he thought that the English Poor Law unjustly burdened and impoverished the middle-income ratepayer while it encouraged ever increasing numbers to apply for assistance and to remain in parishes where they had a "settlement." The parish assessment was particularly ineffective because it required relief from those areas least able to pay because of their poverty (II, 7–8, 17, 63).

Even recent changes in the English system were prime examples of how not to aid the poor. In a heretofore unattributed review of Frederic Morton Eden's massive three-volume *State of the Poor* (1797; reviewed in 1798), Cabanis attacked the Poor Law modifications of 1795 in the southern (Speenhamland) counties. Here wage supplements, indexed to bread prices, were intended to give each worker a minimum real income. In Cabanis' view such "indiscriminate" aid would reward the less productive and establish the bad precedent of supporting able-bodied, employed poor from public funds, which should have been reserved for the unemployed or infirm. Cabanis thought that a proposal presented to the House of Commons in 1795 to allow magistrates to set minimum wages was also unwarranted interference with the natural price of labor. While Cabanis did not blame the poor for their plight, he expected a public assistance program to encourage the incentive to work, rather than to guarantee wages, and to maintain self-respect, rather than to foster dependence.[10]

Much more surprising for a Physiological Ideologist concerned with the

8. *Etat actuel du dépôt de Soissons, précédé d'un essai sur la mendicité* (Soissons, 1789), pp. 2–7, 15–25.

9. L. Duguit and H. Monnier, *Les constitutions . . . depuis 1789* (Paris, 1925), pp. 1–2.

10. *Mercure français,* 29 (20 messidor VI), 272; 32 (20 thermidor VI), 70–71; for attribution of the review to Cabanis, see *Recueil de mémoires sur les établissements d'humanité,* ed. Adrien Duquesnoy (Paris, an VII), VII, 5n.; for sponsorship of this collection by the Interior Minister, François de Neufchâteau, see AN F[17A] 1014.

power of hygiene, Cabanis also shared Eden's distaste for family allowances which offered additional sums for wives and children. This "prodigal" subsidy, he wrote, was "founded on the idea that the poor cannot live without the best white bread, cheese, sugar, and tea."[11] Elsewhere, Cabanis expressed his interest in the ventures of Benjamin Rumford, who claimed that his soup kitchens provided food which was more nutritive and inexpensive than the usual fare of charitable institutions (II, 10). But though Cabanis was prepared to grant considerable effect on sensitivity to regimen in general, he apparently was not willing to provide costly nutritional supplements which would overburden taxpayers and discourage frugality in food budgets.

In 1789, Cabanis' solution to aiding the able-bodied poor had been to provide work for the unemployed at public expense. Cabanis saw several virtues in this policy. First, from the purely economic viewpoint, the productivity of charitable workshops would defray the costs of helping the sick and disabled. Second, there was the traditional argument that "whoever devotes his time and strength to regular occupations lacks the energy to turn his imagination and desires toward objects whose pursuit might disturb public order" (I, 30, 449). Third, there was a sound physiological reason, that activity of faculties was natural and essential. A "sustained occupation, in nourishing the activity of all the organs, including those of the mind, maintains the faculties in equilibrium, a state constituting the health of the brain, as of other parts of the living system." Work prevented the monomania produced by idleness and helped develop independence, enterprise, and a sense of dignity. Finally, in 1789 Cabanis praised charitable workshops for keeping up the general wage level, "a factor of greatest importance for the class that lives by the work of its hands" (II, 57, 22; I, 31).

Between 1789 and 1803, Cabanis came to see assistance to the able-bodied poor as well-intentioned social therapy in principle but almost always pathological in practice. We can probably explain his change in attitude by the continual problems of the National Assembly and the Commune of Paris with outdoor public works for men. Established by Necker in the depths of the subsistence crisis of 1788 and continued by Jean-Sylvain Bailly in 1789, the work projects were suspended and reorganized three times before their definitive dispersal in June 1791, under the watchful eyes of the Marquis de Lafayette's troops. From an ample employment roll of 8600 men earning a respectable eighteen sous a day for unskilled landscaping labor in 1789, the works fitfully expanded by the spring of 1791 to a mammoth labor force of 32,000, engaged in various road repair and canal construction projects. Despite encouragement from the Mendicity Commit-

11. *Mercure français,* 30 (30 messidor VI), 322.

tee, Commune administrators complained about forged enrollment cards, lack of discipline, crime in public parks, illegal sales of cards entitling the bearer to a bonus for leaving Paris, payroll embezzlement by foremen, and, generally, rising expenditures for little useful work. No doubt, the prevailing economic uncertainty, divided authority, and the lack of definite goals produced considerable confusion.[12] The Paris Department official Germain Garnier unequivocally condemned public works in his November 1791 hospitals and public assistance report as a "gross error of beneficence . . . where aid was immodestly asked for and ungratefully accepted," while in the same period private industry languished.[13] Even the Mendicity Committee, which wished to guarantee a job for all able-bodied men, retreated from the "dangerous idea that government can relieve the anxiety and activity necessary for the poor to achieve their subsistence."[14] The entire experience most likely deterred Cabanis from enthusiasm about workshops and work projects.

In addition, Cabanis read Eden's accounts of English workhouses, which were expensive, disorderly, unhealthful, and, by Eden's standards, morally scandalous. Cabanis had previously warned that large hospitals concealed staff corruption; he now warned against the dangers of large, overcrowded public workshops. Eden's reports of the high mortality rate and the baneful influence of the "vile and corrupt" on the honest unemployed suggested to Cabanis that workhouse life was like an indefinite prison term, worse even than the degradation of begging. "Most sensible men," Cabanis concluded, "expect no real improvement in the condition of the poor, unless workhouses are completely destroyed or entirely reorganized."[15]

By 1803, Cabanis was convinced that the large public workhouse usually provided little incentive for industry among the workers or diligence among the supervisors. Contradicting his concern for wage levels in 1789, he now warned that public assistance wages must not tend to raise the normal day laborer's wage above its natural level or artifically attract workers to the public rolls. Otherwise, employers might experience a labor shortage while taxpayers paid for a swollen assistance budget.

Yet Cabanis was still convinced that there was a social obligation to aid the unemployed. To preserve the individual incentive to work and to encourage habits of industry, Cabanis recommended that the public assistance administration pay at piecework rates if possible. Ideally, the state would

12. Yvonne Forado-Cunéo, *Les ateliers de charité de Paris pendant la Révolution française (1789–1791)* (Paris, 1934); Michel Bouchet, *L'assistance publique en France pendant la Révolution* (Paris, 1908), pp. 212–240.

13. Germain Garnier, *Rapport fait au conseil du département de Paris à l'ouverture de la session du 15 novembre 1791* (Bibliothèque Nationale Lb^{40} 183), p. 44.

14. Camille Bloch and Alexandre Tuetey, eds., *Procès-verbaux et rapports du comité de mendicité* (Paris, 1911), pp. 331, 427.

15. *Mercure français,* 30 (30 messidor VI), 328–330.

distribute work to be done at home in accordance with the ability of family members. The state also needed to insure efficient use of funds—buying raw materials from needy artisans would help a flagging industry, manufacturing products in demand would minimize useless efforts, while employing the poor in familiar trades would increase productivity. Cabanis seemed to know enough elementary economics to realize the impracticality of combining these desiderata, but, above all, he wished to avoid the sense of futility of the workhouse. If cottage industry was one approach, another would be entrusting a large works project to a private contractor, who would furnish apprentices with food, lodging, and clothing subject to public inspection. A state-guaranteed rate of return would insure a sufficient number of willing contractors, while the motives of private interest would keep order more effectively (II, 8–9, 25, 29, 30).[16]

The social agitation of 1790–1791 and especially of 1792–1795 had led Cabanis to stress public order and a favorable economic climate for entrepreneurs as much as social justice in his essays on public assistance. No doubt Cabanis subscribed to the liberal illusion that the best way to promote natural equality was to promote natural inequality—to allow the free market to function. But Cabanis certainly did not adhere to the conventional static eighteenth-century doctrines of the "poor are always with us" variety. He was also too realistic to advocate a laissez-faire approach to poverty, if for no other reason than that the poor might be tempted to become the shock troops of aristocratic or royalist reaction.

In the era of the Consulate, public assistance was becoming primarily communal, while private and clerical foundations were assuming a greater share of all aid to the poor. In this context in 1803, Cabanis insisted on a single, centrally administered assistance fund collected from all taxpayers, rather than a municipal or parish poor rate. Such a fund would tap the wealth of the prosperous regions to aid less fortunate areas and would restore the goal of effective public assistance to its urgent priority. Moreover, any call for a central fund was a partial return to the principles of secular and national assistance ineffectually espoused by the National Convention of 1793–1794.[17]

Cabanis' original motivation for writing on public assistance was his professional concern with hospital reform. In the latter years of the Old Regime the polyglot nature and innumerable abuses of hospitals inspired vocal criticism. For physicians the housing of the sick in institutions which were often also insane asylums, workhouses, and homes for the aged and infirm was a fundamental confusion of social roles. For government officials and economists, hospices and hospitals fostered idleness among the able-

16. Ibid., p. 324.

17. *Réimpression de l'ancien moniteur* (Paris, 1847), XV, 748–749, law of 19 March 1793; Jean Imbert, *Le droit hospitalier de la Révolution et de l'Empire* (Paris, 1954), pp. 147–256.

bodied and were economic anachronisms which froze their endowment capital in unproductive use. The testament of a benefactor could also tie the hands of administrators to narrowly confessional charity which would exclude those unable to meet tests of piety or good character.

Enlightenment humanitarianism was as powerful a motive for hospital reform as economic doctrine or the bureaucratic dream of efficiency. The overcrowded, unsanitary conditions of the largest Paris hospitals had long been a concern of physicians and a stimulus to philanthropists. The disastrous Hôtel-Dieu fire of 1772 destroyed one wing and led to a plethora of proposals for hospital renovation or reconstruction. The head of the Maison du roi and the minister for Paris, the Baron de Breteuil, convened a special commission of the Academy of Sciences, including the distinguished surgeon Jacques Tenon and the astronomer Jean-Sylvain Bailly, in 1785 to examine the need for a new hospital.[18]

The commission found the Hôtel-Dieu in every respect insufficient, uncomfortable, and unhealthful. Infamous for beds sometimes containing four to six patients, with no isolation of the contagious or the convalescent, the Hôtel-Dieu had a general mortality rate of nearly 25 percent. The size of the institution (2500 patients) led to routinized, uniformly administered diet and drugs which made a mockery of individualized therapy. Physicians complained that the secondary or new infections which raged inside the wards made proper diagnosis nearly impossible.[19]

One submission to the commission, highly influential on Cabanis, came from the Physiocrat Du Pont de Nemours (1739–1814). Du Pont carried the critique of the large hospital to its logical conclusion. His premise was that the most salutary, the most humane, and the most economical care for the sick would always be at home. Parish officials would supervise domestic food, drug, and fuel subsidies while specially assigned physicians and surgeons would be able to visit at least 3000 patients. Personal care would bring the family closer, facilitate return to work, and boost the morale of all involved.

Where such self-help was impossible, Du Pont recommended a system of small, 100-bed parish hospices, which would minimize complications in disease and surgery. The aged or feeble could be placed in nursing homes

18. Michel Foucault, *Naissance de la clinique* (Paris, 1963), pp. 32–43; George Rosen, "Hospitals, Medical Care and Social Policy in the French Revolution," *Bulletin of the History of Medicine,* 30 (1956), 124–149; on the commission itself, see the excellent articles by Louis S. Greenbaum, "'The Commercial Treaty of Humanity': La tournée des hôpitaux anglais par Jacques Tenon en 1787," *Revue d'histoire des sciences,* 24 (1971), 317–350; "Jean-Sylvain Bailly, the Baron de Breteuil, and the 'Four New Hospitals' of Paris," *Clio medica,* 8 (1973), 261–284; "'Measure of Civilization': The Hospital Thought of Jacques Tenon on the Eve of the French Revolution," *Bulletin of the History of Medicine,* 49 (1975), 43–56.

19. *Extrait des registres de l'Académie royale des sciences du 22 novembre 1786. Rapport des commissaires chargés, par l'Académie, de l'examen d'un Projet d'un nouvel Hôtel-Dieu,* 2d ed. (Paris, 1787), pp. 6–7, 18–21, 117–134.

run by private contractors at a guaranteed rate of return. Competition among them, he added naively, would insure good patient care.[20]

The commission rejected Du Pont's arguments on several grounds. They believed domestic assistance would be medically unsound—the patient would often be in unhealthful quarters without sufficient professional supervision. Family members could often not act as nurses, since they could not afford the loss of income. Moreover, the small parish hospice would not be in a scientifically designed structure, since Du Pont would merely reconvert church buildings. The hospice would lack competent staff to treat rare diseases, to accommodate the insane, or to perform difficult surgery. It could not accommodate the transient and would entail costly administrative duplication.[21]

The commission therefore recommended a compromise proposal, intermediate between the one massive hospital and the many small hospices. Based on Tenon's painstaking fact-finding tour in England and years of clinical experience, the plan recommended four hospitals of 1200 beds each. Ten separate pavilions of 120 beds each would enable isolation and differentiation of disease, and each hospital would have specialized wards, as for the contagious, or for obstetrics, or for the insane. Each patient would be assured a single bed, and medical considerations would be paramount.[22]

The commission was already defunct when Cabanis entered the debate, but the existence of the Committee to Exterminate Mendicity of the Constituent Assembly spurred new hopes for reform. While Cabanis supported the commission's attack on the Hôtel-Dieu, he was convinced that it had not gone far enough. Like Du Pont, he suggested that aid to the sick be as natural as possible, either at home or in the small hospice. Better ventilation, cleaner beds, more sympathetic nurses could be more easily assured in the small hospice. Cabanis attributed the overcrowding, the complications of disease, and the improper administration of food and drugs at the Hôtel-Dieu to its overall size. In a detailed critique of the Academy of Sciences commission's plan, Cabanis argued that only in the small hospice could the physician practice individualized clinical medicine effectively. Only there could the staff keep precise clinical journals and adjust diet and drugs to the medical history, age, and temperament of each patient. The conversion of existing buildings into 100 to 150-bed hospices would at once save capital expenditure and provide better facilities at lower per capita cost (I, 8–27).

A former official of the Civil Hospitals Department in the controller-

20. *Idées sur les secours à donner aux pauvres dans une grande ville* (Paris, 1786); see Louis S. Greenbaum, "Health Care and Hospital-Building in Eighteenth-Century France: Reform Proposals of Du Pont de Nemours and Condorcet" *Studies on Voltaire and the Eighteenth Century*, 152 (1976), 895–930.

21. *Extrait*, pp. 104–116.

22. Reports of 20 June 1787 and 12 March 1788 in *Discours et mémoires, par l'auteur de l'Histoire de l'astronomie* [Jean-Sylvain Bailly] (Paris, 1790), II, 321–340, 341–391.

general's office, Michel-Augustin Thouret, reviewed Cabanis' essay on hospitals favorably in June 1790, in his capacity as adviser to the Committee on Mendicity.[23] Two months later, Thouret himself recommended a small hospice plan with a maximum size of 250 beds in large cities. As the committee prepared its reports, the financial condition of hospitals became disastrous. Revenues from seigneurial dues, tithes, and church property disappeared, the municipal *octroi* was abolished, while hospitals lost their exemption from real estate taxes. The result was a brutal entry into the liberal economic world.

Though the Commune of Paris retained the right to administer its hospitals, the Constituent Assembly allowed the newly created Department of Paris to assert effective jurisdiction over policymaking. For this purpose the department directory appointed a five-man Hospitals Committee, including Cabanis, Thouret, and Montlinot on 11 April 1791. No doubt Cabanis' notoriety as physician to the dying Mirabeau as well as friends such as Siéyès on the Directory helped gain him this position. Here Cabanis acquired practical experience in hospital inspection and wrote several reports, all unfortunately lost, on Hôtel-Dieu mortality, reform of the Hôpital-Général, prison infirmaries, and the foundlings hospital. From the incomplete Hospitals Committee minutes and the report of a department official, we can reconstruct the committee's accomplishments. In the end they dealt more with flagrant abuses than with issues of high policy. Financial constraints did not permit new hospital construction, but only the reform of practices in existing institutions.

The committee first insisted on a careful audit, particularly of the inflated hospital employee food budget. The official's report to the department noted that one might have thought that hospitals existed chiefly for the staff, and that patients "only owe the care they receive to the benevolence of people on the staff." Like the Academy of Sciences commissioners, the committee favored external contracting, with competitive bidding when possible, rather than the costly internal kitchen with its dangerous open fires. Second, the committee reduced the nonmedical budget by drastically cutting the number of priests and choirboys and discouraged nonmedical charity, such as the endowment of masses. Third, the committee ordered wholesale sanitary and medical reform at the Hôtel-Dieu, including one bed per patient, better food, cleaner laundry, a terrace for convalescents, and more medically attentive nursing care. Contagious cases and the insane would be moved into specialized wards, preferably away from the city center.

The committee also began the process of separating the sick from the chronically disabled, and all the "infirm" from the merely poor, in the

23. Bloch and Tuetey, *Procès-verbaux*, pp. 7, 65.

Hôpital-Général. At the Bicêtre asylum, the committee separated the "insane" from the "criminal," increased the bread ration, and, in the measure long identified with the supervising physician Philippe Pinel, substituted the straitjacket for chains in controlling the manic (II, 58). Finally, the committee applied Cabanis' occupational therapy principle to able-bodied boys in the Hôpital-Général. The administrators arranged for private employers to hire apprentice woolspinners aged at least thirteen for a maximum eight-hour workday (an enlightened arrangement for the time). The employers agreed to permit inspection of food, clothing, and lodging given the apprentices. The department officials were convinced they were benefiting the boys as well as reducing their own budget.[24]

Meanwhile, the Committee of Mendicity and the department directory both approved in 1791 a comprehensive plan for Paris hospital reform which managed to reconcile the Cabanis-Thouret small hospice policy with the Tenon-Bailly plan. They insisted on a neighborhood network of fourteen hospices of 175 beds each, including several medically specialized institutions. Convalescents and venereal disease patients would be assured separate wards, while there would be a general separation of the acutely ill from chronic cases. Foundlings, the aged, and incorrigible vagabonds would be housed in distinct institutions. But the report also recommended the creation of two 700- to 800-bed hospitals for "more complete" clinical instruction, since only in them could one assure all the "means of observation . . . collected in less space . . . in a greater variety of subjects.[25] The argument was remarkable for its frank insistence on the importance of the needs of medical research and teaching. Cabanis remained hostile to large hospitals, but in a speech of 1798 to the Council of Five Hundred, he defended the location of medical schools in large cities. Only there, he admitted, would there be "vast hospitals," with men and women from diverse climates, of diverse temperaments, following diverse regimens, so that there could be adequate teaching of both clinical regularities and rare cases (II, 413). Cabanis' "analytic" method in medicine was thus undeniably statistical, requiring large numbers of cases for study, if not large institutions.

In fact, in 1791 the Constituent Assembly implemented none of the recommended hospital reform plans. Nor did its successor, the Legislative Assembly, reach any conclusive decisions on hospital policy. The Paris Hospitals Committee remained in office until 10 August 1792, when the new,

24. Alexandre Tuetey, ed., "Procès-verbaux du comité des hôpitaux, 15 avril–3 octobre 1791," *Bulletin d'histoire économique de la Révolution* (Paris, 1916), 67–153, esp. 97, 115–116, 126–127, 130–131, 139, 147; Garnier, *Rapport,* 27–28, 31–33.

25. A.D. Seine 6 AZ 52 (Imprimé), *Rapport sur la nouvelle distribution des secours proposés dans le Département de Paris par le Comité de Mendicité* (Paris, 1791), 9.

more radical Commune forcefully reasserted full control over the hospitals of the capital.[26]

In the revised Hospitals Committee reports which Cabanis published in 1803, he reiterated the goal of humane treatment of the insane. To eliminate the Old Regime abuses of the *lettre de cachet,* Cabanis insisted on a formal legal act and a formal medical certificate before commitment to an asylum. This confinement would be essentially revocable on medical judgment alone, without slow legal process, since the asylum was an infirmary, not a prison. His premise was that madness itself is often temporary and that confinement is an exceptional violation of individual liberty solely to prevent the patient from harming himself or others. As in his philosophical works, Cabanis stressed the effectiveness of physical agents on the mentally ill. The physical components of Pinel's "moral treatment" avoided preaching and provided better diet, better air circulation, and, here again, occupational therapy (II, 48–59). Even in an insane asylum, there were ways of restoring a semblance of a natural environment.

Cabanis' Hospitals Committee experience also led him to advocate policies for foundlings which would make public assistance as "natural" as possible. The enormous mortality rate and inadequate apprenticeship training at the Paris Foundlings Hospital horrified the committee investigators. Modern studies have corroborated the eighteenth-century opinion that there was an enormous increase in the number of children abandoned in the provinces who were taken to the already overcrowded Paris Hospital.[27] Here Cabanis favored local responsibility for assistance to eliminate the illegal, murderous transport of infants to Paris by unscrupulous traffickers. Cabanis also confidently predicted that the new regime would reduce the chief cause of abandonment—poverty and dissolute morals. This view was more than rhetorical since new legislation allowed the possibility of divorce and gave more tolerance to unwed mothers.

Cabanis favored placement of healthy foundlings in the country with foster parents who would receive a pension indexed to local prices until the child reached age seven. At that time, the foster parents could keep the child as an apprentice until age twenty-one, or return him to public care. When the child reached twenty-one, either the parents or the state would emancipate the youth with an endowment equal to the public pension of the first seven years. Ideally, foster parents would legally adopt abandoned children. The Paris Hospital already placed many more children in the country than it admitted. But Cabanis' seemingly modern approach would eliminate the

26. M. Brièle, *Collection des documents pour servir à l'histoire des hôpitaux de Paris* (Paris, 1883), II, 284–286.

27. Claude Delasselle, "Les enfants abandonnés à Paris au XVIIIe siècle," *Annales,* 30 (1975), 187–218.

risks of the transport as well as the disadvantages of any upbringing in a large public institution (II, 37–43).

Like his contemporaries on the Committee of Mendicity, Cabanis saw a necessary public role in assuring free medical and surgical care for the poor. Just as the physician had to intervene to assure health itself, the state had to intervene to assure a right to health care.[28] Aid to the sick and infirm, unlike aid for the able-bodied, did not raise the question of discouraging salutary industrious habits. But even here, there was a balance between priority for the natural order—giving aid at home or in a homelike hospice—and the needs of clinical medicine, which required large samplings, for which the otherwise necessary evil of the hospital might be useful.

During the Consulate and Empire, members of religious orders were reintroduced as hospital employees, but the local administrative council modeled after the Paris committee of 1791 became general throughout France. The hospital reorganization envisaged in 1791 never occurred as such. But the tie between the hospital and medical instruction, with clinical journal-keeping as its most important characteristic, and the establishment, after 1802, of the competitive medical internship became the strong points of the Paris hospital system. Moreover, the principles of hospital specialization, and of separation of the sick from the disabled and from the able-bodied poor became the foundations of the modern appraoch to hospitals as exclusively medical institutions.[29]

The Revolution was a critical era for schools in part for the same reasons as it was an era of change for hospitals. The schism in the clergy in 1791, the abolition of teaching orders in 1792, the loss of church revenues and the abolition of the corporate status of colleges and faculties all encouraged continual debate on whether education should be public, secular, universal, and compulsory. These debates, already evident under the Old Regime, continued in the constitution committee of the Constituent Assembly.[30]

Before the committee authorized Talleyrand to report its conclusions in September 1791, Cabanis published a "work on public education" allegedly "found in Mirabeau's papers."[31] No one has yet resolved the problem of the

28. Dora Weiner, "Le droit de l'homme à la santé: une belle idée devant l'Assemblée nationale constituante, 1790–1791," *Clio medica,* 5 (1970), 209–223.

29. Erwin Ackerknecht, *Medicine at the Paris Hospital, 1794–1848* (Baltimore, 1967), 17–22.

30. In the vast literature on education during the Revolution, works consulted include James Leith, "Modernization, Mass Education, and Social Mobility in French Thought, 1750–1789," *Studies in the Eighteenth Century, II* (Canberra, 1973), 223–238; Félix Ponteil, *Histoire de l'enseignement primaire en France de la Révolution à la loi Guizot (1789–1833),* (Paris, 1959); Célestin Hippeau, *L'instruction publique en France pendant la Révolution,* 2 vols. (Paris, 1881–1883); Albert Duruy, *L'instruction publique et la Révolution* (Paris, 1882); Louis Liard, *L'enseignement supérieur en France, 1789–1889,* 2 vols. (Paris, 1888–1894); H. C. Barnard, *Education and the French Revolution* (Cambridge, 1969).

31. "Travail sur l'éducation publique trouvé dans les papiers de Mirabeau," *Oeuvres complètes de Cabanis,* ed. François Thurot (Paris, 1823), II, 363–581.

authorship of these four discourses (II, 546–547). We know that Dominique-Joseph Garat and Constantin-François Volney introduced Cabanis to Mirabeau at the National Assembly on 15 July 1789, and that the young physician thereafter entered the veritable workshop of ghostwriters which the great orator usually employed. Cabanis modestly assigned the work to Mirabeau, though his widow and some critics have claimed it for Cabanis. No doubt Mirabeau found the opinions acceptable, even if he had not revised the text for publication. A recent study presents manuscript evidence that the fourth discourse, on education of the dauphin, was probably the work of the future Girondin, the abbé Antoine-Adrien Lamourette. Mirabeau's correspondence also shows that he asked another collaborator, the Genevan pastor Etienne-Salomon Reybaz, to write a discourse on "national education," but there is no record of Reybaz's reply.[32] Passages on the faculty of sensitivity, the relationship of the physical and the mental, the limited role of the state in general education, and the necessity of supervising medical education are all consistent with Cabanis' views and even with his style. If the topics of the discourses depended on Mirabeau, it seems likely that Cabanis provided the actual drafts and possibly the major ideas.

The discourses aligned themselves on the central axis of Cabanis' thought—the desirability of leaving the spontaneous and natural situation to itself, qualified by the indispensable intervention of external authority. The general discourse on public education adopted the familiar agricultural metaphor, previously used by Cabanis and at least as old as Hippocrates, on the power of education. The outcome of good education was comparable to the harvest of a skillful farmer, who, given his knowledge of the terrain, planted and cultivated carefully to improve the "wise dispositions" of nature (I, 76–77). Especially in impressionable childhood years, education had great power to cultivate necessary habits and eliminate destructive ones. At the current juncture, proper education would combat Old Regime prejudices and "lead human inclinations back to nature."[33]

Though the inventors of "laissez faire" theory, the Physiocrats, favored universal free primary education, the author of the discourses followed more closely the views of Adam Smith.[34] Whether or not Cabanis was the author, he explicity admitted later that he had once shared the view of the "Mirabeau" plan that education is a free profession rather than a public obligation. While educators would intervene in natural development, the state would not disturb the free operation of competition and industry among schoolmasters.[35]

But at least three principles forced compromise with this rigorously liberal

32. J. Bénétruy, *L'atelier de Mirabeau* (Geneva, 1962), pp. 315–317, 330, 466–479.
33. Cabanis, "Travail," pp. 371, 384–386, 482–486.
34. Adam Smith, *Wealth of Nations* (1776), bk. V, pt. III, art. 2.
35. Cabanis, "Travail," pp. 437, 481, 490–492, 549.

theory of schooling. First, the state could not remain indifferent to the "urgent need" and "profound ignorance" of the people. Therefore the state must "protect, excite, and reward" parish schoolmasters (still, for the most part, the constitutional clergy) with bonuses for recognized teaching ability, prizes for excellent pupils, and subsidies for authors of elementary textbooks. There would also be department and national education committees to supervise instruction, but no government ministry.[36]

Second, the state would recognize its obligation to some talented, but impoverished children by endowing faculty chairs and allocating one hundred student stipends for a higher-education National Lycée, with chairs in modern languages and literature, the sciences, public economy and ethics, and several modern philosophical subjects. "Universal Method" would include "analysis," the decomposition and recomposition of objects, and classification by analogies and resemblances, and would show the relationships among the sciences. "Universal Grammar" would study the role of signs in fixing ideas and of languages as analytic methods. "Metaphysics" would analyze mental faculties in the fashion of John Locke, Etienne Bonnot de Condillac, Claude-Adrien Helvétius, and Charles Bonnet (all authors later cited favorably by Cabanis) and give rules for the "degree of certainty possible in each subject."[37] Thus, the discourse included a kind of Ideology *avant la lettre,* the kind of psychology, "analysis," grammar, and logic which would later be the subjects of Idéologue treatises and part of the curriculum of the secondary-level "central schools" of the Directory.

Third, the state would insure regulation of those professions (such as grain merchants, goldsmiths, and physicians) where the public needed protection from fraud and quackery. Under any regime the state would have to license health care practitioners. The text paralleled Cabanis' own views on unified education of physicians and surgeons, certificates for pharmacists and midwives, penalties for quacks, and Cabanis' emphasis on clinical observation and teaching in hospitals.[38]

No doubt, the discourses appeared conservative when compared to the more famous Talleyrand report or to the subsequent Condorcet plan of April 1792. Talleyrand supported state-funded primary schools, while Condorcet drew on the public treasury for "higher" primary schools, as well as for secondary and higher education.[39] The Mirabeau-Cabanis plan did not envisage secularization, or compulsory attendance, and hardly encouraged universal primary schooling or any kind of social mobility. It gave the generally conservative departmental directories jurisdiction to inspect secondary and medical schools and to choose pupils for the National Lycée.

36. Ibid., pp. 390–393, 437, 382, 492–493.
37. Ibid., pp. 505, 509–520, 527–536.
38. Ibid., pp. 392–398, 422–431.
39. Hippeau, *L'instruction publique,* I, 33–288.

Still, as the Revolution progressed, Cabanis confessed that he had to refine his views on public education. In the session of the Council of Five Hundred for 1798–1799, Cabanis sponsored a plan for organizing the medical profession (4 messidor VI) and for attaching medical schools to higher-education establishments known as lycées (29 brumaire VII) (II, 388–401, 405, 424). These proposals remained close in spirit to the discourses of 1791. But he also prepared a major speech for the general debate on public education in which he had to recognize that the "natural" competition of private schoolmasters and the "natural" ability of pupils to pay fees could never eradicate Old Regime prejudices (II, 425–450).

Ever since the Le Chapelier Law of 1791, there had been no medical guild in France. Military needs for physicians and surgeons led the Convention to create three new medical schools in 1794 with a single, government-approved curriculum, emphasizing clinical instruction, for both physicians and surgeons. However, in 1798 there was still no medical certifying examination, so that military health officers or others who lacked degrees from the old faculties, largely inactive since 1792, had no formal professional status.[40] Partially in response to Directory messages of concern, Cabanis renewed efforts in June, 1798, to obtain a new legislative statute for medical practitioners. He proposed formal licensing of all physicians and surgeons, with suitable relaxation of examination standards for health officers of proven competence who were long removed from their studies. Yet two physicians, Jean-Marie Calès of Toulouse (in the session of 1797) and Louis Vitet of Lyon, opposed legal penalties for quacks and implied that Cabanis was the spokesman for a privileged elite that wished to suppress the newly acknowledged freedom for anyone to practice any profession.[41]

In December 1798, the public instruction commission of the Council authorized Cabanis and Antoine-François Hardy to present a report on medical education which attempted to maintain rigorous medical standards while opening access to medical education to the less wealthy.[42] Cabanis would not concede any departure from strict regulation. Only success in medical examinations at existing medical schools would allow the right to practice medicine or surgery. Moreover, the single diploma for physicians and surgeons seemed to Cabanis the cornerstone of all medical reform

40. David Vess, *Medical Revolution in France 1789–1796* (Gainesville, Florida, 1975), pp. 71–92, 117–136, 162, 170.

41. *Rapport fait . . . sur un mode provisoire de police médicale présenté par Cabanis* (4 messidor VI) (BN, Le[43] 2075); *Projet de résolution de Calès* (12 prairial V) (Le[43] 1017); *Projet de résolution de Vitet* (17 ventôse VI) (Le[43] 1816), pp. 13, 29–31n.; *Motion d'ordre de Vitet* (4 messidor VI) (Le[43] 2076).

42. *Rapport fait . . . sur l'organisation des écoles de médecine* (29 brumaire VII) (BN, Le[43] 2450); *Rapport fait par Hardy . . .* (1 frimaire VII) (Le[43] 2455); on the 1790 proposals, see Jean-Gabriel Gallot, *Observations sur le projet d'instruction publique lu par M. Talleyrand . . .* (Paris, 1791); Henry Ingrand, *Le comité de salubrité de l'Assemblée Nationale constituante, 1790–1791* (Paris, 1934).

proposals since 1789. But, in line with suggestions already made to the Constituent Assembly in 1790, there would be schools of "elementary" medical instruction and teaching clinics in twenty civil hospitals, directed by the chief health officers of the institution (II, 417–418). Vitet and another physician, Jean-François Barailon, favored a separate diploma and less rigorous standards for surgeons, but at the same time argued that instruction would be incompetent in the civil hospitals. Thus, there was disagreement on the means of assuring a sufficient number of qualified practitioners for rural areas.[43]

Action on medical standards occurred only under the Consulate in April 1803, when Thouret and Antoine-François Fourcroy supported a law establishing two categories of practitioners—physicians and surgeons with regular degrees and health officers certified on the basis of a shorter course of study or years of clinical experience. However, arguments for this law openly avowed a kind of social discrimination with the intention to confide the "extended class" of "industrious and active people" to the less qualified health officer.[44] Such a distinction was never implicit in the defeated Cabanis plan of 1798, which was based on the principle that government certification maintained high professional quality.

Cabanis did not, however, wish to extend government regulation of medical education to the point that all medical schools had equal status. Calès, Barailon, and Vitet all insisted on the same number of professors and the same curriculum at each of the three medical schools. Moreover, the Interior Ministry appointed professors in Paris from a short list of three elected by colleagues, rather than after a competitive examination. All three of Cabanis' opponents in the debates charged that some Paris professors were appointed because of favoritism and failed to fulfill their responsibilities. In part, this attack was personally intended for Cabanis, whose recurrent ill health and political career prevented him from teaching.[45] But both Cabanis and Hardy testified that the Paris School of Medicine was overflowing with pupils and defended its need to have a larger staff. They remained dedicated to the principle of centers of excellence rather than institutional parity. Moreover, while they proposed creation of three new medical schools, Paris was the only one of the three existing institutions to be attached to new

43. *Opinion de Vitet* (23 nivôse VII) (BN, Le^{43} 2683); cf. *Opinion de Barailon* (17 germinal VI) (Le^{43} 1883), p. 9.

44. *Tribunat. Rapport fait . . . par Thouret sur le projet de loi relatif à l'exercice de la médecine* (16 ventôse XI), pp. 13–14; for a contrary interpretation of the social implications of Cabanis' report, see Foucault, *Naissance,* pp. 69–86.

45. BN, Le^{43} 1017, p. 14; Le^{43} 1816, p. 11; Le^{43} 1883, pp. 7–8; Le^{43} 2683, p. 2; for Cabanis' career at the Ecole de Médecine, see AN AJ^{16} A^{P} 1–7 (provisional classification), "Procès-verbaux de l'assemblée des professeurs," I, 308, II, 239, 346, 386, "Registre des procès-verbaux," pp. 65, 247, "Procès-verbaux . . . du comité d'administration," fol. 47v., 51v., 116.

institutions of higher education known as lycées (II, 409).[46] The natural superiority of intellectual life in Paris would remain unchallenged.

Both the medical schools and the lycées were a portion of a comprehensive reform of public education presented by the physics teacher Roger-Martin to the Council of Five Hundred in November, 1798.[47] Cabanis explicitly associated himself with the speakers sponsoring each section. In the process he moved far beyond the cautious innovations proposed in the Mirabeau discourses.

In the most radical phase of the Revolution, Robespierre himself recommended to the Convention a plan requiring compulsory attendance at secular public elementary boarding schools.[48] By the closing days of the Thermidorian Convention, there was a significant retreat not only from the principle of compulsion, never a majority view, but also from the principle of free primary education, acceptable to moderates in 1791 and 1793. The law of 25 October 1795 (3 brumaire IV) authorized creation of only about 5000 primary schools for children aged seven to ten with a curriculum limited to reading, writing, arithmetic, and "republican morality." Teachers received only lodging indemnities, not salaries, while parents had to pay fees, though up to 25 percent per school could be exempted because of poverty. The Convention protected private schools in Article 300 of the Constitution of 1795, which gave any schoolmaster the right to open an institution.

The Public Education Committee of the Convention also curtailed the number of courses in the partially state-funded secondary central schools. Thus, while Cabanis collected several months' salary in 1795 as professor of "hygiene" in the Paris central schools, he never actually taught, since his subject disappeared from the final draft of the law.[49]

For the next four years, neither primary nor central schools functioned as the Convention had planned. At the outset, public primary schools suffered the handicap of lack of buildings and a shortage of qualified teachers which was aggravated by payment in depreciating currency with no fixed minimum salary. Regional teacher-selection juries struggled with the problem of appointing candidates both competent and patriotic. They excluded nonjuring priests and others who refused the oath of hatred to royalty. But

46. AN F^{17} 2273 has Interior Ministry records of petitions for new medical schools.

47. *Rapport général fait par Roger-Martin sur l'organisation de l'instruction publique* (19 brumaire VII) (BN, Le^{43} 2438; redraft, pluviôse VII, Le^{43} 2801).

48. Robert J. Vignery, *The French Revolution and the Schools: Educational Policies of the Mountain, 1792–1794* (Madison, 1965); documents and debates in J. Guillaume, ed., *Procès-verbaux du comité d'instruction publique de la Convention nationale,* 6 vols. (Paris, 1891–1907).

49. Guillaume, *Procès-verbaux,* VI, 869–873, 553, 575; on Cabanis, see AN F^{17} 1344^{27}; on the central schools, see L. Pearce Williams, "Science, Education, and the French Revolution," *Isis,* 44 (1953), 311–330.

the boards did appoint former nuns and priests for lack of other personnel, and high enrollment often correlated with quasi-legal religious instruction. In at least one case armed force had to prevent irate villagers from expelling a schoolmaster who took over the residence of the beloved curé. Peasants who feared exclusion from communion of children studying with godless republican instructors sometimes destroyed government-sanctioned textbooks. Even in the Department of the Seine an estimate of public primary school attendance was one-half to one percent of children of the eligible age group. In the same year there were only fifty-six public primary schools in the Paris region compared to about 2000 private establishments.[50]

In the polarized atmosphere following the coup of fructidor (September 1797), the Directory attempted to regulate by decree what the Legislative Councils refused to approve by law. In November, candidates for public office had to verify attendance of themselves or their children at public schools. In February 1798, the Director François de Neufchâteau required local officials to inspect all schools for teaching of the Rights of Man, use of approved textbooks, and observance of holidays in the revolutionary calendar.[51] While municipalities closed a significant number of schools, surviving private institutions evaded the rules, presented a republican facade to visitors, and retained the vast majority of pupils.

By October 1798, the Interior Ministry persisted in blaming primary school problems on royalists, negligent local administrators, and absence of regulatory power. A Directory message to the Council of Five Hundred recommended several reforms, such as local guarantees of fixed minimum salaries for instructors, intermediate-level schools to fill the gap between elementary and secondary instruction, and exclusion of priests. Official anticlericalism was so triumphant in the post-fructidor Directory that the message declared priests "unfit to educate youth in the principles of purified virtue" and stipulated that "philosophical and universal morality must be the exclusive basis of republican instruction."[52]

The issues in the subsequent debate crystallized Cabanis' concerns about balancing freedom and state intervention, permitting natural inequality, and eradicating artificial inequality. They illustrated the dilemmas of Directory moderates, who in principle favored free enterprise in education. Yet an open society in a time of foreign and latent civil war might be exploited by royalist or clerical enemies of the Revolution. Conversely, arbitrary regu-

50. Jonathan Helmreich, "The Establishment of Primary Schools in France under the Directory," *French Historical Studies,* 2 (1961), 189–208.

51. AN F[17A] 1014.

52. *Procès-verbaux des séances du Conseil des Cinq-Cents* (Paris, brumaire VII), pp. 111–141, esp. 125; see Ernest Allain, *L'oeuvre scolaire de la Révolution* (Paris, 1891), p. 272, and Hippeau, *L'instruction publique,* II, 293.

latory measures would merely alienate some republican support and narrow the power base of the regime.

A group of moderates presented four parts of the Roger-Martin plan—reform of primary and central schools, an ambitious new scheme of higher-education lycées, and measures applying to private schools.[53] Roger-Martin himself repudiated the ideas of Adam Smith in advocating state action to correct the defects of the law of 1795. On the premise that "instruction is the need of all" and "ignorance the worst enemy" he recommended creation of five times the previously authorized number of schools. The republic would guarantee to all instructors, who could not be priests, a minimum salary of 100 to 400 francs, depending on local population and paid by the canton from direct tax revenues. Though the proposal envisaged enormous outlays for higher education, fully 60 percent of educational expenditures would be devoted to two levels of primary schools. To gain additional compensation for instructors, municipalities would collect fees from parents, ranging from twenty-five centimes to one franc per month per pupil, according to income level calculated by four classes of direct taxation. Twenty-five percent of the pupils would still be eligible for exemption from fees, while parents who chose private schools would be forced to pay double the maximum fee.

An especially controversial provision envisaged about 500 "reinforced" primary schools (one per canton), where several instructors would teach children aged eleven to thirteen. Inspired by the ideals of Condorcet, Roger-Martin would make available to children "of comfortable artisans and propertied farmers" an enriched program of basic skills, practical subjects such as surveying, geography, and bookkeeping, and the elements of Latin, French grammar, and literature. Thus those who were not rich enough to attend central schools or who lived too far away would have the opportunity to extend primary education, while others would be better prepared for the academic secondary instruction.[54]

No doubt Cabanis found particularly appealing some of the features of the plan relating to the central schools, including the addition of a chair in "logic and analysis of the understanding." In addition, the Idéologues had long been encouraging a restoration of higher education, which here would take the form of five lycées, each with thirty nonmedical chairs distributed among the mathematical sciences, the physical sciences, the moral and political sciences, and belles-lettres.[55]

53. Deputies presenting sections of the plan were: Heurtault-Lamerville (22 brumaire VII, BN, Le[43] 2440) on primary education: Bonnaire (23 brumaire VII, Le[43] 2444) on central schools; Briot (27 brumaire VII) on lycées; Dulaure (2, 7 frimaire VII, Le[43] 2456) on surveillance of private schools.

54. BN, Le[43] 2438, pp. 10–11, 22, 25–35; Le[43] 2440.

55. BN, Le[43] 2438, pp. 4–6, 13–17; cf. Daunou plan for higher education in 1797 in Liard, *L'enseignement*, II, 419–471.

The harshly worded auxiliary plan to regulate private schooling excluded from state scholarships to central schools anyone who failed state examinations and included extensive powers, especially in the first draft, for Directory commissioners to interrogate pupils to verify their knowledge of republican principles and to write reports on the zeal of the masters.[56]

The polarities in subsequent council debate clarify Cabanis' search for a middle path between state indoctrination and academic freedom.[57] At least one neo-Jacobin deputy resurrected Robespierre's principle of compulsory public boarding schools to insure egalitarian character formation. He threatened loss of civil rights to parents boycotting the public primary school, which he would place in every commune, approximately double the number in the Roger-Martin plan. In his view no cost was too great to prevent a nation "deprived of enlightenment, falling under the aristocracy of the rich, the most odious of all subjection." Education in common would force children of the rich to receive the same instruction and acquire the same opinions as the poor.[58]

At the other extreme was the fiery moderate lawyer from Nancy, Boulay de la Meurthe, later aligned with Siéyès and the brumaire conspirators, who opposed the expense of primary schools in a financial crisis. The great maxim which should direct the government was "laissez faire," the best protection against unnecessary taxes. Moreover, the meddlesome inspection plan ignored the truth that "bayonets cannot destroy the power of habits." The new system would be "more intolerant than Papism."[59] Another moderate lawyer from Grenoble who opposed the curriculum of reinforced primary schools had already warned the previous year against overeducating artisans and farmers who might be attempted to abandon their work and who would slow down the proceedings of primary electoral assemblies. Only wealth and leisure, he argued, fit a man for education, and "human perfectibity" was a "vain speculation."[60]

In addition, Boulay's antipathy to the plan and the opposition of two other speakers stemmed partly from their distrust of the radical Eighten-

56. BN, Le[43] 2456, pp. 18–32.

57. Of thirteen speakers whose opinions were printed, seven favored more primary schools than in the plan (numbers from BN Le[43] series): Boilleau (2687), Duplantier (2688), Scherlock (2689), Bremontier (2704), Joubert (2706), Sonthonax (2807), and Andrieux (2994); Duplantier, Scherlock, and Sonthonax supported compulsory public primary schooling with an eccentric interpretation of the constitutional guarantee of private schools; two deputies opposed expansion of primary schools: Bailleul (2924), and Boulay de la Meurthe, reprinted in *Moniteur universel* (23 and 29 germinal VII), XXIII, 826–828, 851; two deputies opposed "reinforced" primary schools: Sonthonax (2807) and Andrieux (2994); *one* favored them: Challan (2705); three opposed stringent surveillance: Boulay, Pison-du-Galland, reprinted in *Moniteur universel* (14, 15, 16 germinal VII), XXIII, 791–792, 794–796, 798–800, Andrieux (2994).

58. Sonthonax (BN, Le[43] 2807), pp. 5–10.

59. Boulay, pp. 828, 826, 851.

60. Pison-du-Galland (brumaire VI) (BN, Le[43] 28), pp. 4–5;

ment. Despite the explicit references to a Supreme Being in the primary school report, they were scandalized that ethics might be taught without God. One noted that the Directory message seemed to ignore that morality was a question of feeling and authority, not of reason. Impicitly attacking the Idéologues' influence in the central school reform, he caricatured materialism as a system "which, confounding soul and body, would make of us a machine, ruled like the animal by the impulse of fleeting appetites; which makes our will a passive instrument of chance, a blind collision of the elements."[61]

Contrary to the suggestions of textbook histories of education, then, there was no single "bourgeois liberal" attitude toward education in the Council of Five Hundred. Moderates themselves were divided on the desirability and urgency of primary education, the justice of state regulation, and the philosophy of the central schools. The same debate also revealed disagreements among the Idéologues themselves. The literary critic and *Décade* editor, François-Stanislas Andrieux, thought republican festivals more useful than higher primary schools and opposed a curriculum for primary schools prescribed by the state. He thought the surveillance plan would make republicans seem like a "ferocious sect," who "having sown fear will harvest only hatred." Interrogation of instructors would make them automata parroting government doctrine rather than teachers cultivating a love of liberty. In addition, he found the higher education plan much too grandiose.[62]

We also know the ambivalent feelings of Destutt de Tracy about primary education. In his essay of 1801, intended to save the central schools at a time when the Interior Ministry of the Consulate was reorganizing education, Tracy devoted minimal attention to the first level of schooling. Distinguishing between schools for the "working class" and those for the "learned class," he proposed two parallel educational systems—terminal primary schools for sons of workers and a system of several levels for those preparing for the professions. Tracy recommended primary schools only for those communes where parents valued education enough to pay for them. They would also be established gradually as the central schools turned out teachers qualified to staff them.[63]

Cabanis never actually entered the 1799 debate in the Council of Five Hundred, but the manuscript of a speech he prepared was obviously intended for delivery some time in May (after 11 floréal VII).[64] He agreed with

61. Boulay, p. 828; Pison-du-Galland, *Moniteur,* pp. 794–796.

62. BN, Le[43] 2994, pp. 3, 11–12.

63. *Observations sur le système actuel d'instruction publique* (Paris, an IX), pp. 1–5, 64–68.

64. Cabanis' references to Boulay (Lehec, ed., II, 429) date composition of the speech after 21 germinal; remarks on pages 441 and 445 apparently refer to Andrieux's comments on 1 and 11 floréal; several opinions were reprinted 4 prairial, *Procès-verbaux des séances du Conseil des Cinq-Cents* (Paris, prairial VII), p. 96; closure of discussion may have occurred before

Boulay that public education should not stipulate forced attendance at boarding schools. Analogies from ancient Greece or Rome were inapplicable since there the "lowest class of the people; miserably enslaved, was always sacrificed to the upper classes, alone free: their pretended democracies were at bottom violent aristocracies. . . ."

Since the constitution explicitly guaranteed the right to open a legally certified school, it would be unwise to attempt to place the progress of enlightenment at the mercy of the government. He ridiculed the overemphasis of some speakers on the educational role of national festivals. The moral influence of these events among the ancients as well as the moderns, he pointed out, could never be realized by merely "mechanical and vulgar" means—compulsory attendance at ceremonies.

Nevertheless, as a member of the public instruction commission, he gave his full support to all sections of the Roger-Martin plan. Referring to his own agreement with the Mirabeau discourses, Cabanis wrote, "The ideas of Smith thus modified were, I confess, mine for some time; but I declare with the same frankness, that after more mature reflection, I consider them not very solid in general . . . and especially in no way applicable to the circumstances of the French nation." We must now remove, he continued, the "overly great influence of knowledge not by trying to restrict its progress to certain channels, or to halt it, but to spread it in great waves everywhere." Knowledge must not become the exclusive privilege of "classes already favored by fortune" who sometimes oppose the national interest. As Cabanis added in somewhat florid rhetoric, "often the happiest dispositions are hidden and languish under the humble roof of the poor." The minimum salary for primary school instructors was necessary since teachers depending exclusively on fees would be at the mercy of parents. Paternal preferences should not be all-powerful at this juncture, and present conditions in the primary schools showed that republican instructors would not necessarily be the most prosperous (II, 428–430–433–434).

At the same time, neglecting educational levels above the primary would only dry up the source of primary instructors while not helping primary education. Unlike Andrieux and Tracy, Cabanis defended the "reinforced" primary schools of the Roger-Martin plan as important means of overcoming the influence of private schools. Perhaps Cabanis was aware that the drafting course of the central schools had the greatest enrollment of sons of artisans at the secondary level. In a scientific age, Cabanis argued, craftsmen, such as tanners, bleachers, masons, and cutlers, and practical men such as civil and military engineers and sailors, would need to study applied mathematics and the applied sciences—just the kind of subject

---

Cabanis' speech; cf. Joanna Kitchin, *Un journal philosophique: La 'Décade' (1794–1807)* (Paris, 1965), pp. 199–201.

which could be taught in the "reinforced" primary schools. Against the scorn of some colleagues for "demi-savants," he pointed out that educated artisans and ingenious farmers would be precisely those responsible for necessary innovation and increased productivity so vital to the nation (II, 436–437, 443–444).

Aside from the need for technical education, Cabanis noted that in a republic, ignorance among potential public officials was dangerous. The government must teach the elements of ethics and politics, based on the "needs and faculties of human nature," rather than on "certain religious beliefs," which would crumble as reason developed. Public supervision of ethics teaching would not be prejudicial to the rights of parents. As the Mirabeau discourses had already argued, there was a need to enlighten the conscience of the citizen before encouraging him to follow it. Finally, no one need fear the consequences of education for the poor. Ignorance could only mean dependence, rather than happiness or virtue. The diffusion of enlightenment would be the best defense of liberty if for no other reason than to teach the populace to see through the blandishments of demagogues (II, 449 n. 1, 439).

Despite impassioned pleas from several speakers, the Council of Five Hundred once again tabled the Roger-Martin plan. There was no further reform of the central schools and no further aid to primary education under the Directory. But despite the lukewarm attitude of the council majority to education in a time of financial crisis, Cabanis did favor vigorous efforts in primary education to break the monopoly on knowledge of the wealthy.

To assess his difference from the complacent liberalism of the Consulate, one need only cite the arguments of Fourcroy and Pierre-Louis Roederer for the public education law in the changed political atmosphere of the spring of 1802. In a speech to the Tribunate (24 floréal X), Roederer restated the wellworn arguments about overeducating pupils who would then disdain their naturally destined condition. He also noted the uselessness of teaching the "moral and political sciences" at the secondary level. The real hostility of Napoleon Bonaparte to the central schools emerged in Roederer's comments on the "legislative" sciences as immature, unfocused, and of uncertain and contradictory method. Several weeks earlier, Fourcroy noted that state salaries for primary instructors would encourage negligence and inertia, and he explicitly cited Adam Smith's argument that education was best left to private initiative.[65]

Certainly, no one would deny the commitment of Bonaparte to technical education in the lycées and in the Polytechnique or even to the establishment of some arts and crafts schools. But the tenor of educational reform, in 1802

65. Maurice Gontard, *L'enseignement primaire en France de la Révolution à la loi Guizot (1789–1833)* (Paris, 1959), pp. 202–211.

and in the establishment of the Imperial University, was less the diffusion of enlightenment than service to the military and administrative needs of the state. Thus primary education was left to communal initiative, and the religious teaching orders once more assumed a significant role. The revolutionary ideal of secular universal education was postponed for several generations.

Cabanis' participation in or approval of several political coups and his markedly elitist justification of the Constitution of 1799 certainly made clear his misgivings about political rights for the uneducated and unpropertied in the determination of the public interest. Yet he continually adhered to a certain ideal of the moderate revolution which opposed privilege and hoped that economic growth would bring general prosperity.[66] Precisely because he lived in a revolutionary era Cabanis could not advocate merely laissez-faire policies.

The psychophysical engineering of the applied science of man assumed that natural temperament was the necessary point of departure of producing an "acquired temperament." The physician or moralist was in a superior position to determine the best interest of individual health, and with increasing education, everyone would be able to learn sound physical and mental habits. Similarly, enlightened legislators would determine the public interest. As fundamental self-interest became enlightened and fundamental sympathy became cultivated, individuals would turn toward the public good without losing their natural rights.

To this end, unemployment relief, even in the form of cottage industry or projects arranged with private contractors, and hospital care for the sick would meet the obligations of society and express the natural compassion of the more fortunate. But charity which perpetuated dependence and hospitals which complicated the disease were the wrong kind of therapy for social problems. As the revolution became more radical, Cabanis became more skeptical of crowds in relief projects and advised against indiscriminate aid which would destroy the incentive to self-improvement. For Cabanis, the natural situation of a person meeting his own needs could occur only through the independence of work rather than the subordination of public assistance. Independence would ultimately lessen artificial inequality while leaving a salutary residue of natural inequality.

The small hospice would also provide a more "natural" environment for medical diagnosis and therapy, though statistical compilation of records was essential for clinical medicine. For the mentally ill, Cabanis' goal was rehabilitation, not aimless confinement.

66. Analysis of Cabanis' participation in Directory coups and theory of government has been deleted and appears in *Cabanis: Enlightenment and Medical Philosophy in the French Revolution* (Princeton, 1980).

The political circumstances of the Revolution convinced Cabanis more firmly of the need for public primary education to prepare all for citizenship and of flourishing central schools to teach ideology to those destined for professions. Unlike Tracy, he saw no outright division between education for workers and for the learned, and he even favored widespread technical training for artisans. Moreover, he was always convinced that the state could regulate the medical professions and medical education without encroaching upon individual rights.

Doubtless Cabanis' attitude toward the propertyless and allegedly unenlightened was paternalistic. His arsenal of weapons against socially produced inequality did not include progressive taxation, or the minimum wage, and the meager food budgets he prescribed for those on public assistance might certainly have resulted in malnutrition which we now know affects educational potential. Even if one avoids an unhistorical twentieth-century perspective, one must admit that Cabanis wrote little about the concept of social insurance already being formulated by Thomas Paine in the *Rights of Man* (1791) as well as by French colleagues such as Montlinot.

Despite these reservations, one must recognize that Cabanis was not resigned to a static, hierarchical society. Even while he warned against artificial economic controls, he prophesied in 1803 that "a political constitution based on human nature and the eternal rules of justice must in the long run almost completely eradicate the traces of poverty and distribute without upheaval all the means of enjoyment (*moyens de jouissance*) in a more equitable manner. This constitution may very much diminish the number of crimes committed by eliminating both colossal fortunes and extreme poverty" (II, 59). Similarly, in a November 1803 review of Volney's study of American climate and soil, Cabanis remarked that high wage scales across the Atlantic alleviated many causes of dissension prevalent in Europe. He added, "The most serious of these is the enormous quantity of beggars, the rate of wages which is much too low, and as a result, the constant war of the poor and the rich, which must finish by overturning everything, if one does not hasten to favor, everywhere, without violence, a more equitable distribution of the goods of nature and the powers of society."[67]

In the context of early nineteenth-century French history, the passage seems to temper the politics of a Guizot with the social awareness of a Saint-Simon. In some respects Cabanis was the forerunner of the "Enrichissez-vous" outlook of the July Monarchy, but he never would have accepted the platitudes of academic notables of the 1830's who believed that society owes nothing to the poor. The extent of society's debt, the possibility

67. *La décade philosophique, littéraire et politique* (10 frimaire XII), p. 398n.; cf. Kitchin, *Un journal*, p. 196.

of promoting social mobility by education, the viability of extensive welfare and social service programs without reducing work incentives, without overburdening taxpayers, and without bankrupting municipalities, remain at the heart of the dilemmas of modern capitalism. While Cabanis was a staunch defender of individual rights, he was also an Idéologue for whom human perfectibility implied therapy for society as well as for the individual.

III

SCIENTIFIC INSTITUTIONS

SALOMON'S HOUSE EMERGENT: THE EARLY ROYAL SOCIETY AND COOPERATIVE RESEARCH

MARIE BOAS HALL

THAT the Royal Society of London for the Improving of Natural Knowledge drew its inspiration from the methodology of Francis Bacon and his dream of man's ability to understand nature was tacitly taken for granted by thoughtful contemporaries. Thus the virtuoso Joseph Glanvill in 1665 could say "Salomon's House in the *New Atlantis* was a Prophetick Scheam of the Royal Society";[1] Thomas Sprat's *History of the Royal Society*[2] was given a frontispiece in which the president and Bacon acted as heraldic supporters for the bust of Charles II; and the virulent Henry Stubbe cast ridicule upon Bacon's discussion of the sweating sickness as part of his attack upon the Royal Society of the early 1670's. Bacon was the self-proclaimed restorer of the arts necessary for the discovery of nature, and the Royal Society seemed attractively to fit his picture of Salomon's House on the island of Bensalem, where in Utopian peace, plenty, and cautious isolation all men worked for the discovery of truth in an elaborately structured cross between a modern research laboratory and a government bureaucratic agency. For did not the Royal Society undertake cooperative work, and did not the leading spirits set the tone and provide subjects for the investigation of the natural world?

There is here an intriguing paradox. The foundation of the Royal Society in England, as of the Académie Royale des Sciences in France, undoubtedly signalized a new stage in the study of nature, a step on the road to science as it is understood today. Equally certainly Bacon's dream organization was a Renaissance conception of the purest dye, comparable with Thomas More's Utopia, or the Christianopolis of J. V. Andreae.[3] These visions in turn

1. *Scepsis Scientifica* (London, 1665), dedication.
2. Finished in 1663 but finally published in 1667, the delay presumably the result of the effects of plague and fire on the book trade as well as on the Royal Society.
3. Described in *Republicae Christianopolitanae descriptio* (Strasbourg, 1619).

reflected an actual social structure of Renaissance intellectual life, the learned household. The great humanists frequently gathered around themselves students and assistants who lived together patriarchally and devoted their laborious days and nights to the pursuit of learning. A notable case is that of the great sixteenth-century printing family of Paris, the Estiennes or Stephani; it was said that the whole household, babies, compositors, masters, and servants spoke Greek as fluently as French. The first Renaissance academies—the Platonic academy of fifteenth-century Florence, the academies associated with the poets of the Pléiade and the court in sixteenth-century France,[4] the Lyncean Academy of which Galileo was so proud to become a member—there were all associations, more or less close, of learned men who spent much time in debate and discussion and who somehow felt that they had thus created a corporate entity. The scientific organizations of seventeenth-century France to which Harcourt Brown has introduced us so learnedly and lucidly were not very different. Their aim was a social grouping, even a club, in which men shared their intellectual interests in debate and discussion, informally and often confusedly, with no responsibilities, no definite end in view other than intellectual stimulus, no qualification for membership except personal introduction, and no lasting results.

Bacon's Salomon's House differed from these in having a tightly knit hierarchy which directed the investigation of lesser workers who did not see the ends, but labored none the less effectively for the truth. This however was not precisely what its seventeenth-century admirers saw; they thought of Salomon's House as an inspiration for the setting up of an organization for cooperative research by men of diverse capacities and interests who were yet drawn together by a shared desire to discover truth—in this instance, truth about the natural world. Salomon's House was a true aristocracy, with intellectual and political government directed and controlled by a few choice spirits; the later seventeenth-century ideal was an intellectual democracy in which all "curious minds" brought their wits to bear on the problems of nature and all shared in the gradual exposure of truth; yet the holders of this later ideal turned to Bacon as their progenitor.

Seventeenth-century scientists were obsessed with the necessity for communication and the interchange of information to a degree hardly known in earlier centuries. This again has some parallel in the activities of sixteenth-century humanists, whose letters often reveal more about their true intellectual preoccupations than do their books. This new need arose, I believe, because of the new interest in discovery as distinct from the older aim

4. See Frances Yates, *The French Academies of the Sixteenth Century* (London, 1947); Harcourt Brown, *Scientific Organizations in Seventeenth-Century France* (Baltimore, 1934), picks up the story, although only in 1620. For the Lyncean Academy the best description is to be gleaned from Edward Rosen, *The Naming of the Telescope* (New York, 1947).

of understanding. The methods of medieval science, as of medieval learning, had been primarily directed toward the understanding of difficult texts and of difficult problems raised by these texts; these methods were equally suitable for the student or the master, for the student could be expected to have the same difficulty with elementary works that the master might experience with more advanced works. (Similarly, the texts could indifferently concern nature or salvation; the problems of interpretation and comprehension were the same). The revival of the Greek view of progress in science, and the growing belief in the reality of new discoveries brought the need for communication. New discoveries must not be lost, for upon them depended the further progress of science. Moreover, there was urgent need for discussion and debate, for the age had an ardent belief that two heads were better than one. Great things had come from individuals working in isolation—although it must not be forgotten that such different contributors as Tycho Brahe and Galileo had both had large households with students and assistants—but yet greater and more continuous progress might come when men of equal intelligence should match wits together.

The Royal Society, like so many other societies and academies of the first half of the seventeenth century, had its first faint beginnings out of the belief that discussion brought clarity, and that to debate scientific problems was a useful end in itself. This, I think, amply explains the terms in which John Wallis, many years after the event, described the meetings of a group of young men, nearly all at first university graduates, to which he traced the origin and source of the Royal Society. Although it is very familiar, perhaps I may be permitted to quote it yet again, calling attention to Wallis' emphasis on aims, and noting that the latest scholarship, in spite of much debate, has on the whole accepted Wallis' recollection as sound historical fact.

> I take its first ground and foundation to have been in London, about the year 1645, if not sooner, when Dr. Wilkins . . . and others, met weekly at a certain day and hour, under a certain penalty, and a weekly contribution for the charge of experiments, with certain rules agreed amongst us. . . .
>
> Our business was (precluding matters of theology and state-affairs), to discourse and consider of Philosophical Enquiries, and such as related thereunto: as Physick, Anatomy, Geometry, Astronomy, Navigation, Staticks, Magneticks, Chymicks, Mechanicks, and natural Experiments; with the state of these studies, as then cultivated at home and abroad. We then discoursed on the circulation of the blood, the valves in the veins, the venae lacteae, the lymphatick vessels, the Copernican hypothesis, the nature of comets and new stars, the satellites of Jupiter, the oval shape (as it then appeared) of Saturn, the spots in the sun, and its turning on its own axis, the inequalities and selenography of the Moon, the several phases of Venus and Mercury, the improvement of telescopes, and grinding of glasses for that purpose, the weight of air, the possibility, or impossibility of vacuities and nature's abhorrence thereof, the Torricellian experiment in quicksilver, the descent of heavy

> bodies, and the degrees of acceleration therein; and divers other things of a like nature.[5]

This would sound like a pure discussion society, were it not for "the weekly contribution for the charge of experiments" and a fact Wallis also noted, that the meetings were usually held where there was convenient access to instruments, such as telescopes. It was this experimental aspect which constituted the great novelty of this early group, of its Oxford and London continuation, and of the early Royal Society. The need for instruments and the charge for weekly experiments had two aspects. In part they arose from the nature of the new science: men could effectively debate the validity of the Copernican hypothesis without instruments, but they could not dispute about the shape of Saturn, the selenography of the moon, the weight of the air, or "other things of a like nature" without appeal to the senses. No one could properly understand the problems raised by new experiments without repeating them, as investigators were rapidly discovering, and for this reason the Royal Society took as its motto "Nullius in verba," and William Petty was only half flippant in suggesting that St. Thomas should have been their patron saint rather than St. Andrew.[6] Thomas Sprat in his *History of the Royal Society* (written under the supervision of John Wilkins) wrote "In their Method of Inquiring, I will observe, how they have behaved themselves, in things that might be brought within their own Touch and Sight: . . . I shall lay it down, as their Fundamental Law, that whenever they could possibly get to handle the subject, the Experiment was still perform'd by some of the Members themselves."[7]

In conformity with this belief the Royal Society laid great stress on the need for confirmatory experiments all during its early years. Hence when accounts of experiments were brought in to meetings the first demand was always that they be repeated by suitable Fellows, preferably before a meeting of the Society. Examples abound throughout the seventeenth century. Thus at the first meeting after the decision had been taken to form a society "for the promoting of Physico-Mathematicall Experimentall Learning," that is, on 5 December 1660, Christopher Wren was "desired to prepare against the next meeting for the Pendulum Experiment"—although he failed to do so. On 19 June 1661 it was ordered that "Mr Boyle's experiment of

5. The first sentence is taken from Wallis' *A Defence of the Royal Society and the Philosophical Transactions . . . in Answer to the Cavils of Dr. William Holder* (London, 1678); the second from an autobiographical letter written in 1696/7 and published first by Thomas Hearne in the preface to his edition of *Peter Langtoft's Chronicle* (Oxford, 1725), I, cxl-clxx. There are numerous references to the early history of the Society in Wallis' letters; see A. R. Hall and M. B. Hall, eds., *The Correspondence of Henry Oldenburg,* vols. I-IX (Madison, Milwaukee, and London: University of Wisconsin Press, 1965–73); vols. X onwards (London: Mansell, 1975–). See esp. X, 279.

6. As reported by John Aubrey in his "Brief Life" of Petty.

7. P. 83 (pt. II, sect. XII).

the flat marble . . . be tried at the next meeting." Or again, on 7 May 1668, "Mr Oldenburg produced an experiment sent him from Paris, shewing, that when the picture of an object falls just upon the optic-nerve, there is no vision. The experiment itself was made and succeeded."[8]

When the Academia del Cimento sent its long-awaited *Saggi* to the Society in the winter of 1667/78, the experiments—mainly on sound and air—were discussed at great length, and various related experiments suggested. For example at the meeting on 19 March 1667/8: "It being mentioned, that in the said book [the *Saggi*] there was related an experiment of making organ-pipes sound in an exhausted receiver, by blowing them there with bellows; it was ordered, that Mr. Boyle, as the person, who had been the first known to have suggested this experiment, should be desired to make it; or, if he had already made it, to acquaint the society with the success of it." And again, at the meeting of 26 March 1668, "It being mentioned again, that the Florentines had affirmed, that sounds move equally swift against and with the wind, it was suggested by the President, that the experiment might be conveniently enough made between Deal and Dover, and that he would desire the governor of Deal-Castle to take care of it."[9] It should be remembered that Lord Brouncker, the president from 1662 to 1677, was one of the Navy commissioners. (Unfortunately there is no record as to the carrying out of this experiment, although this does not preclude the possibility that it may have been performed, as many of the Fellows who resided in London were friends and in the habit of meeting frequently for business, pleasure, or scientific converse, and often told each other informally of what should have been entered in the Minutes.) When Erasmus Bartholin in 1670 sent a sample of Iceland spar and a description of its peculiar optical properties, the "stone was referred to Mr. Hooke, who was desired to give an account of [its properties] at the next meeting."[10] And so on; the examples are endless.

Indeed, it was in large part for this very reason that the Society from the beginning had in its employ men whose function it was to perform experiments or to make preparations for the performing of experiments at the Society's meetings. The first post to be established was that of "Operator"; this was a menial post at small salary, occupied by a succession of skilled or semiskilled mechanics, whose names are known but who made no independent impact upon the advance of science.[11] The second post was that of "Curator," a position held for many years after 1662 by Robert Hooke,

8. See Thomas Birch, *The History of the Royal Society*, 4 vols. (London, 1756–57), I, 4, 31, II, 2–81.

9. Ibid., II, 257, 261.

10. Ibid., II, 448–49.

11. Richard Shortgrave served from 1663 to 1676; he was then replaced by Henry Hunt, who had been Hooke's "lab boy" since 1673, and who served the Society until his death in 1713 (he was also Keeper of the Library and Repository from 1678).

who was also a Fellow of the Society and soon to have a considerable independent reputation. After Hooke's appointment the operator partly functioned as his assistant, just as the amanuensis assisted the other paid officer of the Society, the "active" secretary, Henry Oldenburg. (There were two secretaries named in the charter, but Oldenburg from 1662 until his death in 1677 was the working secretary, charged with day-to-day administration and the conduct of the Society's correspondence).

This desire to see for themselves, to view all experiments in public or failing that to have one or more members of the Society report independently upon the reliability of the report, is of course by no means the only reason for the Royal Society's insistence upon experiment, any more than it had been for the young scientists of the 1640's and 1650's. Experiment symbolized (though it could not circumscribe) the method by which knowledge was to be advanced upon a rational and secure plan and cooperative experiment temptingly suggested itself as a natural way to secure this admirable end. Here I do not wish to discuss the balance between theory and empiricism;[12] rather I should like to try to examine how the Royal Society in practice proceeded in its endeavor to establish some sort of "Salomon's House" by attempting to harness the best scientific minds of the century to a single broad aim.

The most obvious and direct way was to persuade those with access to information about out-of-the-way areas of the world to provide materials toward "a universal natural history"; to this information about less exotic areas could also usefully contribute. One of the earliest hints of this plan is to be found at that very first real meeting of the Society already mentioned, when it was ordered "that it be referred to lord viscount Brouncker, Mr Boyle, Sir Robert Moray, Dr Petty, and Mr Wren, to prepare some questions, in order to the tryal of the quicksilver experiment upon Teneriffe."[13]

The next step was on 6 February 1660/1, when "A committee was appointed for considering of proper questions to be inquired of in the remotest parts of the world." The business did not then prosper; but finally in August 1664 the Royal Society's "Committee for correspondence met the first time at Mr Povey" when "Generall inquiries were drawn up, serving for all parts of ye world; and Authors were distributed amongst ye members of the Committee, to be perused for ye collecting thence particular inquiries for particular countries."[14] The second part of this plan came to nothing, but the "general inquiries" were often sent to country gentlemen, travelers, foreigners, seamen, diplomats, and others, usually in the emended form

12. I have discussed this question in "Science in the Early Royal Society," in M. P. Crosland, ed., *The Emergence of Science in Western Europe* (London, 1975).

13. Birch, *History*, I, 5.

14. As reported by Oldenburg to Boyle; see *The Correspondence of Henry Oldenburg*, II (1960), 209.

edited by Robert Boyle which was to be published in the *Philosophical Transactions* in 1666. These, as Oldenburg remarked to Boyle, were the more necessary, "most men not knowing, what to inquire after, and how?"[15] In 1665, Hooke had written and showed to some of the Society his own version, which, entitled "Lectures of Things requisite to a Natural History," is in the Royal Society's archives; although Hooke spoke of printing it, he never did so. As early as 1663, Oldenburg had set John Winthrop, Jr., collecting information about New England, and for the next several years no opportunities for acquiring information about the Near and Far East, the North, the West Indies, or even the more familiar parts of Europe and the British Isles were to be neglected. In the winter and spring of 1665–66, as the Plague abated and some members returned to London, Oldenburg could write to Boyle in Oxford, "We have thoughts of engaging as many of ye Society, as are cordiall and have opportunity, to observe and bring in, what is any wayes considerable of Naturall productions in England, Ireland, Scotland; every one his Symbol [contribution], for ye bringing together a Naturall History of what is in yesd Kingdoms, as well as we intend to collect what is abroad, by enlarging our Correspondencies every where."[16] It was indeed to become one of the chief functions of the "country members," virtuosi of, generally speaking, more zeal than scientific ability and originality, to offer accounts of agriculture and arboriculture and of the mineral resources of their counties and districts.

They could and did contribute as well to that very Baconian aspect of the universal natural history, the "history of trades." Thus as early as 19 December 1660,[17] "Dr. Petty and Mr. Wren [were] desired to consider the philosophy of Shipping, and bring in their thoughts to the company about it"; and on 2 January 1660/1, "Dr. Merret was to bring in his history of Refineing," which he did two weeks later. On 9 January, "Mr. Evelyn [was desired] to shew his catalogue of trades," and a week later "an history of engraving and etching: And Dr. Petty to communicate the history of some trade at his own choice"; this proved to be "clothing," in fact, textiles. For the next few years members continued to bring in partial or complete histories of trades: of the manufacture of alum, brass, cerusse, china varnish, copperas and lead, of brewing, cider-making and wine, of calico-printing, earthenware, the coloring of marble and paper, of masonry, mining, hats, potash, salt, soap, and vinegar—the list is very long. Sprat printed those on niter or saltpeter (by Nathaniel Henshaw), dyeing (by Petty), and oyster cultivation; of the rest some were printed in the *Philosophical Transactions* and others remained in the registers of the Society.

The universal natural history also required help from abroad, not always

15. *Correspondence,* III (1966), 33.
16. Ibid., p. 32.
17. For these and the following notes, see Birch, *History,* under the appropriate date.

easy to obtain. In 1666, Oldenburg wrote to Sir John Finch, F.R.S. and English ambassador at the court of the Grand Duke of Tuscany in Florence.

> One thing more there is, I must sollicite; vid. yt it being a part of ye R. Society's dessein, to compose a good Nat. History, to superstruct, in time, a solid and usefull Philosophy upon, and ye compiling of ye natural Histories of particular Contries appearing very conducive to such a dessein, you would use your interest in Italy to excite some able and diligent persons to set upon yt work for yt contry; as I hope, the like will be done, by our importunity, in Spaine, Portugall, France, Germany, Poland, Hungary etc. And it being, of no slight importance, to be furnisht wth pertinent Heads, for ye direction of inquirers, there hath been printed in ye Last of the . . . Phil. Transactions, a list of Generall and comprehensive Articles for yt purpose, of wch I shall endeavor to get some transmitted to you. . . .[18]

Sir John Finch proved on this occasion, as on others (notably his embassy to Constantinople in 1673) too idle to assist the Society, but other, busier, men were more helpful. Thus Johannes Hevelius, already fully occupied with his brewing business, his activities in connection with the administration of the city of Danzig, and his astronomical researches, in 1666 sought and obtained information from friends about the Baltic regions, and especially the effects of their extremes of heat and cold, and through the years offered information on amber. Early in 1668, Oldenburg wrote a letter to the mathematician and churchman of Liège, René François Sluse, ending "I shall in conclusion urge this upon you which I have already requested several times: namely, that you take the trouble to stimulate some ingenious and skillful men to compile a natural history of your region."[19] Sluse responded to this appeal with a general survey, mainly directed to mineral resources, and some months later with an account of fossil shells.[20] Five years later Oldenburg was bombarding Sluse with queries upon mineral waters, taking advantage of Joseph Williamson's appointment to the peace negotiations at Cologne in 1673.[21]

In the summer of 1668, young Dr. Edward Browne, already a Fellow, set off on a fantastic journey of something over a year which took him into the depths of the Carpathians and Balkans. From there he reported on the mineral resources and mining practices of these remote regions, much to the interest and edification of the society when these accounts were read to meetings.[22] In 1671, Oldenburg, having heard from Thomas Hill, an English merchant in Lisbon whose brother had been an officer of the Society, that there was an (unnamed) Jesuit (actually a competent astronomer, F. Valentin Estansen) in Brazil who would be willing to supply information

18. *Correspondence,* III, 87.
19. *Correspondence,* IV (1967), 212.
20. *Correspondence,* IV, 265 f.; V (1968), 90–91.
21. *Correspondence,* IX (1973), 627–631, and X, 446–452.
22. See *Correspondence,* V, VI (1969), and VII (1970).

about that little-explored country, produced a most impressive list of nearly one hundred questions, drawn from the few printed accounts of that country.[23] These touchingly reveal the hunger men felt for knowledge of remote regions; unfortunately postal communication was too broken for a reply ever to be received.

From neighboring France there was little, even though, as Oldenburg reported to Boyle in early 1666, a Parisian correspondent had written "Je parle par tout du beau dessein, que vous avez de faire une Histoire naturelle, à fin de donner de l'Emulation à nos Messieurs, et les exiter à vous imiter."[24] Various provincial Frenchmen sent accounts of "trades," such as salt or vinegar making, but showed little or no enthusiasm for true natural history.[25] Ironically, after the foundation of the Académie Royale des Sciences, French scientists turned to a cooperative endeavor more closely knit and more truly Baconian than anything the Royal Society ever attempted. For these carefully solicited and compiled accounts of which I have been speaking remained in the Royal Society's archives, or at best were individually printed in the *Philosophical Transactions;* the grand design of a universal natural history, although Oldenburg never ceased to mention it to his correspondents, never came to fruition. The French, contrastingly, settled to truly cooperative work on botany and anatomy, producing beautifully illustrated printed books: *Description anatomique d'un cameleon, d'un castor, d'un dromedaire d'un ours et d'une gazelle* in 1669; *Mémoires pour servir à l'histoire naturelle des animaux* in 1671 (edited by Claude Perrault) and *Mémoires pour servir à l'histoire des plantes* in 1676. These were genuinely cooperative works, with individual contributions subdued to anonymity, quite remote from anything the Royal Society ever attempted or even thought of attempting. Nor were the English scientists ever required to publish anonymously or pseudonymously, but were astonished when Jean Picard's *Mesure de la Terre* was published in 1671 without the author's name on the title page.

The Royal Society certainly never advocated anonymity; on the contrary it was a Society composed of individuals of strong personality, who did not wish to submerge that personality in corporate enterprise. The nearest it came to corporate endeavor was the stimulating of men to take up investigations begun by others, with links maintained either by discussion at the Society's meetings; or by correspondence carefully conducted by the secretary; or by publication in the *Philosophical Transactions* (from 1665 to 1677 not a publication of the Society, but a private, though privileged,

23. *Correspondence,* VIII (1971), 220–251.

24. *Correspondence,* III, 68.

25. Notably Elie Richard, a physician of St. Martin on the Ile de Ré at La Rochelle; see *Correspondence,* VI, 82–85, 198–200, and 200–202. These were printed in the *Philosophical Transactions* in translation.

venture by its secretary, Henry Oldenburg). A somewhat dull example is the prolonged discussion in 1669 and 1670 about the "bleeding" of trees such as willows and sycamores when cut in the winter.[26] This was a roughly three-cornered (and almost totally inconclusive) effort, much of which was published in the *Philosophical Transactions,* between Israel Tonge (then resident in Kent, later to acquire notoriety for his association with Titus Oates and the Popish Plot), Francis Willughby assisted by John Ray in Warwickshire, and Martin Lister in York. A more stimulating example, and on an international scale, is the interest in blood transfusion which led to several daring and puzzling experiments in 1667.[27] The initial work had been done at Oxford as injection experiments conducted by Robert Boyle and the young Christopher Wren, about 1656 and 1657. When in 1666 reports were received of similar attempts at the injection of medicines abroad, Richard Lower at Oxford (with Boyle's encouragement) tried experiments of the injection of blood from one dog to another—that is, transfusion. These experiments upon dogs were repeated in London and, word of this reaching France through the *Philosophical Transactions,* the experiment was tried upon a madman with apparent success. The experiment of transfusing sheep's blood into a madman's veins was, also successfully, repeated in London upon a man named Arthur Coga. He was subjected to only one attempt and lived; he was evidently of a strong constitution, and very possibly less than the ten or eleven ounces thought to have been transfused actually were introduced into his vein. The French subject was unlucky; he survived one transfusion, but having relapsed into frenzy he received one or more further transfusions. Either from these (as his wife claimed) or from poison administered by his wife (as the surgeons claimed) he died some weeks later. The result was a public legal inquiry, and no more such experiments were conducted in either England or France.

There is an analogous case in the experiments on the "blood-staunching liquor" or styptic fluid publicized by Jean Denis in Paris and London in 1673, and tried by surgeons of the English Fleet at the command of Charles II, who was impressed by accounts at its efficacy on dogs, even when an artery had been severed.[28] Martin Lister claimed to have as effective a fluid of his own invention, which he tried (successfully, as he reported) in his medical practice at York;[29] others using their own or others' recipes reported no better success than with the use of clean compresses.

The whole affair of transfusion, in spite of its inauspicious conclusion which might have led all to dissociate themselves from the experiments, in

26. See *Correspondence,* VI and VII.
27. See *Correspondence,* III (esp. p. 611), and IV.
28. See *Correspondence,* X *passim.*
29. See *Correspondence,* X, esp. pp. 177, 222.

fact left behind a legacy of mistrust and suspicion, as well as unsettled claims for priority.

Controversies over priority indeed were many in the seventeenth century, and they were an important factor working against cooperative research. While the Royal Society as a body might feel, as it did, that secrecy was undesirable and that individuals should in all cases be encouraged to publish; since publication could stimulate the efforts of others, these same individuals were often jealous to maintain their rights of priority, and either kept their discoveries secret, or wrangled acrimoniously with fellow scientists. Robert Hooke's tendencies toward both secrecy and subsequent wrangling are well known, and I pass them by. A great aristocrat such as Christiaan Huygens (Original Fellow of the Royal Society as well as one of the first members of the Académie Royale des Sciences) could wrangle as bitterly as Hooke—and he did so from 1667 to 1669 with the Scotch mathematician James Gregory. (Ultimately, in spite of patriotism, most English mathematicians absolved Huygens either of plagiarism or of being anticipated. But Huygens did not always have the right of it in every such dispute and Wallis in 1673 after the publication of *Horologium oscillatorium* was to accuse Huygens of having become so Frenchified as to have lost his sense of fairness in these matters.)[30] Much sillier controversies of this nature were often at least quasi-patriotic: thus Dr. Timothy Clarke wrangled tediously and ungraciously with the young Dutch anatomist Regnier De Graaf, claiming both that De Graaf's discoveries were not true, and that anyway they had been anticipated by English anatomists. More profitable was the way in which certain other English anatomists such as Edmund King and William Croone were stimulated to repetition and possible expansion of De Graaf's discoveries.

Other futile controversies involved the Oxford mathematician John Wallis, ever ready to correct less mathematical heads at great length, until even the patient Oldenburg seemed to find Wallis' effusions wearisome, and ceased to publish them. There was a prolonged exchange between Wallis and a minor French mathematician, François Du Laurens, in 1668; and there was an even longer exchange—begun in 1660, before the foundation of the Royal Society, and continued intermittently over a dozen years—between Wallis and the aged Thomas Hobbes, who for all his intellectual keeness failed to grasp the basic postulates of Euclidean geometry, as Wallis repeatedly explained at very great length. Wallis appears in better light in 1668, when he used the same pertinacity of argument to correct and improve the ideas in mechanics of his former pupil William Neile, a gifted mathematician who died young "at his father's house at White Waltham" as

30. See *Correspondence,* X, 42, 526.

the *Dictionary of National Biography* touchingly puts it.[31] Wallis in 1673 firmly and more attractively advanced Neile's priority in certain aspects of mathematics against Leibniz and others.[32]

Wallis' controversy with Neile in fact illustrates another form of cooperative enterprise working within the Society usually engineered or at least guided by its able secretary, Henry Oldenburg. It was often thought good in the seventeenth century to stimulate men to intellectual activity by the somewhat drastic method of telling A what B had just done in a field in which A was expert, and fanning the resultant sparks of indignation into the flames of discovery. For although the seventeenth century was an age in which men were jealous of their discoveries and careful to claim priority, they were also often little inclined, or sometimes not able, to publish. Isaac Newton was by no means unique in keeping important scientific and mathematical ideas to himself, and even Newton would have published more of his early work on fluxions, for example, if booksellers could have expected a better market for mathematical books than they did, in fact, experience. Huygens delayed long in the publication of his ideas on mechanics: the content of *Horologium oscillatorium* (1673) had mostly been in notes for ten years or more, while his *Dioptrics,* which Oldenburg often urged him to publish,[33] stayed in manuscript form until after his death, although parts were utilized for the *Traité de la lumière* of 1690. Marcello Malpighi would never have published his *Anatomy of Plants* or his work on embryology without the Royal Society's encouragement, as communicated by Oldenburg. (To the honor of the Society it put justice before patriotism in the case of Malpighi, rightly judging that his work on plant anatomy preceded that of Nehemiah Grew.) To these examples could be added others, but these should suffice to demonstrate the point that there was good reason for someone like Oldenburg to crack men's heads together in the hope of generating new ideas.

The *Philosophical Transactions* in themselves served this function, and the more so as throughout Oldenburg's term of editorship (1665–1677) most of the direct communications were in the form of letters, although these were supplemented by reports from foreign journals and book reviews. The subtitle of the journal describes its aim as "giving some accompt of the present Undertakings, Studies, and Labours of the Ingenious in many considerable parts of the World." It served to inform, but also to stimulate men to work on particular topics, both at home and abroad, although there were always those who demurred at free communication to foreigners, and favorable reporting of their work. But this was at best random stimulation.

31. See *Correspondence,* IV and V.
32. See *Correspondence,* X, esp. 276–283, 291–293.
33. See, e.g., V, 464.

Oldenburg was able to do more by careful personal selection of his correspondents.

An early example is the controversy, encouraged and stimulated by Oldenburg, between the French astronomer Adrien Auzout and Robert Hooke in 1665, over selenography.[34] Oldenburg not only translated for Hooke's benefit various letters about the moon from Auzout, but in at least one case made marginal comments obviously designed to stimulate Hooke to reply, ranging from a mere "NB" to "What say you to this?" "A handsom sting again will be necessary," or "Me thinks, here you may tosse railleries wth him." Hooke declined to reply. Insofar as Auzout was on the whole in the right he was correct, but I suspect that part of his refusal was grounded in Auzout's urgent appeal for Hooke to communicate his alleged discoveries for the improvement of telescopes. One cannot but agree with Auzout's remark "If M. Hook will impart to us his Invention, we shall be obliged to him; and I wish I had a secret in the matter of Telescopes to encourage him to communicate it."[35] Had Hooke been a more open man, or less inclined to announce as an invention what was really only the idea for one, this exchange might have produced the kind of results that Oldenburg was constantly seeking.

In many other cases he was more successful, even though his methods might produce temporary resentment. Thus in October 1668, Hooke suggested at a meeting of the Society that "there might be made experiments to discover ye nature & laws of motion, as ye foundation of Philosophie and all Philosophical discourse" whereupon Brouncker remarked that both Christopher Wren and Huygens "had considered that subject more yn many others, & probably found out a Theorie to explicate all sorts of experiments to be made of that nature," as Oldenburg duly wrote to Wren and Huygens.[36] Now both Wren and Huygens were notorious nonpublishers, but this particular appeal had a good effect. Wren promptly replied, reporting that he had "looked out those papers of the Experiments that concerned the Lawes of Motion arising from collision of hard bodies," but "found them somewhat indigested as I left them at first. & I could be glad you would give me a little time to examine them."[37] An ominous remark, usually a prelude to procrastination. But in this case, helped on by encouragement from Oldenburg; Wren actually produced a finished paper; the original remains in the archives of the Royal Society and it was printed in the *Philosophical Transactions* for 11 January 1668/9 along with papers by Neile and Wallis. This is one of the very few contributions which Wren

34. See *Correspondence,* II.
35. Ibid., p. 473.
36. *Correspondence,* V, 118.
37. Ibid., p. 125.

made to science in his mature years, and his agreeing to have it printed was a great tribute to the efforts of the Royal Society.

With Huygens things took a more complex turn. Oldenburg had written him an account of the Society's meetings in terms similar to those he used to Wren, asking him to "inform" the Society "when [he] would bring [his] speculations and observations on this subject before the public,"[38] and if they were not yet in publishable state to communicate his "hypothesis of motion" to be registered in the Society's books. Huygens replied with enthusiasm, professing himself delighted at still being regarded as an ordinary Fellow of the Society (it was some years since he had attended a meeting); he expressed willingness to have a summary of his ideas registered, but thought he had nothing yet ready for publication.[39] The Society was satisfied with this, as Oldenburg informed Huygens;[40] Huygens then sent his paper, which was duly registered along with those of Wren and Neile. In return, Oldenburg sent a copy of Wren's paper, and informed Huygens that the Society intended to ask all those Fellows who had been working on the problem of collision to consider Huygens' ideas. He added, "We hope that, since there are at this time so many fine minds engaged with so fine a subject, this matter will at last be solidly digested and perfectly established; and we hope that all those who apply themselves to it will bring to the business a disposition to compare the efforts of others with their own, amiably and without prejudice, in order to get rid of everything which will be found contrary to good sense and to recognize and embrace the truth, whatever it may appear."[41] This is a splendid statement of the aim which lay behind the industrious dissemination of ideas and discoveries to which Oldenburg dedicated so much of his working life. The various ideas of Huygens, Wren, Neile, Wallis, Croone, Willughby, and others were discussed at meetings, and experiments were suggested to confirm or deny the hypotheses. All went very smoothly indeed until, early in February 1668/9, Oldenburg sent to Huygens a copy of the *Philosophical Transactions* containing papers by the English contributors to the discussion, but not that of Huygens. Huygens was decidedly offended, and did not reply to several letters by Oldenburg until, toward the end of March, he could send Oldenburg a copy of the *Journal des sçavans* in which was printed his own summary, with a claim for priority (or at least independence) over Wren. Oldenburg insisted that he had never intended to offend Huygens, that he thought he was not authorized to print Huygens' communication, and that in any case Huygens had the *Journal des sçavans* in which to publish. He made amends by reprinting Huygens' paper in *Philosophical Transactions* as being taken from the

38. Ibid., p. 104.
39. Ibid., p. 127.
40. Ibid., p. 177.
41. Ibid., p. 332.

Journal.[42] By his adroit handling of the situation Oldenburg managed to assuage Huygens' temporary pique, and in the end Huygens could write "Ie n'ay aucun suject de me plaindre, mais au contraire je suis tres satisfait de la maniere que vous en avez usé en ce qui regarde nos loix du Mouvement."[43] The episode amply indicates the grave difficulty of trying to produce cooperation by submitting the work of one man to another's consideration and criticism; it was a truly dangerous method of procedure, but one very capable, as in this case, of producing publication by those normally reluctant to make their thoughts public.

Later work by Huygens provoked more amiable exchanges. Thus in 1672, Huygens took up his old interest in experiments on the air, and in July sent a paper for publication to the *Journal des sçavans* about the possible cause for a phenomenon he had discovered a dozen years earlier, the anomalous suspension of mercury, or, as Oldenburg's English translation describes it, "The Experiment is briefly this; That a Tube, being, after the Torricellian way, filled with Mercury, and before inversion perfectly purged of Air, doth, when inverted, remain top full, even to the height of 75 inches."[44] Huygens explained this odd phenomenon by recourse to the supposed pressure of the Cartesian *matière subtile* or aether, which he thought to be strong enough to support the mercury when acting only on one side, as it may do until the tube is tapped or a bubble of air is introduced, when the aether is forced to be in contact with the mercury in several directions, and can no longer support the weight. The Royal Society was not sitting during the summer, but some members continued meeting informally, and they discussed the matter at some length. Further, Oldenburg promptly published a translation in the August *Philosophical Transactions* (few of the Fellows read French fluently). It became clear at the Society's meeting of 30 October 1672, the first since the summer vacation, that Hooke, Wallis, and Brouncker had been especially interested; this further appears from a series of letters from Wallis to Oldenburg, later converted into a formal paper for publication in the *Philosophical Transactions*.[45] Wallis rejected the subtle matter, and after considerable debate, opted for the cause being the "want of spring" (elasticity) of water or mercury compared with the natural elasticity of any included air. Hooke read a paper containing his own views to the meeting of 6 November 1672; though it was well received, he failed to hand it in to be registered, so there is no certainty as to what his theory was. He did however produce a supplementary experiment on 20 November, when "an experiment was made to shew, that water in a tube, open at both ends,

42. Ibid., pp. 465–466.
43. Ibid., p. 554.
44. *Phil. Trans.*, no. 86 (19 August 1672), 5027–5028.
45. *Correspondence,* IX, 258–262, and 278–280; also *Phil. Trans.*, no. 91 (24 February 1672/3), 5160–5170.

will, when lifted up, stand at eight inches, before it begins to fall. Mr. Hooke was desired to describe the contrivance of this experiment, to be registered."[46] Here the reporting of an experiment and the offering of explanations proceeded in international amity, even if without any certain conclusion.

It is hardly necessary to mention that astronomy provided a very constant form of cooperation. Every announcement of the viewing of an eclipse, a conjunction, a new star, or a comet was an invitation to the sharing of data determined by others. Hevelius in Danzig never failed to send Oldenburg accounts of his observations of all interesting celestial phenomena, most of which were printed in the *Philosophical Transactions.* When Saturn's "anses" became visible after a period in 1670, astronomers everywhere contributed their observations, as they had done with the comet of 1664/5, the various novae of the 1660's, or the brilliant sunspot display of 1671. John Flamsteed indeed established his reputation on the basis of careful tables of lunar motion and predictions for the coming year which were published in *Philosophical Transactions* and invariably proved more accurate than those of older astronomers, although his telescopes, and possibly his observational powers, were far inferior to those of Cassini. In the summer of 1673, Oldenburg was able to establish a profitable correspondence between the two men, in which Flamsteed generously acknowledged Cassini's abilities and his wider facilities.[47] Here cooperation was often restricted by weather, and only observations, not theory, were being made.

The most famous case of cooperation by correspondence through Oldenburg as intermediary centers around Isaac Newton's first communications of his doctrine of light and colors. Although well known, this affair is not well understood, since historians have often neglected to look at the whole network of correspondence. It is therefore worthwhile to consider the matter in some detail. It really begins with the communication to the Society (by means not so far known) of Newton's invention of a reflecting telescope. The Society was much delighted with the idea; Oldenburg was instructed to get Newton to authorize a description for communication abroad; this done, Oldenburg sent a copy to Huygens, and the account was subsequently published in the *Philosophical Transactions,* in the *Journal des sçavans* and in Jean Denis' newly founded *Memoires concernant les arts et les sciences,* in all of which forms it attracted much attention. The avowed aim of publication was to establish Newton's priority, and this aim was achieved.

Oldenburg was then instructed to try to get more material from Newton; the result was Newton's famous letter of 6 February 1671/2 on the nature of white light and the phenomenon of dispersion, printed in the *Philosophical*

46. Birch, *History,* III, 61.
47. Mainly printed in *Correspondence,* X.

Transactions for 19 February with Newton's promptly given permission.[48] There were criticisms from Hooke (made immediately after the reading of Newton's paper to the Society on 8 February); and, as a consequence of the printing of the paper in the *Philosophical Transactions,* from Huygens (who had already produced some criticisms of Newton's reflecting telescope, although he upheld Newton's priority), and the French Jesuit physicist Gaston-Ignace Pardies. Hooke's criticisms were decidedly ill-natured, and an embarrassment to the Society which at its meeting had publicly commended Newton's work; it was decided not to publish Hooke's comments until Newton had had an opportunity to reply, which he did finally some months later in June. Oldenburg, acting mainly at the direction of the more senior members of the Society, conducted the affair with all the tact he could consistent with getting Newton to reply, but it was a constant irritant to Newton. After all, he was a virtually unknown Cambridge professor of mathematics, while Hooke, the senior man, was Curator of the Royal Society, Gresham Professor, Cutlerian Lecturer, author of two books, with an established reputation. While the rest of the Society urged him to produce more, Hooke was hostile and intolerantly so; what was Newton to believe? It is no wonder that the controversy with Hooke tended to make him withdraw into his shell.

Newton's disagreement with Huygens produced no such unfortunate consequences. Huygens remained unconvinced by some of Newton's arguments, but politely and at a purely scientific level, unmarked by personal acrimony. The publication of some part of their disagreement in the *Philosophical Transactions* proceeded amicably. Even more amiable was the disagreement between Pardies and Newton. Both wrote to Oldenburg, sending replies to the papers for which Oldenburg acted as epistolary intermediary; both agreed to publication; and in the end Pardies was convinced, having elicited from Newton both the admission that his famous "crucial experiment" was not clear in the absence of a diagram (a criticism also raised by Huygens), and a fuller explanation. In this case, enlightenment had resulted from an epistolary exchange, and both Newton and Pardies were happily aware of it. It was a vindication of the method of stimulating men creatively by exposing them to the criticisms and comments of others.

All these examples show how the Royal Society strove to establish a working cooperation among its highly individual members. But it was no Salomon's House, for there was no hierarchy of investigators—rather a loosely knit association of men of differing capabilities, but free and independent minds. No one in the Society ever seems to have even thought of

48. As all this correspondence was published in *The Correspondence of Isaac Newton,* I, ed. H. W. Turnbull (Cambridge, 1959), only summaries of Newton's letters were printed in the Oldenburg *Correspondence,* though the letters to and from Oldenburg for correspondents other than Newton are given in full. See *Correspondence,* VIII and IX.

trying to make use of the contributed work of others in a grand synthesis.[49] There was no such anonymity of contribution as characterized much published work of the Académie des Sciences, or the *Saggi* of the Academia del Cimento. In so far cooperation meant discussion, the Fellows of the Society welcomed it; indeed, discussion in an orderly and friendly atmosphere was the true purpose of the Society's meetings. Fellows were ever willing to turn their minds to particular topics at the suggestion of other Fellows, or of foreign correspondents. They could be stimulated, but not commanded; urged, but not compelled. In the last analysis the Society acquired its reputation not because it was a research institute directed by a brilliant autocrat, but because it was an association of independent equals who willingly shared with others the discoveries and ideas achieved independently in the solitude of their own studies.

49. Oldenburg often appealed to correspondents over the years to contribute to what he described as a universal natural history, but no attempt was ever made to do more than make the material available to individuals either by printing it in *Phil. Trans.*, or by reading it at meetings.

GUY DE LA BROSSE AND THE JARDIN DES PLANTES IN PARIS

RIO C. HOWARD

THE Muséum National d'Histoire Naturelle in Paris is an institution of long and distinguished history. Before the French Revolution it was known as the Jardin des Plantes or the Jardin du Roi and in the late seventeenth and eighteenth centuries it was one of the principal scientific centers in France. Here Georges Louis Leclerc, comte de Buffon and his collaborators prepared the famous volumes of the *Histoire naturelle* and here scientific figures such as G.-F. Rouelle gave public lectures to which not only the *beau monde,* but also young scientists such as Antoine Lavoisier came.[1]

Yet the origins of the Jardin des Plantes are only obscurely known. Its founder was one of Louis XIII's physicians, Guy de La Brosse, of whom there is even less record than of the early years of his garden. Two royal edicts in 1626 and 1635 established the Jardin des Plantes, but beyond these bare facts, the histories of the period give various and often contradictory accounts of those involved and of their actions in the realization of the garden.[2]

Earlier versions of this paper were read to the Western Society for French History, November 1974, and to the Metropolitan Section, History of Science Society, December 1974. I am indebted to Henry Guerlac for first suggesting the early history of the Jardin des Plantes to me as a topic of research and to Professors Jerry Stannard and John Bosher for a critical reading of this paper.

1. On the Jardin des Plantes in the eighteenth century, see Yves Laissus, "Le Jardin du Roi," in *Enseignement et diffusion des sciences en France au XVIIIe siècle,* ed. René Taton (Paris: Hermann, 1964), pp. 287–341. The following abbreviations are used below: BN, Bibliothèque Nationale, Paris; AN, Archives Nationales, Paris; MHN, Muséum National d'Histoire Naturelle, Paris.

2. An indication of the extent to which confusion reigns in the accounts of the early Jardin des Plantes is the fact that any number of persons are credited with the actual founding of the garden including, in one instance, the son of Jean Riolan II, the famous Parisian anatomist. Cf. *Mémoires du Cardinal de Richelieu, publiés d'après les manuscrits originaux pour la société de l'histoire de France (Série antérieure à 1789) avec le concours de l'Institut de France—*

It is a story worth elucidating, for it forms an important part of a botanical renaissance which characterized the sixteenth and seventeenth centuries, a renaissance which laid the ground for the eventual emergence of botany as an independent science and the work of men like Linnaeus.

Botany had traditionally been a medical subject and from classical times the greatest interest in plants in Europe was occasioned by their medicinal properties. For the Western Middle Ages the herbal of Dioscorides, in different editions and adaptations, was one of the standard reference works on medical plants.[3] Yet, especially during the early Middle Ages, the study of botany became a largely literary pursuit as the iconographical tradition in medieval manuscripts indicates. The illustrations which accompanied medieval botanical treatises were copied so frequently that they often became stylized beyond recognition. Occasionally the same figure even did service for several plant descriptions.[4]

The sixteenth century saw a great change in botanical study, spurred perhaps in part by the importation of plants from beyond Europe, and in part by a new interest in direct observation.[5] This interest was revealed by the production of splendid herbals, in which the illustrations, newly drawn from life, far outstripped the old texts that accompanied them in accuracy of description. Such are the herbals of Otto von Brunfels, Hieronymus Bock, and Leonhard Fuchs.[6] Moreover, as plants began to be observed more closely, several difficulties with the traditional botany of Dioscorides presented themselves. It was only slowly discovered, for example, that the plants described by the great Greek herbalist were those of the Mediterra-

---

Académie Française (Fondations Debrousse et Gas) (Paris: Librairie Ancienne Honoré Champion, 1929), IX (1629) publié sous la direction de M. Lacour-Gayet par Robert Lavollée, p. 76, n. 2. The best history of the early Jardin des Plantes is still the account of A. L. Jussieu, "Notice historique sur le Muséum d'histoire naturelle," *Annales du Muséum d'Histoire Naturelle,* I-IV (1802–1804).

3. On the editions of Dioscorides, cf. John M. Riddle, "Dioscorides" in the *Dictionary of Scientific Biography,* 4 (New York: Scribner, 1971), 119–123, and the accompanying bibliography. Cf. also Charles Singer, "Greek Biology and Its Relation to the Rise of Modern Biology," in *Studies in the History and Method of Science,* ed. Charles Singer (Oxford: Clarendon Press, 1921), II, 1–101, especially pp. 56–78.

4. Cf. Agnes Arber, "From Medieval Herbalism to the Birth of Modern Botany," in *Science, Medicine and History, Essays on the Evolution of Scientific Thought and Medical Practice Written in Honour of Charles Singer,* ed. E. Ashworth Underwood (London: Oxford University Press, 1953), I, 317–336.

5. On the sixteenth-century renaissance in botany, especially on the influence of the non-European world on Europe, cf. Emile Callot, *La renaissance des sciences de la vie* (Paris: Presses Universitaires de France, 1951), pp. 29–42. See also Agnes Arber, *Herbals, Their Origin and Evolution. A Chapter in the History of Botany, 1470–1670,* 2d ed. (Cambridge: The University Press, 1953).

6. Otto von Brunfels (Otto Brunfelsius), *Herbarum vivae eicones . . . ,* Strasbourg apud Joannem Schottum, 1530; Hieronymus Bock (Hieronymus Tragus), *New Kreutterbuch von underscheydt, würchung und namen der Kreutter, . . .* zu Strassburg, durch Wendel Rihel, 1539; Leonhard Fuchs (Leonhardus Fuchsius), *De historia stirpium . . .* , Basle, in officina Isingriniana . . . , 1542. On these herbals, see the references above, notes 3, 4, and 5.

nean region. Those of more northern climates were often only generally similar and sometimes quite unrelated to those known in traditional botany.

Along with the renewed interest in observational botany in the sixteenth century came the appearance of large botanical gardens intended for research and teaching. There had always been botanical gardens of one kind or another in Europe, of course. They were attached to great houses or monasteries, or kept by apothecaries for the supply of medicinal plants. But the public research gardens of the sixteenth century were a new development. These are the gardens which Linnaeus, in his *Bibliotheca botanica,* calls "horti academici" to distinguish them from strictly medicinal gardens and from the grand, royal or aristocratic ornamental collecting enterprises which became more common in the seventeenth and eighteenth centuries.[7] The gardens intended for research and teaching appeared first in Italy. As a rule, they were attached to university medical faculties, and thus the first of these gardens were founded in the mid-sixteenth century at Padua, Pisa, and Bologna. The idea spread rapidly north, and what was to become one of the greatest botanical gardens in Europe was founded at Leyden in 1577.[8]

In France the first such botanical garden did not appear until quite late, in spite of the urgings of distinguished men, who cited the example of the Italian gardens and recommended similar establishments in France. Pierre Belon, as early as 1558 in his *De neglecta plantarum cultura,* commended the benefits of botanical gardens for the teaching of medicine and suggested to the Faculty of Medicine: "that it establish a public place in the city, in which various kinds of plants will be maintained, the care of which will be given to a man experienced and properly trained in their knowledge, in order to increase the appreciation of these objects and the knowledge of the learned of the universities, and, at the same time, to establish and advance [their] literature.[9] Petrus Ramus included the organized study of "herbs, plants and all kinds of simples, in meadows, gardens and woods" in his

7. Carolus Linnaeus, *Bibliotheca botanica recensens Libros plus mille de plantis huc usque editors, secundum Systema Auctorum Naturale in Classes, Ordines, Genera & Species disposito, additis Editionis Loco, Tempore, Forma, Lingua & c. cum explicatione Fundamentorum Botanicorum pars prima,* reprint of the Amsterdam edition of 1736 (Munich: Werner Fritsch, 1968), p. 65. Linnaeus cites the garden in Paris as among the best of the Academic gardens.

8. There is considerable disagreement among authorities as to the founding dates for these public gardens, most often because different dates in the history of an institution are taken as critical by different authorities. Louis Dulieu gives dates of 1545, 1546, and 1568 for Padua, Pisa, and Bologna respectively. Louis Dulieu, "Pierre Richer de Belleval," *Monspeliensis Hippocrates,* no. 40 (Summer 1968), p. 8. On the botanical garden at Leyden, cf. W. T. Stearn, "The Influence of Leyden on Botany in the Seventeenth and Eighteenth Centuries," *The British Journal for the History of Science,* 1, pt. 2, no. 2 (December 1962), 137–158.

9. *Petri Bellonii Cenomani Medici De neglecta Plantarum Cultura, atque earum cognitione Libellus; Edocens qua ratione Silvestres arbores cicurari & mitescere queant. Carolus Clusius Atrebas ante aliquot annos è Gallico Latinum faciebat, & nunc denuo recensebat.* Printed in Clusius' *Atrebatis, Aulae Caesareae quondam Familiaris, Exoticorum Libri Decem: quibus animalium, plantarum, aromatum, aliorumque peregrinorum fructuum historiae de-*

suggestions for the reform of medical teaching at the University of Paris in 1562.[10] Nevertheless, it was not until 1593 when Pierre Richer de Belleval persuaded Henri IV to subsidize a garden to go with a newly created professorial chair in the medical faculty at Montpellier that France could claim a public botanical garden like those in Italy.

Richer de Belleval is almost as obscure a figure as Guy de La Brosse.[11] Although he was trained in the medical faculty at Montpellier, he took his doctoral degree at nearby Avignon. He convinced Henri IV to create a fifth royal chair for him in the medical faculty at Montpellier, and to attach the intendance of a botanical garden to that chair. He was a committed botanist. His colleagues complained that, among other things, Richer de Belleval devoted so much time to plant expeditions away from Montpellier that he rarely fulfilled his teaching duties in the university.

Richer de Belleval's plan for the botanical garden was an elaborate one. The garden in Montpellier was initially divided into three large areas called the Jardin du Roi, the Jardin de la Reine, and the Carré du Roi. These last two were reserved for mountainous plants and those of a purely research interest, respectively. The first, the Jardin du Roi, was the most important and included a section called the "Ecole de Médecine" in which plants of medicinal value were arranged in parallel rows with walkways in between so that students might have ready access to them individually. In 1622 this garden, which was located outside the walls in Montpellier, was destroyed by Catholic forces attacking the city. Richer de Belleval set to work immediately to rebuild his garden, using his own resources when recalcitrant authorities would not advance him funds.[12] The Jardin des Plantes in Montpellier, however, never really regained the glory of its early days. When its founder died in 1632, he was succeeded by a nephew who showed himself but little devoted to the garden.

Richer de Belleval had intended the Jardin des Plantes in Montpellier to serve not only medical students, but also all those interested in botany for other reasons, whether economic or scientific. The garden of Guy de La Brosse in Paris, the next royally patronized botanical garden in France, was

---

scribuntur . . . , (Antwerp, ex officinà Plantinianà Raphelengii, 1605), p. 239. This translation and those which follow are by the author.

10. Pierre de la Ramée, *Advertissements sur la reformation de l'Université de Paris Au Roy.*, (1562) reprinted in *Archives curieuses de l'histoire de France,* ed. L. Cimber and F. Danjou (Paris: Beauvais, 1835), lère série, V, 115–164.

11. On Richer de Belleval, cf. Louis Dulieu, "Pierre Richer," and Louise Giraud, *Le premier Jardin des Plantes français. Création et restauration du Jardin du Roy à Montpellier par Pierre Richer de Belleval* (*1593–1632*) (Montpellier: Imprimérie Roumégous et Déhan, 1911).

12. This account is taken largely from Dulieu. At the time of his death, Richer de Belleval claimed some 100,000 livres from the king for expenses incurred in the reconstruction of the Jardin des Plantes. He left provision in his will for the use of these monies, should they ever be repaid, to found a college of botany, whose students would maintain the botanical garden in return for room and board (Dulieu, pp. 16–17).

also established with these wider aims in mind. The Paris garden, however, was not under the aegis of a medical faculty.

It is this second aspect of the founding of the Jardin des Plantes in Paris which makes it a phenomenon interesting to political as well as to scientific history. From medieval times the Faculty of Medicine in Paris had claimed the right to control all matters related to the teaching and practice of medicine in the city. It even had national, or international, pretensions to the same monopoly, asserting that it was the most important of the medical faculties in France, the only one whose doctors were permitted to practice anywhere, even outside the Paris region, *hic et ubique terrarum,* as its ceremonies proclaimed.

The medical faculty did not view with equanimity challenges to its authority in medical matters. It carried on a running fight with the lesser medical corporations of surgeons and apothecaries in Paris regarding the limits of training and practice for these professions and it prosecuted unlicensed practitioners of all kinds.[13]

In the 1620's, a particularly exacerbated phase of the faculty's relations with the surgeons and apothecaries led to the publication of a series of books which explained simple operations and the concoction of remedies. The aim of these books was to undermine the professional pretensions of the lesser corporations by describing their skills in public manuals. The treatises were written by various members of the faculty under the covering name of one of their number, Philibert Guybert. The books were provided instruction in medical techniques, so that these could be practiced in the home without professional aid except, of course, for that of the doctor himself. As such, these manuals represented the attempt of the faculty to maintain its claim to the control of medical practice and teaching in Paris even if, as in this instance, it were necessary to destroy the business of the other corporations to compel their submission to the faculty.[14]

The struggles of the faculty to police its jurisdiction in this period were complicated by the growing use of chemical methods and remedies in the practice of medicine. The faculty condemned this approach as contrary to the orthodox medicine of the Greek tradition and dangerous, as well.[15] The

13. Pearl Kibre, "The Faculty of Medicine at Paris, Charlatanism, and Unlicensed Medical Practices in the Later Middle Ages," *Bulletin of the History of Medicine,* 27 (1953), 1–20.

14. A number of the books in Guybert's series, at least one of which was probably written by Guy Patin, were published in 1629 as *Les oeuvres charitables de Philibert Guybert . . . Sçauoir, Le medecin charitable, Le prix et valeur des medicaments. L'apothiqvaire charitable. La maniere d'embavmer les corps morts, et les tromperies dv Bezoard descovvertes,* (Paris, chez Iean Iost). For the ancient controversy between the faculty and the lesser medical corporations, cf. C. A. E. Wickersheimer, *La médecine et les médecins en France, à l'époque de la Renaissance* (Paris: A. Maloine, 1906), chap. 3, sect. 1.

15. There is an extensive literature on the introduction of chemical techniques and remedies into medicine in this period. Cf. Walter Pagel, *Paracelsus. An Introduction to Philosophical Medicine in the Era of the Renaissance* (Basel: S. Karger, 1958); Allen G. Debus, *The English*

battle between the faculty and its chemical opponents ranged along many fronts. It included a skirmish of more than a hundred years' duration concerning the legitimate use of antimony as an internal remedy, which was only settled by the cure of the king himself in 1658.[16] Then, in the early years of the seventeenth century, Cardinal Richelieu patronized a doctor sympathetic to chemical remedies, Théophraste Renaudot, over the outraged protests of the faculty. Renaudot, who is perhaps best remembered as the founder of the *Gazette de France,* established what he called a Bureau d'adresse in the late 1620's in Paris. Along with a variety of other public services, free medical consultations and remedies for the poor were soon provided. Renaudot also sponsored a series of weekly conferences whose topics of discussion included medical subjects. He even obtained royal permission to construct furnaces and equipment for the purpose of producing medicines. The faculty was unable to prevail against Renaudot until the death of his patron Richelieu in 1642.[17]

The founding of the Jardin des Plantes presented yet another challenge to the Faculty of Medicine. It not only breached the teaching monopoly of that group, but its founder La Brosse, like Renaudot, was an avowed adherent of chemical techniques in medicine and openly advocated the necessity of chemistry in medical training. In light of these circumstances, therefore, the phenomenon of La Brosse's success in founding the garden becomes the more interesting not only for the place the Jardin des Plantes holds as one of the earliest research gardens in France, but for the insight it provides into the politics of science in the period.

La Brosse was by no means the only person to feel the necessity of a botanical garden for the teaching of medical students in Paris. The faculty itself had made sporadic attempts to maintain small gardens for teaching purposes. Its most recent effort dated from 1597 when the faculty registers for 30 October record that the doctors paid Jean Robin some 73 livres plus 36 livres as a stipend to establish and maintain a garden for them.[18]

Paracelsians (New York: Franklin Watts, 1966), and Robert Multhauf, *The Origins of Chemistry* (New York: Franklin Watts, 1967). Cf. also Henry Guerlac, "John Mayow and the Aerial Nitre, Studies in the Chemistry of John Mayow—I," *Actes du VIIe Congrès international d'histoire des sciences,* Jerusalem (4–12 August 1953) (Paris: Académie internationale d'histoire des sciences and Hermann et Cie, Editeurs, 1953), pp. 332–349, and "The Poet's Nitre, Studies in the Chemistry of John Mayow—II," *Isis,* 45 (1954), 243–255.

16. On the "antimony war," cf. Pascal Pilpoul, *La querelle de l'antimoine* (Paris: Librairie Louis Arnette, 1928).

17. On Renaudot, cf. Howard M. Solomon, *Public Welfare, Science and Propaganda in Seventeenth-Century France: The Innovations of Théophraste Renaudot* (Princeton: Princeton University Press, 1972).

18. *Commentaires,* Faculty of Medicine, Paris, IX, fol. 77 verso. Paul Delaunay, *La vie médicale au XVIe, XVIIe* et *XVIIIe siècles* (Paris: Editions Hippocrate, 1935), 71, says that the Faculty of Medicine had maintained a botanical garden since 1506. This garden must have fallen into disuse by the time Ramus undertook to advise the Faculty of Medicine on the reform of its teaching.

Jean Robin was perhaps the best known gardener in Paris. He was by trade an apothecary, and for some time before his faculty appointment, had had a garden on the western end of the Ile de la Cité.[19] He profited from the fashion for colorful flowers which the ladies of the court liked to design in embroidery, and it was perhaps through Pierre Vallet, embroiderer to the king, that Robin came to the notice of the court. Vallet frequented Robin's garden looking for new flowers to design, and it may have been for this reason that Robin was given a pension for his garden by the king and the title "botaniste et simpliciste du Roy."[20]

It is not certain whether Robin kept a separate garden for the Faculty of Medécine after 1597. The *Catalogus stirpium* which he published in 1601 and dedicated to the faculty would certainly seem to indicate the attempt of its author to provide a demonstration of his competence and the fulfillment of his commission, even if the garden described was his own.[21]

Robin was later joined by his son Vespasien,[22] and together they pub-

19. M. Bouvet, "Les anciens jardins botaniques médicaux de Paris," *Revue d'histoire de la pharmacie,* 35, no. 119 (December 1947), p. 226, describes Robin's garden as having been "at the western tip of the Cité, in the neighborhood of the present Place Dauphine." After the Place Dauphine was constructed in 1607, it seems probable that Robin's garden had to be relocated, but where or when this took place is not known. However, Robin is also described as having kept a garden in the Louvre and he may have moved his plants there from the *cité* when construction forced him out. Since by then he enjoyed royal patronage, the move to the Louvre may have been connected with Robin's functions at court. Cf. Jeanne Pronteau, "Etude sur le Jardin royal des plantes médicinales à Paris (1626–1788)," *Annuaire 1974–1975* of the Ecole Pratique des Hautes Etudes, IVe section (Paris, 1975), pp. 651–669.

20. The date of this appointment is uncertain. It was before 1601 since Robin identifies himself in a catalogue of that year as "botanicus regius." Tournefort, in the *Isagoge in Rem Herbarem,* which he added to the Latin edition of his *Elémens de Botanique* in 1700, traced royal patronage of Robin as far back as 1570, which seems early, and of dubious accuracy since Tournefort cities the king as Henri IV: "Henry IV, most powerful King of France and Navarre, about the year 1570, publicly gave his garden filled with rare plants to Joannes Robin, second to none in the identification and cultivation of plants." Joseph Pitton de Tournefort, *Institutiones Rei Herbariae . . . ,* editio altera, Paris, Typ. regia, 1700, p. 48. In 1608, Vallet and Robin published a collection of engravings of some of the more "exotic" flowers in the latter's garden, called *Le Jardin du Roy tres chrestien Henry IV Roy de France et de Navarre.* Séguier, *Bibliotheca botanica,* p. 168, gives the following title for the work, which suggests that Vallet was responsible for the plates and Robin for the text: *Le Jardin du Roi Henri IV. ou recueil de fleurs gravées par Pierre Vallet brodeur du Roi, & décrites par J. Robin . . .,* Paris, 1608. Jean-François Séguier, *Bibliotheca botanica, sive Catalogus auctorum et librorum omnium qui de re botanica, de medicamentis & de horticultura tractant a Joanne-Francisco Seguierio Nemausense digestus. Accessit Bibliotheca botanica Jo. Ant. Bumaldi, seu potius Ovidii Montalbani Bononiensis.* (The Hague, apud J. Nealume, 1740).

21. The full title of Robin's catalogue is *Catalogus Stirpium tam indigenarum quam exoticarum quae Lutetiae coluntur à I. Robino botanico regio, & Iatrici horti celeberrimae schole Parisiensis Curatore,* (Paris, ex typographia Philippi à Prato, 1601). André Bourde, *Agronomie et agronomes en France au XVIIIe siècle* (Paris: S.E.V.P.E.N., 1967), I, 109, says that Robin did keep a separate garden for the faculty which he describes as "attached to the College de Clermont" and that Robin "on the other hand, had his own garden."

22. There is some doubt as to whether Vespasien was the son, nephew, or brother of Jean Robin, although he was most probably the son. Bouvet, "Les anciens jardins," p. 227, so identifies him, and says that he had two brothers, Jacques and Etienne. On Vespasien Robin, cf.

lished a second catalogue in 1624, probably written by Vespasien and entitled *Enchiridion Isagogicum ad Facilem Notitiam Stirpium, tam indigenarum quam exoticarum quae coluntur in horto D. D. Ioannis & Vespasiani Robin....*[23] Vespasien was later to be named as the first "underdemonstrator" of the Jardin des Plantes in 1635 and he pursued a distinguished career as a botanist in his own right.

It seems to have been generally agreed in the Faculty of Medicine that the small garden which had been subsidized under the care of Jean Robin had not fulfilled its purpose with sufficient effect. Something more was needed. There appeared in 1618 a small treatise entitled *Requeste au Roy Pour l'establissement d'un Jardin Royal en l'université de Paris.* The author was Jean Riolan II, himself a member of the Paris medical faculty and a distinguished anatomist.[24] The *Requeste au Roy* begins with some pages of exalted rhetoric addressed to the king, but comes abruptly to its point: "The school of medicine, founded in your university of Paris, has need of a royal garden, in order to place in it all the plants of the world which can be obtained. Henry the Great... caused a magnificent one to be erected at Montpellier. By his death, he left for you the honor and the benedictions which you will receive in causing a similar garden to be built in the university of Paris, which is, in your kingdom, the city of letters...."[25] Riolan notes that as one of the first acts of his reign, Louis XIII had laid the cornerstone for what was to be the building of the Collège Royal and he suggests that a fitting sequel to this "proof of your paternal affection" would be the erection alongside the new edifice of a Jardin Royal. To match the magnificence of its patron, this garden would be no small undertaking; it should contain "all the plants which the earth brings forth, which number some three thousand or thereabout...." Perhaps encouraged by this modest estimate of the bounty of Nature, Riolan added that the cost should not be prohibitively expensive since it had cost Aristotle but 800 talents or "four hundred and eighty thousand écus of our money" to write the natural histories of plants and animals.

Guy de La Brosse conceived his own plan for a Jardin des Plantes, according to his own testimony, as early as 1614–1618.[26] It is difficult to verify

especially Philippe Tamizey de Larroque, *Deux jardiniers émérites. Peiresc et Vespasien Robin* (Aix: Ve. J. Remondet, 1896). On Vespasien's correspondence with Peiresc, cf. E. T. Hamy, "Vespasien Robin, Arboriste du Roy, premier sous-démonstrateur de botanique du Jardin royal des plantes (1635–1662)," *Nouvelles archives du Muséum d'Histoire Naturelle,* 3e serie, 8 (1896), 1–24.

23. Paris, apud Petrum de Breschem, 1624.

24. On Riolan, see Jerome J. Bylebyl, "Jean Riolan Jr.," *DSB,* 9 (1975), 466–468, with bibliography.

25. Jean Riolan II, *Requeste au Roy Pour l'establissement d'un Jardin Royal en l'université de Paris* (1618), MS copy AN: AJ^{15} 501 no. 8.

26. La Brosse, *Description du jardin royal des plantes medecinales estably par le Roy Louis le Iuste à Paris contenant le catalogue des plantes qui y sont de present cultivées, ensemble le*

these dates, for little is known of the family or early life of La Brosse apart from the written works he left behind. Virtually the only certain date we possess is that of his death in the night of 30–31 August 1641.[27] The parish register at St. Médard recorded his age at death as fifty-five which suggests a birthdate around 1586.[28]

La Brosse's father, Isaie Vireneau, sieur de La Brosse, was a doctor at the court of Henri IV, and an uncle, Hierosme de La Brosse, served the Comte de Soissons as a secretary.[29] La Brosse describes himself as having spent summers as a young man collecting plants around Paris, but beyond this, nothing is known of his youth.[30] From the evidence of the title pages of his published works and scattered family documents we know that he was successively doctor to Henri II de Bourbon, Prince de Condé, and *médecin ordinaire* to Louis XIII, but it has not been possible to determine where he took a medical degree or, indeed, if he took one at all.[31] Apart from his published appeals for the garden, he left a *Traicté de la peste* (1623), a *Traicté contre la mesdisance* (1624), and his major work, *De la nature, vertu et utilité des plantes* (1628).[32] In this last work, La Brosse revealed

plan du iardin, (Paris, 1636), p. 18. Hereafter cited as *Description du jardin royal* (1636). Cf. also *L'ouverture du jardin royal de Paris, pour la demonstration des plantes medecinales,* à Paris, par Iacques Dugast, 1640, in which La Brosse says (p. 15) that the garden was the fruit of twenty-four years of work, "eighteen of endeavor and six of cultivation." *Catalogue des Plantes Cultivées a present au Iardin royal des plantes medecinales, estably par Louis le Juste, a Paris, . . .* (Paris, chez Jacques Dugast, 1641), p. 1, hereafter *Catalogue* (1641).

27. Guy Patin, *Lettres,* ed. J.-H. Reveillé-Parise (Paris: J.-B. Baillière, 1846), I, 81–82. See also E. T. Hamy, "Quelques notes sur la mort et la succession de Guy de La Brosse," *Bulletin du Muséum d'Histoire Naturelle,* 3, no. 5 (1897), 152–154.

28. Hamy, "Mort de La Brosse," p. 152.

29. René Pintard, *Le libertinage érudit dans la première moitié du XVIIe siècle* (Paris: Boivin, 1943), p. 195 and accompanying notes. On the life and work of La Brosse, cf. Henry Guerlac, "Guy de La Brosse," *DSB,* 7 (1973), 536–541. See also, E. T. Hamy, "La famille de Guy de La Brosse," *Bulletin du Muséum d'Histoire Naturelle,* 6, no. 1 (1900), 13–16. New information on the family of Guy de La Brosse appears in R. C. Howard, "The Family of Guy de La Brosse," *Proceedings of the 25th International Congress in the History of Medicine* (Quebec, August 1976).

30. La Brosse, *De la nature, vertu et utilité des plantes,* (Paris, chez Rollin Baragnes, 1628), p. 75. Hereafter cited as La Brosse, *De la nature des plantes* (1628). Cf. also Hamy, "Famille," pp. 13–14.

31. Although La Brosse is often described as a doctor of Montpellier, there is no La Brosse listed as a student at Montpellier in Marcel Gouron, *La matricule de l'Université de Médecine de Montpellier* (*1503–1599*) (Geneva: E. Droz, 1957). According to Louis Dulieu, there is no record of Guy de La Brosse in the medical records at Montpellier. Cf. on this, Henry Guerlac, "Guy de La Brosse and the French Paracelsians," *Science, Medicine, and Society in the Renaissance, Essays to Honor Walter Pagel,* ed. Allen G. Debus (New York: Science History Publications, 1972), I, 195, n. 26. La Brosse is described as "médecin ordinaire de M. le Prince" in the marriage certificate of his cousin, Françoise de La Brosse in 1619. "Contrat de marriage de Pierre Morin et de Françoise de La Brosse, May 14, 1619." MS copy by E. T. Hamy in AN: AJ[15] 507 no. 130. The title page of La Brosse's *De la nature des plantes* (1628) qualified him as "médecin ordinaire du roy."

32. *Traicté de la peste . . . auec les remedes preseruatifs,* (Paris, chez Ieremie & Christophe Perier . . . , 1623). *Traicté contre la mesdisance* (Paris, chez Ieremie & Christophe Perier,

himself as a follower of Paracelsus and called for the use of chemical techniques in the analysis of the medical properties of plants.[33]

It is the activity of La Brosse in connection with the Jardin des Plantes which is most fully documented both in his own writings and in public records. He turned for help in the realization of his plans to the royal government and it was his shrewd and assiduous cultivation of patrons which insured his success in founding the Jardin des Plantes and protecting it against the faculty of Medicine. La Brosse solicited patronage in the form of public letters to the king and to various royal officials describing the garden he wished to found and urging its support for a number of reasons.[34]

A public letter to Richelieu, which bears no date but which appeared first in 1628 bound with *De la nature des plantes,* is remarkable in many ways. La Brosse suggests in support of the establishment of a botanical garden that one of the most practical uses of plants is to prolong life. He even says obliquely that with the proper knowledge of how plant "virtues" operate, eternal life may be possible for man on earth. The reference is only passing, for such an opinion would surely be blasphemous. It is certain, says La Brosse, that God would hardly have seen fit to number man's days if the length of life could be changed at will through the use of plants. But the truth of the matter is that man was originally created to live forever (why should it have been otherwise?) and only through the Fall was he condemned to die. Our desire for a long life is thus nothing but the desire to return to grace, what he calls an "image of the future."[35] The attempt to lengthen one's life can hardly be impious in such circumstances. After all, God has already, in the Ten Commandments, promised long life to him who honors his parents. If this is so, surely there must be other means by which God has enabled man to extend the length of his life. The use of plants as remedies should be chief among them.[36]

It seems probable that Richelieu favored the establishment of the Jardin

---

1624). On the *Traicté contre la mesdisance,* cf. Antoine Adam, *Théophile de Viau et la libre pensée française en 1620* (Paris: Librairie E. Droz, 1935), p. 379, and Pintard, *Le libertinage,* p. 195.

33. On La Brosse's Paracelsianism, cf. Henry Guerlac, "Le Brosse and the French Paracelsians." The more traditional botanical thought of La Brosse is discussed by Agnes Arber in "The Botanical Philosophy of Guy de la Brosse: A Study in Seventeenth-Century Thought," *Isis,* 1 (1913), 359–369.

34. The letter to the king is not dated. Louis Denise describes it as reprinted in *De la nature des plantes* (1628), *Bibliographie historique et iconographique du Jardin des Plantes* (Paris: R. Daragon, 1903), p. 21.

35. La Brosse, *A Monseigneur le tres-illustre et le tres reverend Cardinal, Monseigneur le Cardinal de Richelieu,* published with *De la nature des plantes* (1628), p. 706. *De la nature des plantes* is also dedicated to Richelieu.

36. The letter concludes: "Et que la posterité sache, qu'un tres-illustre a reverend Cardinal de Richelieu, luy a procuré le richelieu des tresors de la santé & de la longue vie . . . ," the cardinal should support the project. Louis Denise, *Bibliographie,* p. 23, refers cautiously to the "curious theological reasons in favor of the enterprise" which are adduced by La Brosse in this letter. *De la nature des plantes* (1628), pp. 729–738.

des Plantes for many of the same reasons that he encouraged and patronized Renaudot's Bureau d'adresse. Renaudot and La Brosse shared a great many aims in common. Both men stressed the social utility of what they wanted to do, both were interested in providing better and cheaper medical care for the poor, and both wanted to open the closed orthodoxy of the medical schools. The Bureau d'adresse and the Jardin des Plantes offered the crown a means of dealing with some of the problems of public relief, problems which in the seventeenth century were only beginning to receive the concentrated attention of the royal government as legitimate objects of its concern.

Then, too, the Jardin des Plantes and Renaudot's Bureau provided means by which the monopolistic pretensions of the Paris Faculty of Medicine might be undermined. These pretensions showed no signs of diminishing in the early seventeenth century. The Paris doctors held as tenaciously to what they took to be their corporate rights as did any trade guild. The Jardin des Plantes was an assault, launched with royal approval, on the claims of Paris doctors. Its foundation suggested that medical teaching and practice might be at least partially organized under royal and central control, rather than that which was corporate and local.[37] There seems every reason to think that Richelieu favored these aims applied to medicine even as he sought to realize them in other spheres of French life.

But, although it seems likely that Richelieu knew of La Brosse's plans and favored them, I can find no direct evidence of the Cardinal's interest in the Jardin des Plantes. La Brosse derived more immediate help from lesser royal officials, especially from the two men who were successively chief physician to Louis XIII, Jean Héroard and Charles Bouvard.

The king's chief physician headed a medical establishment at court whose

37. There had been earlier attempts, most notably on the part of the king's doctors, to centralize various aspects of the teaching and practice of medicine under some sort of royal control, generally that of the chief physician to the king. In the last years of the sixteenth century, Jean Ribit, sieur de la Rivière, Henri IV's chief physician, had convinced the king to place him in charge of the practice of surgery and pharmacy throughout France. La Rivière was disabused of his pretensions by the Paris Faculty of Medicine, which obtained a sentence from the Parlement of Paris in 1601 nullifying the king's original decree. Nothing daunted, Henri IV in the same year gave La Rivière the superintendence of all the mineral water sources in France, which were used for medicinal purposes. Cf. A. Chereau, "Les médecins de Henri IV," *L'union medicale,* 22 (1864): nos. 49 (26 April), 161–166; 50 (28 April), 177–183; and 53 (5 May), 225–236. La Rivière died in 1605, but on his inspiration Henri IV charged his chief physician in 1606 with the appointment of one or two surgeons in each city of the kingdom to assist in the local practice of forensic medicine. Cf. Roger Vaultier, "Henri IV et la Faculté," *La presse médicale,* année 61 (31 October 1953), p. 1414. On La Rivière, who has been commonly confused with another of Henri IV's doctors, Roch le Baillif, sieur de la Rivière, cf. Hugh Trevor-Roper, "The Sieur de la Rivière, Paracelsian Physician of Henri IV," in Debus, ed., *Pagel,* II, 227–250.

Jean Héroard, when he became *premier médecin* in his turn, attempted to subject the practice of medicine itself to the control of the king's doctor, making an exception only for the Paris faculty. The Paris doctors, however, alerted their provincial colleagues to the danger and in 1611 the Grand Conseil ruled against Héroard. These attempts continued throughout the seventeenth century. Cf. Paul Guillon, *La mort de Louis XIII. Étude d'histoire médicale d'après de nouveaux documents* (Paris: A. Fontemoing, 1897), pp. 117–118.

doctors were often "foreign," that is, they held their medical degrees from schools other than Paris, both inside and outside France. They were permitted to practice in Paris on the bare sufferance of the Paris Faculty of Medicine. Throughout these years a rivalry existed between the royal doctors, sometimes referred to as the "Faculté du Roi," and the Paris faculty. This competition manifested itself in many connections, from doctrinal fights to precedence and legitimacy quarrels. To embittered defenders of the faculty's prerogatives such as Guy Patin, it seemed that the "Faculté du Roi" was not only a threat to medical practice, but that it sheltered a number of "chemical" doctors as well.[38] In soliciting the patronage of the king's chief physician, La Brosse may have hoped that the longstanding rivalry between the king's doctors and the faculty, together with the preference of some of the former group for chemical medicine would help to insure his garden against what were certain to be the assaults of the Faculty of Medicine on the new institution.

It was Jean Héroard who probably first brought La Brosse's idea to the king's attention. Héroard, who was made doctor to Louis XIII at the latter's birth in 1601, was trained by the chief medical rival of the Paris school in this period, Montpellier. He is perhaps best remembered for his daily record of his royal patient's health and activities, *Journal de la santé du Roi,* which was continued by his successors in the post of chief physician.[39] La Brosse, in a public letter to Héroard, urged him to support the project of the garden, not for his own personal glory but for the public good, adding directly, however, that his name would be on the plaque over the entrance to the garden: "For who will be able to prevent that on the front of its portico, one read . . . Louis XIII, the Just and the Victorious, on the advice of N. [*sic*] Erouard, his Counselor and chief physician, caused this garden of medicinal plants to be built."[40] Héroard makes no mention of La Brosse's plans in his journal, but his own name is prominent in the edict of 1626 which first

38. For particularly splenetic outbursts against "the chemists," cf. Patin, *Lettres,* especially I, 446–448, and II, 562–564.

39. *Ludovicotrophie ou journal de toutes actions de la santé de Louis, Dauphin de France qui fut ensuite Louis XIII depuis le moment de sa naissance, 1601 jusque à 1628.* After Héroard's death, this account was continued by his successors as *Journal de la santé du Roy.* It was published by Eud. Soulié and Ed. de Barthélemy as *Journal de Jean Héroard sur l'enfance et la jeunesse de Louis XIII* (*1601–1628*) (Paris: Firmin Didot, 1868).

Héroard's other major works included a treatise on horses, *Hippostologie, c'est à dire, Discours des os du cheval* (Paris: M. Patisson, 1599), and another work on the education of princes, *De l'Institution du prince* (Paris: J. Jannon, 1609). A recent interesting article on Héroard's relationship with his royal patient is Elizabeth Wirth Marvick, "The Character of Louis XIII: The Role of His Physician," *Journal of Interdisciplinary History,* 4 (1974), 347–374. Cf. also E. T. Hamy, "Jean Héroard, premier surintendant du Jardin Royal des Plantes médicinales (1626–1628), notice iconographique," *Bulletin du Muséum d'Histoire Naturelle* no. 5, (1896), pp. 1–5.

40. La Brosse, *A Monsieur Erouard, premier médecin du Roy,* not paginated, BN: Sp 12320. Cf. Denise, *Bibliographie,* p. 21.

authorized the idea of a royal botanical garden: "We have availed ourselves with affection of the advice and proposals made to us by our well beloved and faithful Counselor and chief physician, the Sieur Heroüard, for the establishment and construction, in one of the faubourgs of our good city of Paris, of a Royal Garden of Medicinal Plants, since these latter are the best instruments that Nature has produced for the cure of the sick."[41] The edict gave the superintendence of the garden to Héroard as chief physician and to his successors in that post, together with the power to appoint personnel for the garden including an intendant. On 7 August of the same year, Héroard nominated La Brosse for the latter post, and on the eighth the nomination was confirmed by the King.

On 8 February 1628, Héroard died and Charles Bouvard was eventually appointed to succeed him. The competition for the favor of the new chief physician must have been formidable if one may judge from the number of books dedicated to him in and around that year. Bouvard was himself a doctor of the Paris faculty and the brother-in-law of Riolan. It was, therefore, with some reason that the faculty looked to him for support. The dedication of the most recent of their instructional manuals on medicine written against the surgeons and apothecaries, was made to Charles Bouvard, "Conseiller du Roi." The book was entitled *Les tromperies du Bezoard* and was published in 1629, most probably just before Bouvard became chief physician but when the likelihood of his appointment was known.[42]

For all his connections to the Paris faculty, Bouvard took the cause of a Jardin des Plantes to heart. The frequency of the mention of his name in the next major edict concerning the garden, that of 1635, indicates that his was a primary role in helping La Brosse to obtain the firm establishment of the project. In fact, Bouvard became so devoted to the garden that he wrote a small tract in 1639 against the Faculty of Medicine and its attempts to control the newly established institution.[43]

In addition to the chief physicians as patrons, La Brosse received particular help from the superintendents of finance. To these men he looked for the provision of specific funds to cover the purchase of land and the construction and maintenance of the garden. The edict of 1626 had merely provided

41. "Edit dv Roy, Pour l'establissement d'vn Iardin des Plantes Medecinales," 6 January 1626. Printed copy in *Actes royaux* no. 8321 (1625–1628), II, BN: F23610 (793). The date of January 6 is one established by E. T. Hamy. Cf. "Les débuts de l'anthropologie et de l'anatomie humaine au Jardin des plantes. M. Cureau de la Chambre et P. Dionis (1635–1680)," in *L'anthropologie,* V (Paris: G. Masson, 1894), 3.

42. *Les tromperies du Bezoard* was published with *Les oeuvres charitables de Philibert Guybert.* La Brosse also addressed a public letter "A Monsieur Bouvard, Conseiller du Roy en ses Conseils & son premier Medecin," BN: S6751 (3). Cf. Denise, *Bibliographie,* p. 28.

43. *Mémoire pour le Jardin Royal des Plantes par Charles Bouvard* (1639). BN: MS fr. no. 17308, fols. 150–151. On Bouvard, cf. Guillon, *La mort de Louis XIII,* pp. 107–119.

in a vague way for "such a sum of deniers as shall be judged necessary," and this may have been one reason why nothing immediate was done on the garden; no specific funds had been allocated. By 1635 and the second edict, not only had the salaries of all personnel and operating expenses been itemized, but the specific tax base from which these costs were to be met was included in the edict. Moreover, by 1635, La Brosse had succeeded in negotiating the purchase of a piece of land for the garden.

Another of the public letters which appeared in 1628 with *De la nature des plantes* was addressed to "Monseigneur le Surintendant des finances." It was most probably written to the Marquis d'Effiat, who held that office after 1626.[44] The letter is as straightforward in its appeal as a business proposition, quite unlike the letter to Richelieu. La Brosse says merely, think what posterity will say of you. Your support of the garden will place you among the benefactors of a greater monument than any Sully left behind. Nothing could be of more public good than a medical garden.

This letter produced no visible change in the status of funds to establish the garden, however. It was not until the death of d'Effiat in 1632, as he was about to embark on a military campaign, that La Brosse found a more sympathetic audience. Richelieu caused two of his supporters to be appointed as superintendents of finance: Claude de Bullion and Claude le Bouthillier. It was to the good offices of these two men that La Brosse owed the success of his plans. They were particularly instrumental in securing the purchase of the land for the garden in 1633.[45]

Bullion was one of the most powerful ministers in the government. He was a close friend and patient of La Brosse and often dined with him. He is reputed to have enjoyed good living and the contemporary chronicler, Tallemant des Réaux, suggests that it was La Brosse's recommendation of the vines which grew on a hill in the Faubourg Saint-Victor which obtained the superintendent's active support for the purchase of land to establish a garden on that site. Of Bullion, Tallement des Réaux says, "the good fellow amused himself. He went often to the home of La Brosse, his doctor, whom he had established in the Jardin des Plantes in the faubourg Saint-Victor; there he had his favorites and indulged himself entirely at his ease."[46]

These, then, were Guy de La Brosse's patrons, the men whose help and influence he used to realize the idea of a Jardin des Plantes in Paris.

44. On Antoine Coeffier, dit Ruzé, Marquis d'Effiat et de Chilly, baron de Macy et de Longjumeau, cf. Gédéon Tallemant des Réaux, *Les historiettes,* ed. Antoine Adam, Bibliothèque de la Pléiade (Paris: Librairie Gallimard, 1960), I, 294–295.

45. The *Description du Jardin royal* (1636) is dedicated to "Mgr. de Bullion, Surintendant des Finances," and the *Catalogue* (1641) to "Claude Bouthillier, chevalier conseiller du Roy en ses Conseils, Commandant Et grand Tresorier de ses ordres Et Surintendant des finances de France." On Bullion and le Bouthillier cf. Orest A. Ranum, *Richelieu and the Councillors of Louis XIII. A Study of the Secretaries of State and Superintendents of Finance in the Ministry of Richelieu 1635–1642* (Oxford: Clarendon Press, 1963), chaps. 7 and 8.

46. Tallemant des Réaux, *Les historiettes,* I, 301. The *Historiettes* include a vignette of Bullion: I, 300–304.

La Brosse had a highly developed conception of what he wanted to be included in his garden. In the course of various published appeals, the details of his plan were described, and it is evident that the garden as he wanted it to be was quite different from other related institutions which had preceded it. La Brossa wrote a number of short treatises in which he discussed his project. Principal among them are: 1. *Advis defensif du iardin royal des plantes medecinales à Paris,* Paris [before 1628]. 2. *Ordre du dessein du iardin royal des plantes medecinales,* Paris [before 1628]. 3. *Dessein d'un iardin royal pour la cvltvre des plantes medecinales a Paris,* A Paris, chez Roolin Baraigne, 1628. 4. *Advis pour le iardin royal des plantes medecinales que le Roy veut establir à Paris. Presenté à Nosseigneurs du Parlement,* A Paris, De l'imprimerie de Iacques Dugast, 1631. 5. *Pour parfaictement accomplir le dessein de la construction du iardin royal, Pour la culture des plantes medecinales,* n.p., n.d.. 6. *A Monseigneur le Chancelier,* n.p., n.d.. 7. *Description du jardin royal des plantes medecinales estably par le Roy Louis le Iuste, à Paris,* 1636.[47] In the first two of these treatises, which appeared with *De la nature des plantes* in 1628, are contained the lengthiest and most detailed discussions by La Brosse both of specific proposals for the building of the garden, and of those aims which he felt justified the founding of such a new institution.

The *Advis defensif* was later published again as the body of the letter to the Paris Parlement in 1631, and still later under the title *Pour parfaictement accomplir le dessein de la construction du iardin royal pour la culture des plantes.*[48] It includes a summary of La Brosse's views on the development of medicine, on the problems of medicine in his own day, on Paracelsianism, and finally, on the fundamental reasons for which he thought a major botanical garden should be established in Paris.

As they are outlined in the *Advis defensif,* the functions of the garden were to be three, all contributing to the improvement of society. They were: "1. The instruction of apprentices in medicine, even those most advanced in its practice, in the knowledge of the principal instruments of their art, long neglected. 2. That the art be more sincerely and readily practiced. 3. And that the poor, overwhelmed by necessity and weakness, find help there, charitably, according to their needs."[49] Robin's garden was a charming idea according to La Brosse, but it was far too small and was maintained on a

47. On the dating of these pamphlets cf. Denise, *Bibliographie,* pp. 24–30. The *Dessein d'un jardin royal,* bound with *De la nature des plantes* (1628), included the letters to the king, to Cardinal Richelieu, to the keeper of the seals, and to the superintendent of finance as well as the *Advis defensif* and the *Ordre du dessein,* all of which Denise describes as being "reprinted" in *De la nature des plantes* under this rubric, together with the edict of 1626 establishing the garden, the letters of nomination for La Brosse as intendant and a *Mémoire des plantes vsagères & de leurs parties que l'on doit trouuer à toutes occurrences, soit recentes ou seches, selon la saison; au Iardin Royal des Plantes Medecinales; Ensemble les sucs, les eaux simples distillées, les sels & les essences.*

48. Cf. Denise, *Bibliographie,* p. 28.

49. La Brosse, *Advis defensif,* in *De la nature des plantes,* p. 756.

ridiculously small pension. It was hardly sufficient to grow more than some two hundred plants, and of those, only one of each kind. In *Dessein d'un jardin royal,* La Brosse proposes a very different project, designed to do much more than Robin's garden had attempted, and on a much larger scale: "The garden which I propose must occupy a space of fifty *arpents* or more; the plants will not only be there singly for teaching purposes, but in great numbers for use and to serve for experimentation. . . ."[50] The garden would serve first as a teaching institution for medical students, since plants were the most important remedy which the doctor commanded. Second, the garden, by raising quantities of plants and processing them, through drying, distilling, and the like, would supply herbs to apothecaries, and would thus hopefully provide some sort of quality control for the simples sold in druggists' shops in Paris. La Brosse laments that lack of such control and suggests that the garden might even supply remedies directly to the poor who could not otherwise afford them. It is here that the resemblance between the plans of La Brosse and those of Renaudot is most evident.

Part of the fulfillment of these functions was dependent upon research into the different qualities of plants. Although La Brosse says in *De la nature des plantes* that he considers chemistry to be the primary tool in such research, he does not mention this point in the *Dessein,* but justifies the construction of a chemical laboratory merely by the fact that it will be necessary to prepare the waters and syrups from plants that they may be used as remedies. A course must therefore be taught in this art, that the students may learn how such preparations are made: "Having assured your majesty that the waters, juices, essences and salts of plants will be kept, of which three are the works of fire, it is very appropriate and necessary to explain their action. To this end, I promise to teach a course in the distillatory art, and to demonstrate all its operations to those desiring to learn."[51]

All this could be done for 200,000 livres initial cost and maintained for 20,000 livres a year. The initial cost would cover the purchase of the fifty arpents of land, "their enclosure, buildings, the collection of plants, both domestic and foreign, the purchase of necessary and proper vessels and utensils."[52] The 20,000 livres maintenance cost per year would pay twelve men necessary to staff the garden and would be used "to maintain the vessels and other utensils necessary for the operations proposed. Six of these men will be employed in distant provinces for the recovery of plants; four of the six others will devote themselves to the cultivation of the garden, and the two others remaining will be engaged in the harvesting of plants, the distillations of waters and essences, and in other works of fire."[53]

50. La Brosse, *Dessein d'un jardin royal,* in *De la nature des plantes,* p. 693. One arpent is equivalent to approximately one and one half English acres.

51. Ibid., p. 696 (misnumbered 196).

52. Ibid., p. 697.

53. Ibid., p. 698.

Such a garden would be better than that at Montpellier and would cost less. As his final argument, La Brosse subdivided his proposed budget even further. As can be seen, he had worked out the plans for the garden in considerable detail:

To buy fifty arpents of land	50,000 livres
For the "enclosure, at one thousand toises around and two toises high, including three feet for the foundation, which makes two thousand toises, at nine livres the toise, to be made of limestone and sand in rows of cut stone; the two thousand toises will cost . . ."	18,000 livres
For the gallery for drying and keeping plants and for distillations "being fifty toises long, four wide and six high, at the end [of which will be] a pavilion to lodge the workers, the interior of the gallery [being] filled with closets for drying the plants, will cost, as much for the masonry of six hundred and forty eight toises, as for the compartments, floors, carpentry and woodwork of the doors, windows and closets, more than . . ."	20,000 livres
"For the principal lodging, consisting of two pavilions connected by a main building, in which will be the rooms for teaching, [with] cellars beneath, [and] on the sides, the stables and other places, as much for lodging the horses used in the garden for the transport of earth and other necessary things, as for the barrows and carts, and two small pavilions at the entrance, for individual lodgings . . ."	60,000 livres
Finally, "to raise a mountain in the midst of the garden, being four arpents wide, and nine or ten toises high, to lay out the parterres, to excavate the fish ponds, to buy the plants which must be in great numbers, to furnish vessels and all the utensils necessary to the distillatory art,"[54]	50,000 livres

In the *Ordre du dessein du jardin royal des plantes medecinales* which also appeared in 1628, La Brosse speaks of the manner in which the plants would be arranged in the garden and what kinds of plants would be cultivated. They should be arranged as nearly as possible in the conditions in which they grow naturally. It is for this reason that the mountain was to be constructed, and various parts of the garden sectioned off for different kinds

54. Ibid., pp. 698–700. One toise is approximately six feet.

of plants. La Brosse proposes to keep domestic as well as wild plants, French as well as foreign, and he asks for the six salaried assistants to gather them.

In 1631, the first appeals had been unsuccessful. La Brosse thereupon devised a scheme for financing the garden which he proposed to the Parlement of Paris. Since all normal sources of royal funds seemed to be unavailable to him, he proposed the creation of a new source of revenue:

> His Majesty, being advised that in many provinces of his kindgom, there were several unused wastelands, damaged woods, islands, islets, alluvial and abandoned places near rivers or along the coast of the ocean, marshes, grazing lands, common pasturages, pools, ponds and trenches for fish in his domaine, from which he derives no benefit, has found it appropriate to aid in the accomplishment of this admirable plan to which he has only given his sufferance, and without demanding anything of his people, to sell such lands, woods, and similar properties . . . [to provide funds for] the purchase of a site necessary and convenient for the establishment of this undertaking and for the plants that must be cultivated there.[55]

Leaving nothing to chance, La Brosse concluded: "That if, until now, such a project has not been begun, it seems to me that it has not been proposed to you [the members of the Parlement]; nor [have] the means of its execution [been] at hand."

Such euphonious phrases seem not to have moved the members of the Paris Parlement. A third version of the *Advis defensif, pour parfaictement accomplir le dessein . . . du Jardin Royal . . .* , was published most probably shortly after the letter to the Parlement, since the request for land has shrunk to 25 to 30 arpents. It is in this pamphlet that La Brosse also mentions for the first time the possibility that astronomy should be taught in the garden: "in the middle of the main building, on the staircase, a platform will be built in the form of a square tower, three and a half *toises* wide, of the same length and about ten *toises* high, to set up astronomical instruments for the measuring of the heavens."[56] This may simply have been an afterthought, since La Brosse had already stressed the importance of a knowledge of astrology to medicine in the *Dessein* and had proposed that a course be taught in this discipline.

The final appeal for funds was made to the chancellor. In a letter, *A Monseigneur le Chancelier,* La Brosse has reduced the amount of land he needs to 20 to 25 arpents, and his proposed budget has also been sharply curtailed; he refers to "the sum of one hundred and ten thousand livres necessary for the purchase of twenty or twenty five arpents of land, for the enclosure, buildings, recovery of plants, purchase of vessels and utensils necessary for the operations, and twelve thousand livres of annual revenue for the maintenance of twelve men, six of whom will attend to the search for

55. La Brosse, *Advis pour le jardin royal* (1631), prefatory letter.
56. La Brosse, *Pour parfaictement accomplir le dessein* (n.d.), p. 4.

plants in the countryside and in foreign lands; four of the other six will cultivate the Garden and the two remaining will be employed in distillations."[57]

In 1632, Richelieu appointed Claude de Bullion and Claude le Bouthillier as superintendents of finance and their names are mentioned prominently in the act recording the purchase of land for the garden on 21 February 1633.

The purchase agreement states that the 67,000 livres tournois, "to which price His said Majesty has himself agreed," were paid "by the verbal command of His Majesty," to Jean Richer, "tutor of the minor children of *maître* Daniel Voisin, deceased,"[58] who had been the owner of the property until his death.

The king had purchased in 1633: "a house located in the faubourg Saint-Victor of this city of Paris, having two entrances on the main street of the said faubourg, consisting of several main buildings, courts, cellar, presshouses, gardens, woods and hillocks, planted with vines, cypresses, fruit trees and others, enclosed by a wall."[59] The agreement expressly stated that the land was to be used for the construction of a royal garden of medicinal plants. The land destined for the new garden was known locally as *le clos Coypeau* from the fact that the corporation of *Coupeaux,* or butchers, practiced its trade nearby.[60]

The doctors of the Faculty of Medicine, upon hearing of the land purchase, immediately complained that they would not permit La Brosse, whom they qualified as a "foreign empiric (*empiric étranger*)," to teach botany in such a garden. Their attack was concentrated on the fitness of La Brosse as a teacher rather than on the question of the garden as a teaching institution outside their control. The reason is not far to seek, since they proposed to take over the entire idea themselves. Through the offices of Richelieu's physician, François Citois, the faculty took its case to the cardi-

57. La Brosse, *A Monseigneur le Chancelier*. BN: printed copy Sp 12314, p. 14. In his speech for the public opening of the Jardin des Plantes in 1640, La Brosse acknowledged not only Richelieu, but the Chancellor Séguier among his patrons, describing the latter as "cet incomparable administrateur de la Iustice." The letter I have quoted here, although it is undated, would seem to have been written before Séguier was made chancellor in 1635, since by then La Brosse already had land for his garden. It was probably addressed to Etienne d'Aligre, chancellor from 1624 to 1635, or to his predecessor, Nicolas Brûlart de Sillery, chancellor from 1607 to 1624. One of Séguier's doctors, Marin Cureau de la Chambre, was among the first demonstrators appointed to the new Jardin des Plantes in 1635 (see below), but apart from this fact, no evidence of Séguier's interest in the project has survived. Like the connection of Richelieu with the efforts of La Brosse, that of Séguier can be established only through La Brosse's mention of him as a patron. Cf. *L'ouverture* (1640), pp. 12–13.

58. "Achat par le roi, dans le fauxbourg St. Victor, d'une maison et d'un terrain destinés à la construction d'un jardin Royal des Plantes Medecinales," 21 February 1633. I have quoted from Hamy's copy which is in AN: AJ[15] 512 no. 521 of original in AN: S1520.

59. Ibid.

60. This is the explanation which Paul Lemoine gives in *Le Muséum d'histoire naturelle, extrait du volume du tricentenaire du Muséum national d'histoire naturelle.*, 6e série (Paris: Masson, 1935), XII, 4.

nal himself, writing to him on 26 April of their reasons for opposing the new garden:

> The first is that this garden, which can only be established and laid out at great expense, cannot, moreover, be useful and suitable to the public in any way, since the greatest number of medicinal plants, not having the strength or virtue of sun necessary to them, can thus be cultivated only with great difficulty; further, that the site proposed is so far from the College of the Faculty of Medicine, that the students, who are always busy with the disputations and anatomies which normally take place, will be diverted rather from these activities, so useful and necessary, and not instructed in the knowledge of simples, which needs to be taught in a garden adjoining the schools of the said Faculty, which from the earliest times has maintained [just] such a one, in which those medicinal plants most necessary were carefully cultivated and taught.[61]

According to A. L. Jussieu, writing a history of the garden in 1802, the doctors asked Richelieu for an alternative garden of their own and, as well, for a new building and an amphitheater. Apparently nothing daunted, Richelieu took an interest in their proposals to the point of having plans drawn up which suggested the Ile Saint Louis as a site for the new establishment. From there, however, the plans were put aside and forgotten.[62]

The faculty next named a commission of doctors to visit the chief physician. On 29 May the commissioners went to see Bouvard.[63] They suggested that the faculty should present Bouvard with the names of three or four qualified professors from their own ranks from whom he should choose all the teachers at the garden. All appointments should be renewed every four years "in order that all might be able to participate in this honor."[64] Jussieu relates that it was Bouvard who, while acquiescing in the demand of the faculty that the demonstrators would be drawn from Paris doctors, added that he wanted one of the demonstrators to teach also "the composition of medicaments." This was the first mention of any explicit intention to teach chemistry to be done in the garden, although La Brosse had proposed the idea as part of his plans.[65]

La Brosse in the interim had lost no time in setting about the construction of the garden itself. He says that he began work before the end of 1633,[66]

61. Quoted by A. Chereau, "Encore Charles Bouvard . . . et le Jardin des Plantes de Paris," *Gazette hébdomadaire de médecine et de chirurgie,* 21, no. 31 (31 July 1874), p. 497.

62. Jussieu, "Notice historique," I, 6.

63. Cf. Chereau, "Encore Bouvard," p. 495.

64. Jussieu, "Notice historique," I, 6.

65. Ibid. Bouvard, however, failed to act on his promise to the satisfaction of the faculty. This may well have been due to his involvement in another quarrel with the faculty. Cf. A. Chereau, "Une tempête dans un verre d'eau . . . minerale," *Gazette hébdomadaire de médecine et de chirurgie,* 21, no. 29, pp. 457–464.

66. La Brosse, *Catalogue* (1641). A copy of this catalogue in the hand of Sébastien Vaillant (MHN: MS 1325) is entitled: *Catalogue des plantes qui ont esté cultivées au Jardin du Roy estably à Paris par Louis XIII. l'an 1633 iusqu'en L'an 1641. Et qui ont Esté demontrées par Guy de la Brosse Medecin ordinaire du Roy.*

and in April of the following spring, when he presented a map of the new garden to the king at Fontainebleau, he had already collected some 1500 kinds of plants.[67] However, the work was somewhat more difficult than he had anticipated. In a catalogue of the plants contained in the garden, which was published in 1641, La Brosse explained what progress had been made in the early years of the garden: "We have properly ordered the number of its parts, as we described them in the plan which we wrote before, but not in the same sequence, not having found the site as precisely prepared as we would have hoped."[68] In the first year the ground was cleared and one parterre made, which was, according to Jussieu, 45 toises long and 35 wide. Many of the first plants which were cultivated there came from the garden of Jean Robin. The problem of a water supply was so acute that La Brosse reports that 75 livres a week alone were spent to pay men to water the plants. The king was petitioned eventually to allow water to be piped in from Rungis. Agents, including Vespasien Robin, were sent throughout France and Europe to search for plants. In addition, La Brosse says that he sent at his own expense agents to the two Indies: "We have also sent to both the Indies, at our own expense, persons astute enough in herbal knowledge, to search out for us and bring back all the seeds they could recover, which has not been totally unfruitful."[69]

The first catalogue of the young garden was published in 1636 with an explanation of the project under the title, *Description du jardin royal des plantes medecinales estably par le Roy Louis le Iuste à Paris contenant le catalogue des plantes qui y sont de present cultivées, ensemble le plan du jardin.*[70] The garden at this time contained more than 1800 different plants. By 1641 when the second catalogue was published, the number of plants included had increased to 2360. La Brosse also contracted with the engraver Abraham Bosse to have a thousand copper plates made of the rarest plants in the garden, together with two perspective views of the garden and a frontispiece for one of La Brosse's books. Bosse was to receive 25 livres for each plate completed and 2000 livres for the two maps and the frontispiece.[71]

La Brosse pressed Bouvard for an edict from the king which would ratify the land purchase, clearly define the functions of the demonstrators, and provide money for the maintenance of the garden. The Edict of May 1635 did all this and more. Unlike the general Edict of 1626 which had really done nothing more than appoint Héroard to be the superintendent of a good

67. *Gazette de France,* No. 54 (2 May 1634), p. 223.
68. La Brosse, *Catalogue* (1641), p. 1.
69. Ibid., p. 4.
70. The *Catalogue* section is subtitled: "In the two and a half years since it was prepared." For a discussion of this *Catalogue,* cf. Ernest Roze, "Le Jardin des plantes en 1636," *Journal de botanique,* 2, no. 11 (1888), pp. 191–196; no. 12, pp. 210–212, and no. 13, pp. 218–220.
71. "Touchant la graveure de mille desseins de plantes." BN: MS fonds français 18967.

idea, the Edict of 1635 spelled out what the garden was to include and how it should be financed. It was the first real constitution of the institution.

The first pages of this edict are devoted to a confirmation of the purchase of land in the faubourg Saint-Victor to be used for the garden. The edict then proceeds to the naming of the demonstrators, who are particularly charged with the teaching of pharmacy, or the "interior of plants" as well as with the "exterior," since this essential subject is taught to medical students nowhere in the kingdom:

> Furthermore, since students are not taught the operations of pharmacy anywhere in Paris, or in the other schools of medicine in the kingdom, from which proceeds a multitude of errors on the part of doctors in their practice and in their prescriptions, and of frequent abuses by apothecaries and their assistants in the execution of these latter, to the ruin of the health and life of our subjects. . . . It pleases us to decree that three doctors, chosen by him [the chief physician] will provide the demonstration of the interior of plants and of all medicaments, both simple and composite, which shall consist in the teaching of their essences, properties and usages, and in performing manually all pharmaceutical operations, the choice, preparation and composition of all kinds of drugs, both by ordinary and common means, and by chemical, in the presence of the students [and] in having [the students themselves] perform these [operations].[72]

The provisions for the teaching of chemistry could hardly have been more explicit, nor more blatantly designed to draw the wrath of the faculty. La Brosse was named to demonstrate the "exterior" of plants with the aid of an under-demonstrator. Both he and Bouvard were confirmed in their offices as intendant and superintendent respectively. Michel Bouvard, the son of the chief physician, was to succeed La Brosse as intendant upon the death of the latter. With the nomination of three doctors to fill the posts of the demonstrators, the chemical aspect of the teaching was once more reiterated. Bouvard was almost as good as his word to the Paris Faculty of Medicine. Two of the three men designated to be demonstrators were Paris doctors, and the third, Marin Cureau de La Chambre, who was identified as a doctor of Montpellier, was specifically qualified as an exception. He was nominated for the post, "although the provision of the said de la Chambre shall not give rise to any consequence, nor detract from the present edict, to the effect that there shall be admitted to the said three offices only Doctors in the Faculty of Paris and [that] on the nomination of our said chief physician."[73]

Bouvard had taken care to nominate good friends and acquaintances,

72. "Edit en forme de declaration, pour l'establissement du Iardin Royal estably au Fauxbourg S. Victor les Paris, du mois de May 1635." BN: Mémoires et documents, fonds français, vol. 1590, fols. 175–178. The date of the edict is given only as May 1635, although it was published on 15 May 1635 and registered in the chancellery. An older copy is signed by Bouvard and dated 21 February 1635 (AN: AJ^{15} 501 no. 12).

73. Ibid.

notwithstanding his promise to the faculty. Cureau de la Chambre was one of his own protégés, and of the Chancellor Séguier, and had just become physician to the latter. Of the other two men named as demonstrators, Urbain Baudinot, a Paris doctor, worked often with Bouvard, while Jacques Cousinot was his son-in-law.

The edict of 1635 further specified that a cabinet was to be kept in the garden containing samples of all the kinds of drugs. The keys to this cabinet were to be in the hands of La Brosse, and he was to be assisted in this, as in his other tasks, by the under-demonstrator, Vespasien Robin.

There followed an annual budget for the operation of the garden and the salaries of the personnel:

Superintendent	3000 livres
"Counselors, Demonstrators, and Pharmaceutical Operators" each 1500 *livres*	4500 livres
Intendant	6000 livres
Under-demonstrator	1200 livres
Maintenance of garden: workers and equipment	4000 livres
Materials and drugs for laboratory	400 livres
Boys to help in laboratory	400 livres
"Receiver and payer of the above said officers of the said garden"	600 livres
	20,100 livres

A total of 21,000 livres was allotted to cover these expenses every year.

In 1626 when the king had first confirmed La Brosse's appointment as intendant of the garden, he had instructed the "Presidents and Treasurers General of France at Paris" to pay to the new intendant such funds as would be allotted for his use for "the maintenance and wages of the said officers of the said garden each year henceforth according to the accustomed terms and manner." The use of these funds was to have been checked for the first year only by the Chambre des Comptes.

The edict of 1635 was much more specific in its provision for the control of the 21,000 livres allotted for the garden. An office of receiver and payer was created to deal with the expenses of the garden and payments were to be made four times a year. Any surplus funds in the budget were to be dispensed by the receiver, on the request of the intendant, for unforseen expenses and maintenance of the buildings in the garden.

E. T. Hamy discovered a separate document, also dating from May 1635, which is now in the Archives des Affaires Etrangères in Paris and which further specifies the staff of the new garden:

> Three principal gardeners
> To wit, one for the grand parterre, which is at present one hundred and ten toises long, who has two men under him;
> Another for the mountain, which contains five arpents, partly in vines and partly in medicinal plants, [who] has another man under him;
> One for the woods, the kitchen garden, and the big open walks, [who] has two men under him and to whom it is necessary to add others, from time to time, when there is pressing work;
> And still all this without the porter, the carter for the manure and the women who gather the weeds.[74]

In June of the same year yet another ordinance laid down that Cureau de la Chambre was to concern himself with the teaching of surgery at the garden and "to provide visual and manual demonstrations of each and every operation of surgery, whatever their nature may be." As Hamy points out in an article on the origins of the teaching of anatomy in the garden, it was through this decree that surgery and soon thereafter, human anatomy, were introduced as subjects to be taught in the garden. This document is known only through its mention in a later decree concerning the garden of 1673.[75]

The intent as well as the language of these decrees which organized the garden was clearly opposed to the Faculty of Medicine in Paris and to its claim to a monopoly on the teaching of medicine and the licensing of doctors in the capital. It was well known that the Paris doctors were opposed to the teaching of chemistry as a part of the medical curriculum. Yet the royal edicts proposed to allow chemistry to be taught to medical students and this in an institution which directly challenged the faculty's role in the training of doctors, since surgery and anatomy would be taught in the garden as well.

There is no record of faculty opposition to the edicts of 1635 until late the following year. However, it is to be imagined that they made their complaints heard, since by 29 November 1636 when they wrote their first

74. "Officiers du Jardin des plantes medecinales du faulxbourg Sainct Victor," May 1635. I have quoted from a copy made by E. T. Hamy, AN: AJ[15] 510 no. 340.

75. Cf. "Déclaration du Roi pour faire continuer les exercises au Jardin Royal des Plantes, registrée au Parlement et Chambre des Comptes le 23 mars 1673." Printed copy in AN: AJ[15] 501 no. 17. Cf. also E. T. Hamy, "Recherches sur les origines de l'enseignement de l'anatomie humaine et de l'anthropologie au Jardin des plantes," *Nouvelles archives du Muséum d'Histoire Naturelle,* 3e série, 7 (1895), 5. The decree concerning Cureau de la Chambre is also referred to under the title, "Lettres patentes pour l'établissement de l'anatomie et Chirurgie," of June 1635, in "Anciennes nottes dans lesquelles il en fait mention des titres du jardin Royal des plantes dont les suivants n'ont pù étre retrouués et qui sont probablement dans les dépôts de la maison du Roi." AN: AJ[15] 501 no. 3. Cf. the same report of a missing document in an inventory of papers pertaining to the garden made by R. A. F. de Réaumur in the eighteenth century. AN: AJ[15] 501 no. 4.

protest to the Chambre des Comptes, the edict of May 1635 had still not been registered in that court.

The Chambre des Comptes decreed that all parties opposed to the registration should be given copies of the edict that they might show cause why the court should not proceed to registration as it had been instructed by the king. In a letter of 29 November, the faculty complained to the court that it had been unable to obtain a copy of the edict. On 9 December 1636, the Paris doctors recorded their opposition to La Brosse and his garden in their own registers, referring to the new intendant as an "empiric."[76] Finally, on 20 December 1636, they submitted to the Chambre des Comptes the list of their objections to the edict of 1635.[77]

Their complaints were lengthy. They began by asserting that La Brosse was unqualified to be the intendant of the garden "given his notorious incompetence." This was evidenced, they felt, by the fact, "that he is very poorly versed in the knowledge of plants and that he has no degree in medicine." Moreover, the provision of Michel Bouvard to succeed La Brosse was but the addition of insult to injury, since Bouvard *fils* had not even studied medicine, let alone taken a degree: "two such would simply demonstrate that favor is of more importance than capacity. [The young Bouvard] can only take up the duties of intendant as a doctor in the Faculty of Paris."

The doctors further objected to the appointment of Cureau de la Chambre as "demonstrateur et operateur pharmaceutique" qualifying him as "so-called doctor of Montpellier," and saying that he certainly was not Parisian. The complaints of the Faculty then turned to administrative matters: "moreover . . . [we ask] that the said Sieur de la Brosse not be allowed to dispose of the lodgings, except for what shall be newly built, to provide for the instruction of students, for a laboratory and for a cabinet to keep rarities, since this would contravene the intention of his Majesty, who wishes that the said students be promptly instructed, as they are able to be from this moment in the most commodious rooms already built, without waiting for new ones to be built, which will happen maybe within the century." That the intendant of the garden should be accountable each year for the manner in which he spent the 4000 livres allotted to him for the maintenance of the garden, as should be the demonstrators for the 400 livres they spent on drugs, was eminently reasonable, but the doctors insisted that such accounting should be made to the faculty rather than to the superintendent of the garden. Indeed, further, the court was asked to consider that: "When the offices created for the said garden shall be vacant, it is entirely appropriate and very reasonable that they be filled by competent persons on

76. *Commentaires* of the Faculty of Medicine, Paris, XIII, fol. 13 recto.

77. This document is reproduced in the *Commentaires* of the faculty, XIII, fols. 16 verso–17 recto.

the nomination and presentation of the superintendent after the said faculty shall have first assembled regarding this matter and elected three persons in the customary manner from whom the superintendent shall have the option of choosing such as he may wish, without however being able to nominate any [who are] not of the Faculty of Paris; and this latter to avoid venality of the posts of professor."

The faculty placed itself again on record as formally opposed "to the teaching of chemistry in Paris, as it has been forbidden and censured by decree of the Parlement of Paris for good and just cause." It was not true, they said, that pharmacy was not taught in the medical schools; there were two professors so charged in the Paris faculty, and if they had but half the wages promised to the teachers in the garden they could produce far better courses than would ever be given to students in the new institution.

Last, the faculty requested that the court see to it that the edict was registered only on the provision that the superintendent of the garden be a Paris doctor, elected by the faculty, to serve for two years. They suggested that candidates for the office might be drawn from former deans of the faculty itself.

The Chambre des Comptes still did not ratify the edict of 1635, since there were parties other than the faculty who objected to its provisions. The officers in charge of dispensing funds for the maintenance of the various royal establishments objected to the creation of a particular receiver to handle the funds destined for the garden, especially the 4000 livres intended for maintenance. The new institution, they asserted, was a royally patronized undertaking and as such should be managed in a manner similar to that used for other royal properties, from funds dispensed through the treasurers for the royal buildings. On 11 September 1638, an agreement was made in which Guy de la Brosse consented to all the conditions which had been specified for the handling of money for the garden in the edict of 1635, but in which the treasurers for the royal buildings were substituted for the receiver.[78] The intendant and superintendent might still dispose of any surplus in the annual budget in the manner specified in the original edict.

The Chambre des Comptes published its ratification of the 1635 edict on 12 October 1638. Several important modifications were introduced which reflected the objections raised by the faculty and the financial officers. La Brosse was confirmed as intendant but he was to provide "from this moment a suitable place for the instruction of students." The demonstrators

78. "Accord fait entre Guy de la Brosse Intendant du Jardin du Roi et la chambre des comptes, pour que les fonds nécessaires à la réparation des Batimens, l'achat des drogues et autres acquisitions utiles dudit Jardin, soient remis entre les mains du trésorier des Batiments du Roi qui en fera l'emploi d'après les ordres de l'Intendant dudit Jardin." 11 September 1638. AN: AJ^{15} 501 no. 13.

were to be paid for their services only upon certification that they had performed their duties and this certification was to be signed by the dean and two other doctors of the Paris faculty. They were to be replaced by appointment of the king from three nominations proposed by the faculty. Michel de Bouvard, son of the chief physician, might succeed La Brosse as intendant, but only on condition "that he become a doctor." Finally, the court provided that the funds for the garden should be handled by the treasurers for the royal buildings. In previous edicts, financial officers had been instructed to deliver funds to authorized personnel "for a simple receipt." Now there was to be a thrice yearly audit by representatives from the Chambre des Comptes: "Let two Counselors and Masters of the said Court . . . betake themselves to the said Garden at least three times a year, and prepare their *procès verbaux,* which shall be reported to the account of the royal buildings for that part of the four thousand livres [for maintenance] used in the name of the intendant of the said garden. The intendant shall bring, at the end of each year, to the clerk of the court, a statement listing all the plants of the said garden, in which shall be a separate chapter on those plants which have been newly added in the said year, signed and certified by him."[79] To further insure the control of the Chambre des Comptes, the superintendent and the intendant were to take an oath in the court itself regarding these conditions.

The modifications could hardly have been expected to pass without comment by those who championed the establishment of the garden. On this occasion, however, it was not La Brosse but Bouvard who took up the cause of the garden against his colleagues in the faculty. In a short piece entitled *Mémoire pour le Jardin des Plantes,* Bouvard requested the king to promulgate *lettres de jussion* requiring the court to register the original edict unchanged, or to pass a new decree which would reaffirm the conditions of the edict, unmodified. Bouvard took issue with all the alterations made by the Chambre des Comptes in its decree of registration and he held the Faculty of Medicine responsible for the better part of these.

The right of nomination and certification of the officers in the garden had been removed from the chief physician and given to the Faculty. Self-righteously ironic, Bouvard asked:

> The chief physician, having observed throughout France the errors, disorders and mistakes in the practice of medicine which are due to the fact that the doctors have not been taught in the Faculty of Paris, nor elsewhere, anything of the operations of pharmacy or of its subject matter, was himself the only person to deal with it by proposing to the king the establishment of this garden as a new school absolutely necessary to the perfection of doctors; Is it reason-

79. "Arrest de Verification en la Chambre des Comptes." 12 October 1638, reprinted in Actes royaux no. 8321, II, (1625–1628), BN: F23610 (793).

> able to defraud him of this honor and of this right in order to adorn the faculty with it, [that faculty which is] itself the cause of these errors and disorders and which has opposed itself directly to this establishment?[80]

Furthermore, the provision that the intendant, as chief administrator of the garden, should be a doctor, was absurd. There were already three doctors connected with the garden who could deal with the medical aspects of teaching to any extent necessary. The intendant's job was quite another; he was to devote himself completely to the culture and the gathering of plants. For this he hardly needed a medical degree. Bouvard was here addressing the provision of the faculty and the court that his son Michel should take a medical degree before being permitted to assume the intendancy. This was arrant hypocrisy on the part of the faculty, said Bouvard. When it had been a question of their own garden, they had hardly seen fit to be so rigorous: "The faculty . . . has never charged any of its doctors with the care of its garden, having always contented itself with Vespasien Robin and his sires, herbalists who were not by any means learned nor doctors. The gentlemen [of the court] should therefore also order that Robin become a doctor and make a profession of medicine, rather than the administrator who is more of a gardener than he is in this garden."[81] Finally Bouvard protested that the institution of a receiver to handle funds for the garden was not without precedent, since the king had done this before and he argued that the office should be left as the king had decreed it.

Work progressed in any case on the garden. La Brosse made his own home there and, on 20 December 1639, he obtained permission from the archbishop of Paris to build a small chapel in the grounds of the garden where mass might be celebrated on all Sundays and holidays except Easter. He included a sepulchre in which he intended to be buried at his death.[82] In 1640, the garden was officially opened and La Brosse published *L'ouverture du Jardin royal de Paris, pour la demonstration des plantes medecinales.* In this pamphlet he specifically thanks his chief patrons: Héroard, but more emphatically Bouvard, as chief physicians; and "those two very prudent dispensers of finances, my lords de Bullion and Bouthillier." A summary of the importance of plants is included together with a list of rules to be

80. *Memoire pour le Jardin Royal des Plantes par Charles Bouvard* (1639). Cf. note 43.

81. Ibid.

82. This incident was uncovered in the archepiscopal records of the diocese of Paris by the Abbé Lebeuf, who gives the following account of it in his *Histoire de la ville et de tout le diocèse de Paris* (Paris: Librairie de Féchoz et Letouzey, 1883), I, 261: "I discovered that Guy de La Brosse, doctor [and] Intendant of this Garden, seeing that he was at some distance from Saint Médard, had it [the chapel] built and obtained from the archbishop of Paris, on 20 December 1639, [the right] to have Sundays and holidays celebrated there, except for Easter, with burials reserved for the curé, even that of the founder, who had chosen there his sepulchre in a vault; on condition however, that on Easter day, at the parish mass, a white taper worth one livre and a gold écu would be offered on behalf of this Intendant.

observed by those attending the public demonstrations. That teaching was already being done in the garden is suggested by a list of students included by La Brosse in another catalogue of plants, published the following year, 1641. La Brosse himself most probably demonstrated the plants and their uses himself.[83]

La Brosse did not live long to enjoy the pleasures of his garden for he died in the following year on 31 August 1641. It was most probably his sister Louise who saw to it that he was buried in the chapel he had had built on the grounds of the garden where it was perhaps also she who wrote in charcoal on the wall above the tomb this touching epitaph:

> Guy de la Brosse
> Whose death overwhelms me with grief
> Though his body be covered with earth
> His name, I hope,
> Will be
> Will be forgotten never.[84]

It was not until 1893 that the remains of Guy de La Brosse were transferred to their present site in the garden in a ceremony presided over by A. Milne-Edwards, the director of the Muséum.[85]

The years immediately following the death of La Brosse were not prosperous ones for the Jardin des Plantes.The competition for the patronage and income involved in the official posts at the garden had a deleterious effect on the program of teaching envisioned by its founder. Nor does an effective and continuing program of research seem to have been set up at the garden either during the lifetime of La Brosse or in the years following his death. Only at the end of the century did the nephew of Guy de La Brosse, Guy-Crescent Fagon, as chief physician to Louis XIV, succeed in transforming the Jardin des Plantes into the teaching and research institution his uncle had planned. In his patronage of such men as Joseph Pitton de Tournefort and the surgeon Pierre Dionis, Fagon inaugurated one of the greatest

83. "Liste des estudians a la connoissance des plantes au Jardin royal de Paris, & aux operations de la Medecine, qui s'y font l'an 1641," *Catalogue* (1641). MHN: 8° Rés. 77. That La Brosse himself demonstrated the plants in the garden is indicated by another copy of the *Catalogue* (1641) in the hand of Vaillant (MHN: MS 1325). Cf. note 66. It is difficult to determine the extent to which any regular program of teaching was established or carried out in the garden before Guy-Crescent Fagon became intendant in 1693. Cf. however the following sources on teaching in the Jardin des Plantes in the seventeenth century: A. L. Jussieu, "Notice historique"; Laissus, "Le Jardin du Roi"; Jean-Paul Contant, *Contribution à l'histoire de l'enseignement de la pharmacie: l'enseignement de la chimie au Jardin royal des plantes* (Cahors: Imprimerie A. Coueslant, 1962); Paul Crestois, *L'enseignement de la botanique au Jardin royal des plantes* (Cahors, 1953).

84. A. Milne-Edwards, "Translation et inhumation des restes de Guy de la Brosse et de Victor Jacquemont. Faites au Muséum d'histoire naturelle le 29 Novembre 1893," *Nouvelles archives du Muséum d'Histoire Naturelle,* 3e série, 6 (1894), vii.

85. Ibid.

periods in the history of the Jardin des Plantes, an era culminating in the work of Buffon and his successors, in which the Jardin des Plantes became one of the great scientific centers in France.

The greatest monument of Guy de La Brosse is the Jardin des Plantes. The history of its founding deserves to be better known, not only for its importance to the history of science and to that of botany and medicine in particular, but for what it reveals about the nature of early royal patronage of scientific institutions in France. The Jardin des Plantes was in fact the first such institution to be so patronized and as such may be regarded as a precursor of the later Académie des Sciences. Its place in the administrative history of France is as important as its place in the history of science.

THE LIBERTIES OF THE PARIS ACADEMY OF SCIENCES, 1716–1785

RHODA RAPPAPORT

THE Duc de Choiseul, consulted in 1768 about a forthcoming election in the Academy of Sciences, offered his advice to the Academy's secretary and then declared: "Furthermore you understand that this observation has no bearing whatever on the other candidates or on the votes in the Academy which must retain all their freedom and integrity."[1] The idea that the Academy possessed freedom of choice was expressed from time to time throughout the century by royal ministers whose duties included "protection" of the several royal academies. Thus, in 1726, the abbé Bignon replied to a solicitation for membership in the Academy of Sciences that he would do all he could to preserve freedom of choice in elections. Another such solicitation in 1784 was transmitted to the Academy's president with the ministerial reminder that it was "essential not to constrict" freedom of choice.[2] These sentiments were shared by many academicians, for reasons like those of chemist P.-J. Macquer: "I have suffered on certain occasions from seeing conniving mediocrity win out over talent unsupported by patronage.... I have often feared... that if the liberty of elections were hampered, this dangerous mediocrity would prevail and ultimately bring about a shameful ruin of the Academy."[3] A retrospective assertion of academic liberties was eventually to be used in defense of the institution on the eve of its abolition, when Lavoisier declared that "even under the old regime, ... a kind of respect... guaranteed the sanctuary of the sciences against the invasion of despotism."[4]

1. Choiseul to Fouchy, 21 May 1768, in Archives of the Academy of Sciences (hereafter AAS), Paris, dossier Bézout.

2. Bibliothèque nationale (BN), MS fr 22234, fol. 28. Archives nationales (AN), O1*495, fol. 186. Also, BN, MS fr 22231, fols. 82, 104; 22234, fol. 25; and Roger Hahn, *The Anatomy of a Scientific Institution: The Paris Academy of Sciences, 1666–1803* (Berkeley, 1971), p. 81.

3. Hahn, *Anatomy,* p. 82.

4. *Oeuvres de Lavoisier,* 6 vols. (Paris, 1862–93), IV, 618.

However earnest their statements, ministers and savants well knew that academic elections were often engineered. Bignon, for example, more than once informed interested parties that he was "working at" a particular election and hoped in time to sway the minds of the voters. Similarly, Grandjean de Fouchy, during his years as Perpetual Secretary, more than once offered advice to colleagues and ministers about how posts in the Academy might be juggled in order to achieve desired elections or promotions. And more than one academician wrote directly to the Minister of the Maison du roi in efforts to obtain a prior commitment for an election or a promotion.[5] In short, while both sides claimed that the Academy had and ought to have certain liberties, both commonly took part in the manipulation of elections.

Conflicting statements about the Academy as an institution—its liberties in principle, the intrigues in practice—have their parallel in descriptions of the tenor of academic life and the behavior of academicians. Early in the century, Bernard de Fontenelle had repeatedly eulogized the objective, detached, unworldly savant, whose very profession raised him above pettiness.[6] More often, however, academicians analyzed their colleagues in less flattering terms. Not only are judgments fallible, declared C.-G. de Malesherbes, but voting academicians can also be swayed by "authority" and even by "nepotism." Professional bias can sometimes blind us to merit, Antoine-Laurent Lavoisier admitted, while Joseph Dombey more bluntly observed that men pursuing similar careers are naturally jealous of one another. Rivalries, jealousies, and competitive place-seeking all had a stimulating effect upon the ego and prompted J.-L. Lagrange to remark that pretensions seemed always inversely proportional to merit.[7] And yet, Lavoisier assures us, bias is not pervasive because to be a scientist requires "the habitual use of reason."[8]

Like these academicians, modern historians have reached no consensus about the nature of academic life or the way in which the institution functioned. There has been, for example, some tendency to minimize in-

5. H. Omont, ed., *Lettres de J.-N. Delisle au comte de Maurepas et à l'abbé Bignon* (Paris, 1919), p. 33. Anonymous account of the activities of Bignon and others, n.d. [1725], in AAS, dossier P. LeMonnier. Jean Torlais, *Réaumur* (Paris, [1936]), p. 212. Fouchy, MS "Etat Present de la classe anatomique," AAS, dossier 1758. Draft from Fouchy to [St. Florentin?], n.d., AAS, dossier 27 June 1770. Letters to LeRoy, Arcy, and Ayen in AN, O¹*412, fols. 323–4, 356; 414, fol. 1107.

6. *Oeuvres de Monsieur de Fontenelle,* 5 vols. (Paris, 1825), esp. I, 83, 96–97, 173–175; II, 100, 250. Hahn, *Anatomy,* pp. 42, 56.

7. Malesherbes to d'Alembert, 1769, in C. Henry, ed., "Correspondance inédite de d'Alembert," *Bullettino di bibliografia e di storia delle scienze matematiche e fisiche,* 18 (1885), 569. Lavoisier MS, AAS, dossier 28 July 1784. E. T. Hamy, *Joseph Dombey* (Paris, 1905), pp. 86–87. *Oeuvres de Lagrange,* ed. J.-A. Serret, G. Darboux, L. Lalanne, 14 vols. (Paris, 1867–92), XIII, 206.

8. Above, n. 7. He was addressing his colleagues.

trigue and to focus instead on merit as the most important criterion in academic elections. At the same time, while one historian claims that merit was also a criterion for promotion within the ranks, another describes promotion as the automatic concomitant of seniority.[9] As for the relationship between crown and Academy, one historian tells us that royal ministers often set aside the Academy's judgment and sometimes appointed a new member not nominated by the voters. Elsewhere we are told that royal influence was confined largely to the choice of members in certain ranks and categories: *honoraires, associés libres,* or *surnuméraires.*[10] On the whole, historians agree that the Academy did not possess complete freedom of choice, but few have recognized that such liberty was thought to exist and was deemed essential by ministers and savants.[11]

The purpose of this study is to examine elections and promotions in the Academy in order to determine what degree of internal freedom that body possessed. Consideration will be given to electoral practices and the extent to which the voters followed rules prescribed for the company in 1699 and 1716. Ministerial influence, before and after balloting, will also be analyzed: how often was such influence exercised, for what reasons, and with what response from the academicians? Given the frequent claims to liberty, and the equally frequent admission that voting was manipulated, some effort will be made to determine what ministers and savants had in mind when they invoked academic liberties.

Certain limitations are imposed on this study by the nature of the subject and the available materials. Discussion will be confined to the "working ranks" in the Academy: assistants, associates, and pensioners (*adjoints, associés, pensionnaires*). Since these were the only ranks required to be productive, the quality of this part of the membership determined the scientific quality of the institution; indeed, assertions of liberty were generally made in behalf of the working ranks. In addition, examples will be drawn almost exclusively from the period 1716–1785, a period in which these ranks and the rules governing procedures underwent no change.

Evidence for this inquiry is both plentiful and severely limited. The Archives of the Academy provide lists of nominees for every post, but they

9. Jean Torlais, *L'abbé Nollet, 1700–1770* (Paris, [1954]), pp. 52–53. Hahn, *Anatomy,* p. 80.

10. Hahn, *Anatomy,* pp. 80–81, 98, 98n. S. L. Chapin, "Les associés libres de l'Académie royale des Sciences: Un projet inédit pour la modification de leurs statuts (1788)," *Revue d'histoire des sciences,* 18 (1965), 7–8. E. Maindron, *L'ancienne Académie des sciences: Les académiciens, 1666–1793* (Paris, 1895), pp. 4–5.

11. Those who discuss the matter usually refer not to ministers but to specific crises which provoked an assertion of liberties by academicians. Cf. K. M. Baker, "Les débuts de Condorcet au secrétariat de l'Académie royale des sciences (1773–1776)," *Revue d'histoire des sciences,* 20 (1967), 229–280; W. C. Ahlers, "Un chimiste du XVIIIe siècle. Pierre-Joseph Macquer (1718–1784): Aspects de sa vie et de son oeuvre" (Thesis, Ecole Pratique des Hautes Études, 1969); Hahn, *Anatomy,* pp. 81–82.

rarely contain the numerical results of balloting, summaries of debates on candidates, or explanations of decisions by the voters and the king. Private correspondence, although abundant, too rarely fills in such gaps. Academicians and ministers alike were, after all, usually in or near Paris, and delicate matters such as academic politics were better reserved for conversation.[12]

Gaps in the evidence mean that certain questions are historically unprofitable. Even in well documented cases, it remains impossible to identify voting blocs in the Academy; at best, one can discover the views of only a handful of the thirty or more voters in a given election.[13] As a result, there is at present no way to answer such questions as: What influence was exerted by the *honoraires,* the pensioners, the savants in particular disciplines? How often did scientific issues sway the voters? What was the role of personal friendships or enmities, of philosophical outlook, of patronage? Partial answers to these questions are sometimes available and are used here when evidence permits; but the major focus in these pages is on patterns and policies rather than on factions, and on the Academy's relations with the crown rather than on the intricate networks of influence and protection.

The *règlement* of 1699 spelled out for the first time the structure of the Academy, rules governing the members, and procedures for elections and promotions; subsequent modifications were then incorporated into letters-patent in 1716, after which date no change was to affect the working ranks until 1785.[14] Six classes were established (geometry, astronomy, mechanics, anatomy, botany, chemistry), each to consist of seven members, the seven distributed in three ranks: two assistants, two associates, and three pensioners. All, but especially the pensioners, were to present periodic proof of research, and all had to reside in Paris, a precaution meant to ensure regular attendance at meetings and frequent communication among members. Before an election, the class concerned was to draw up a list of several names; the list was then offered to all the pensioners and honorary members whose vote determined the two or three names to be submitted to the king. Tallying of votes was done in secret, and the Academy and king alike were merely presented with names listed preferentially. Final choice among the nominees was made by the king, or, rather, in his name. It is worth noting that each list of candidates for promotion had to include the name of at least

12. I am also under no illusion that I have exhausted all available materials. My reliance on the Archives of the Academy has been facilitated and made pleasurable by Mme Pierre Gauja and her staff, to whom I can but inadequately express my gratitude.

13. Cf. Baker, "Débuts de Condorcet," pp. 235–236.

14. Léon Aucoc, *Institut de France. Lois, statuts et règlements concernant les anciennes académies et l'Institut, de 1635 à 1889* (Paris, 1889), pp. lxxxiv–xcv. A discussion of the application of the rules through about 1753 is in Jean Hellot, MS "Collection de . . . Reglemens et deliberations par ordre de Matieres," AAS, dossier Règlements. There are two such MSS, one in Hellot's hand and the other a fair copy; I have followed the pagination of the latter.

one nonmember; the Academy was thereby enjoined to consider outsiders who might be more talented than the members.

After more than thirty years without defined rules and procedures, the membership was understandably pleased to be guaranteed a certain permanence and freedom from royal or ministerial caprice. The architect of the *renouvellement* of 1699, Jean-Paul Bignon, was appointed President of the Academy for that year, and the otherwise unemotional pages of the *procès-verbaux* record that this news was received "with great joy."[15] To be sure, the period 1699–1716 saw a certain amount of internal adjustment taking place. A few members were expelled for prolonged absence from meetings, while others, who held posts requiring absence from Paris, were made to choose between membership in the Academy and the posts in question. Furthermore, when the letters-patent of 1716 abolished the rank of *élève* and created six posts at the assistant level, those *élèves* in excess of six were permitted to stay on as "supernumerary" assistants; during the next few years, the normal process of attrition reduced the working ranks to the statutory forty-two.[16]

With a constitution in hand, the Academy in 1716 presumably had only to follow the detailed rules, while the royal protector had only to watch that this was done. Since the following pages deal with the relationship between Academy and crown, it is appropriate now to introduce the several men who served as intermediaries between the king and the savants. For the years 1716–1749 these men were Jean-Paul Bignon (d. 1743) and Jean-Frédéric Phélypeaux, Comte de Maurepas. Contemporaries and historians agree that both were vitally concerned with progress in the sciences. In addition, both were kept well informed about academic life, Bignon by his friends Fontenelle and R.-A. F. de Réaumur, Maurepas by Bignon and perhaps by such protégés as Pierre Bouguer, Buffon, and H.-L. Duhamel du Monceau. Since Bignon and Maurepas consulted each other constantly, it would be difficult to determine which of the two bore the greater responsibility in matters affecting the Academy.[17] After the fall of

15. AAS, MS "Procès-verbaux des séances" (hereafter PV), 4 February 1699. The *règlement* was read to the Academy that day. See Fontenelle, *Oeuvres,* I, 63, 75.

16. The supernumeraries of 1716 were Bomie, Bragelongne, G. Delisle, Deslandes, and Winslow. Cf. *Index biographique des membres et correspondants de l'Académie des sciences* (Paris, 1954 and other editions). Also, Aucoc, *Institut,* pp. xciii–xciv, and Hellot MS, pp. 14–16, 34–35.

17. Neither man has been studied adequately. For Bignon, see J.-J. Dortous de Mairan, *Eloges des académiciens* (Paris, 1747), pp. 288–313; Torlais, *Réaumur,* pp. 204–215; Fontenelle, *Oeuvres,* I, 62; AAS, dossier Bignon. For Maurepas, M. Filion, *Maurepas ministre de Louis XV (1715–1749)* (Montreal, [1967]); A. Picciola, "L'activité littéraire du comte de Maurepas," *Dix-huitième siècle,* 3 (1971), 265–296; *Oeuvres de Condorcet,* ed. A. C. O'Connor and M. F. Arago, 12 vols. (Paris, 1847–49), II, 167, 466–498. For his relations with Bignon, see esp. BN, MS fr 22231, fol. 138; 22234. fol. 1. The later reputation of Maurepas, when recalled to power in 1774, differs from that of his earlier ministry.

Maurepas in 1749, the ministers charged with academic affairs were the Comte d'Argenson (1749–1757), the Comte de St. Florentin, later Duc de La Vrillière (1757–1775), Malesherbes (1775–1776), Amelot de Chaillou (1776–1783), and the Baron de Breteuil (1783–1787). Intellectually, only Argenson and Malesherbes cannot be called nonentities, but neither played as important a role in the Academy's life as did their less impressive ministerial colleagues. If little of relevance is known about St. Florentin, Amelot, and Breteuil, a contemporary assessment suggests that about St. Florentin, at least, there is little to be known: "He is accessible, he answers letters punctiliously, he listens goodnaturedly, he is neither hard nor unfair by nature, . . . [but] one could not be more limited than he is."[18]

On what basis did the Academy draw up its recommendations for the election of new members? How often was the Academy's preference ignored by the ministry, and why? What evidence is there of governmental pressure applied before an election took place? How did the Academy respond to ministerial intervention?

In the seven decades after 1716, the Academy held elections to fill approximately one hundred vacancies at the lowest rank. The *règlements* specified that candidates had to make themselves known to the membership by "some work of their composition," reported on favorably by an ad hoc committee of members.[19] According to Bignon, this meant that candidates of proven merit were to be preferred to those of future promise.[20] That proof of merit was desired by the membership, too, is suggested by J.-N. Delisle in discussing with Bignon the candidacy of Delisle's own assistant at the Observatoire du Luxembourg. Louis Godin, he explained, was a competent astronomer but had not yet demonstrated his ability to do independent research; the then vacant post in the Academy ought not to be filled until a more tried candidate could be found. The post remained vacant for several years.[21]

Rules and policies notwithstanding, the minutes of meetings until about mid-century seldom record the presentation of memoirs by candidates or reports on them by the membership. The existence of such lacunae, however, may mean only that records are incomplete or that candidates offered few or many memoirs depending upon how well known they were to the members.[22] In later decades, the presentation of memoirs seems to have

18. *Journal de l'abbé de Véri,* ed. Jehan de Witt, 2 vols. (Paris, 1928–30), I, 112. P. Grosclaude, *Malesherbes, témoin et interprète de son temps* (Paris, [1961]). Fouchy, "Eloge de M. le Comte d'Argenson," *Histoire de l'Académie royale des sciences,* 1764 (1767), 187–197. A. Doria, *Le comte de St.-Florentin: Son peintre et son graveur* (Paris, 1933), pp. 15–19.

19. Aucoc, *Institut,* p. xciv. Numbers of vacancies and certain other large figures are approximate because of some gaps and anomalies in the records.

20. BN, MS fr 22234, fol. 1.

21. Archives of the Observatory of Paris (hereafter AOP), B.1.2, fol. 42. The vacancy was unfilled from late 1719 until July 1725; Delisle's recommendation dated from 1722.

22. Abbé Nollet, in MS "Réponses aux observations de Mr. l'abbé Nollet," fol. [5v], AAS, dossier d'Alembert.

occurred more regularly, but still not invariably. Even when candidates did follow the rules, and the records are at their fullest, the historian can still be left in doubt about the influence of scientific merit upon the balloting. Selected examples will illustrate this problem.

In 1725, Pierre Maloet and F.-J. Hunauld were the declared candidates for a vacancy in the *class of anatomy,* and both offered memoirs to be reported on by academicians. The scheduled report on Hunauld's work either was not prepared or has been lost, while that on Maloet was equivocal: he seemed to have abilities, but had tackled subjects too complex for his powers. Maloet was the Academy's first choice for the vacancy. Within three weeks, Hunauld was again a candidate, this time for a vacancy in the *class of chemistry.* Bignon believed that Hunauld had proven his worth, and Hunauld's eulogist later asserted that he had been given a post in chemistry, rather than anatomy, because the Academy wished to elect a man of such talent even if he had to be placed in an inappropriate class.[23] In short, why was Maloet elected? We do not know, but are faced with the possibility that his talents played a less than decisive role.

At a later date, botanists Jean Descemet and Jean-Baptiste Lamarck stood for election after both had presented several memoirs and Lamarck the manuscript of his *Flore française.* All the surviving reports are eulogistic, recommending publication under the Academy's auspices. Why was Descemet first on the list of nominees for the vacancy? Again, we do not know, but there is some likelihood that his earlier appearances on such lists had given him a claim to "seniority" among the eligible candidates.[24]

The election of Lavoisier in 1768 provides another example of the difficulties inherent in discovering the criteria used by voters. In retrospect it is obvious that Lavoisier had greater talent than Gabriel Jars, but this could not then have been apparent to many members of the Academy. While his abilities were appreciated by Macquer and J.-E. Guettard, and his memoirs praised by the *rapporteurs,* Lavoisier was but an interesting novice when compared with Jars who had behind him several years as a metallurgist and government consultant. If proven merit was to be preferred, Jars probably had the stronger credentials in 1768, but Lavoisier was the Academy's first choice.[25]

Preelection campaigning within the Academy was doubtless a normal fact of life, and such evidence as we have again suggests both that criteria other than merit were important and that the motives of the voters are often undetectable. In 1752, for example, chemist Jean Hellot admitted that he favored the candidacy of C.-F. Geoffroy because he had admired Geoffroy's

23. PV, 30 June, 28 July, 11 & 29 August 1725. BN, MS fr 22234, fol. 1. Mairan, "Eloge de M. Hunauld," *Hist. Acad.,* 1742 (1745), 207. Hunauld in 1728 shifted to the class of anatomy.

24. PV, 6 & 24 February, 24 & 28 April, 5 May 1779. The seniority of nonmembers is discussed below.

25. PV, 11, 14, & 18 May, 1 June 1768.

father and would thus do all he could to help the son.[26] At a later date, the aid of Jean d'Alembert was recruited in the election of Nicolas Desmarest. What kind of persuasion d'Alembert employed, however, is unknown. Personally unable to judge the quality of Desmarest's geological work, he was convinced that Desmarest was enlightened and that his research would create distress in the corridors of the Sorbonne. To advance the career of a fellow philosophe always loomed large in d'Alembert's mind, but there is no reason to suppose that such arguments would have carried weight in the Academy.[27]

While voters might be influenced by matters other than scientific competence, it is worth noting that the hierarchical social structure of the Old Regime had at most a negligible effect upon voting. Or, more accurately, votes might be swayed by the knowledge that a candidate had influential patrons, but the rank or status of the candidate himself was relatively unimportant. Early in the century there was, to be sure, some discussion of whether men of rank, possessing scientific ability, could with dignity occupy a post of assistant or associate.[28] There is no evidence, however, that such concerns persisted among academicians. The special category of *honoraires,* created in 1699, was available for titled savants who cared about these distinctions. Although honorary members were usually men of rank or status (*gens en place*) and, at best, scientific amateurs, the existence of this category may well account for the apparent absence of social pressure in elections to the working ranks. At the other end of the social scale, humble birth did not distress the voters, as long as candidates were of "honorable estate." One aspirant, J.-B. LeRoy, thus felt obliged to reassure his future colleagues that he would utterly renounce his career as a clockmaker.[29]

Ministerial interference, before and after balloting, supplies another ingredient in an already murky picture. Perhaps the simplest—certainly the most easily identified—form of intervention was the ministry's occasional decision not to abide by the expressed wishes of the voters. During the Bignon-Maurepas period, in six of about fifty elections the Academy's first

26. BN, MS fr 12305, fol. 374. Argenson, too, is here said to have been swayed by "nepotism." This was the case later referred to by Malesherbes (above, n. 7).

27. Henry, "Correspondance inédite de d'Alembert," pp. 529, 570. *Voltaire's Correspondence,* ed. T. Besterman, 107 vols. (Geneva, 1953–65), LV, 141 (no. 11129). The limited influence of the philosophes in the Academy is treated by A. Birembaut, "L'Académie royale des sciences en 1780 vue par l'astronome suédois Lexell (1740–1784)," *Revue d'histoire des sciences,* 10 (1957), 149–150.

28. Delisle-Louville letters of 1718 and 1724–25, AOP, B.1.1, fols. 81–88; B.1.2, fols. 134, 138. Cf. C.-P. Duclos, "Honoraire," in Diderot and d'Alembert, *Encyclopédie* (Paris, 1751–1780), VIII, 291–292.

29. LeRoy to [Mairan?], 17 August 1751, AAS, dossier LeRoy; quoted in J. Bertrand, *L'Académie des sciences et les académiciens de 1666 à 1793* (Paris, 1869), pp. 74–75. Also, Macquer, in Hahn, *Anatomy,* p. 82, and A. Doyon and L. Liaigre, *Jacques Vaucanson, mécanicien de génie* (Paris, 1966), pp. 219–220, 440.

choice was not confirmed by the king; after 1749 this was to occur only once. Although Bignon and Maurepas wanted to maintain a high level of proven talent in the Academy, their rejections of the voters' wishes did not stem from any independent evaluation of merit. Instead, some degree of merit and a variety of pressures influenced these decisions. In 1725, for example, the younger brother of astronomer J.-N. Delisle, Delisle de la Croyère, was the Academy's second choice and was confirmed by the king. He and Godin, the Academy's first nominee, had equivalent credentials, both having served as assistants to the elder Delisle. But the Delisles were on the eve of a scientific mission to Russia, and the elder brother had convinced Maurepas that regular communication with the Academy would be best ensured if both voyagers were members.[30] Far different was the situation in 1729 when the younger Saurin, son of a member, was twice the Academy's first nominee and was twice denied confirmation. Ministerial approval was first given to Pierre Mahieu who, it was said, had no desire to enter the Academy and was astonished to find himself in its ranks. The second effort of Saurin *fils* resulted in the vacancy remaining unfilled for two years, at which time A.-C. Clairaut, the Academy's second choice in 1729, was confirmed. This odd series of events suggests that antagonism toward one or both of the Saurins was the motive behind ministerial actions.[31] The confirmation of the abbé Nollet in 1739 and of Lamarck in 1779 probably were cases of influence in high places. Nollet, unlike his competitor Mignot de Montigny, was well known to such influential academicians as Réaumur and C.-F. de Cisternay Dufay, and his teaching of experimental physics was already earning the admiration of the court. Lamarck's principal protector was Buffon whose recommendations carried great weight with the Minister of the Maison du roi.[32]

After mid-century the case of Lamarck was unique, and the evidence suggests that successive ministries were confirming both the first and second nominees when votes were close or when Bignon or Maurepas might have chosen only the second candidate. The election of chemists H.-T. Baron and

30. AOP, B.1.2, fols. 153, 158–160. Delisle actually convinced Cardinal Fleury and others to persuade Maurepas, who had the final voice in such matters. Also, J. Marchand, "Le départ en mission de l'astronome J.-N. Delisle pour la Russie (1721–1726)," *Revue d'histoire diplomatique,* October–December 1929.

31. PV, 3 & 7 September 1729. Fouchy's later account (AAS, dossier 28 July 1784) contains some inaccuracies, but stresses the offensive behavior of Saurin *père.* Maurepas had been willing to confirm Clairaut as early as 1728, if the Academy so desired (BN, MS fr 22231, fol. 82). For Mahieu, see the unpublished study by E. Bonnardet in AAS, dossier Mahieu. Saurin *fils* (Bernard-Joseph) became a lawyer, poet, and dramatist.

32. Torlais, *Nollet,* pp. 26–31, and *Réaumur,* pp. 81–83. Also, AN, O¹*384, fol. 46. For Lamarck, see Hahn, *Anatomy,* p. 81; E. T. Hamy, *Les débuts de Lamarck* (Paris, [1908]), p. 27; Buffon correspondence of 1777, in AN, O¹*488, passim; and Joseph Laissus, "La succession de LeMonnier au Jardin du roi," *Comptes rendus du 91ᵉ congrès national des sociétés savantes,* Rennes, 1966 (Paris, 1967), Section des sciences, t. I, Histoire des sciences, pp. 146–147.

C.-F. Geoffroy in 1752, astronomers Jean-Sylvain Bailly and E.-S. Jeaurat in 1763, and Lavoisier and Jars in 1768 saw very close results in the balloting; all six became members, three as "supernumeraries." A fourth example of this kind dates from 1778 when both C.-M. Cornette and P.-E. Fontanieu received ministeral confirmation. At that time, however, the Academy chose to question the appointment of Fontanieu, and its reasons for doing so are discussed below in the context of the events of the 1770's. Here it is sufficient to note that ministerial confirmation of two names, when votes were close, was said by one academician to be the normal expectation in 1768.[33]

Governmental pressure and persuasion before an election were probably of regular occurrence, although in this area the evidence is fragmentary. In 1723, for example, Bignon and Réaumur seem to have been responsible for the election of Dufay; indeed, before he had attended a single meeting of the Academy, efforts were being made to promote him. But why Dufay was so esteemed at that stage in his career is not known; nor is it known how the voters were persuaded in his behalf.[34] When Buffon was elected in 1733, one of his friends remarked: "I have known for a long time that a post in the Academy of Sciences was being arranged for him."[35] Unlike Dufay, Buffon had presented two well-received memoirs to the Academy and had two nonentities for his competitors. At the same time, he was engaged in research of interest to Maurepas. These various factors all doubtless played a part in his election, but the nature and importance of ministerial pressure in his behalf remain obscure.[36]

If preelection pressures too often elude our grasp, two instances, both dating from 1770, show that subtle pressures were at times replaced by virtual orders to elect certain candidates. Two vacancies occurred among the astronomers in less than one month, and the Academy nominated, to fill the first post, Charles Messier and J.-D. Cassini (Cassini IV). In the letter confirming the election of Messier, St. Florentin announced that the king "would have simultaneously confirmed Cassini [for the second post, but] ... His Majesty had not wanted to disrupt regular procedures or

33. BN, MS n.a.f. 21015, fols. 198–199. The context here is a particular promotion, not an election.

34. "Lettres de Dufay à Réaumur," *Correspondance historique et archéologique,* V (1898), 306–309. Two of the four letters published here are now in AAS, dossier Dufay; all lack the year, but clearly concern his election and possible promotion. The latter did not take place in 1723, but the post remained vacant for a year, after which Dufay obtained it. For assessments of his abilities as scientist and courtier, see Fontenelle, *Oeuvres,* II, 386–7, 392, 396–7; BN, MS n.a.f. 21015, fols. 6–9; P. Brunet, "L'oeuvre scientifique de Charles François Du Fay (1698–1739)," *Petrus Nonius,* 3 (1940), 77–95.

35. Quoted in H. Monod-Cassidy, *Un voyageur-philosophe au XVIIIe siècle: L'abbé Jean-Bernard Le Blanc* (Cambridge, Mass., 1941), p. 478n.

36. Lesley Hanks, *Buffon avant l'"Histoire naturelle"* (Paris, 1966), pp. 39–42. Despite Buffon's claim, Fouchy was not hastily transferred to another class to make room for him in the Academy; cf. BN, MS fr 22229, fols. 219–220.

undermine in the slightest way the rules of the Academy, which [should] itself elect Cassini." A week later Cassini received an overwhelming vote.[37] Equally forceful were efforts in behalf of chemist Balthazar-Georges Sage, who is said to have attracted the interest of Louis XV himself. Sage's abilities and reputation were called to the attention of the Academy by one of the minister's underlings, who repeatedly urged the Academy to hasten its procedures and elect Sage before disbanding for its annual vacation. Although no unseemly or unlawful haste followed, Sage was indeed elected.[38]

The foregoing discussion of elections suggests that more evidence would merely disclose more intrigue. Until the 1770's, however, pressure applied by the ministry seems to have excited little anxiety among academicians, even when, as in the case of Sage, the candidate was thought to be mediocre.[39] Similarly, intrigue within the Academy itself was to arouse misgivings only at an equally late date.

The one electoral problem to attract attention as early as mid-century involved the judgment of merit. To seek men of merit had always been an acknowledged principle, but whether the principle was consistently applied in practice was a question called to the Academy's attention in 1759 by the Chevalier d'Arcy, by d'Alembert in 1769, by J.-C. Borda in 1770, by d'Alembert, Arcy, and Montigny in 1778, and by Borda and Condorcet in 1784. Although the proposals by these academicians treated several issues, all explicitly or implicitly raised questions about how and by whom merit was to be judged. So many talented men were cultivating the sciences, d'Alembert argued, that the Academy could afford to be more selective in its choices. The Academy, he added, had never demanded sufficient proof of talent, and it therefore stood in need of strengthened rules and new voting procedures.[40] Voting reforms were also advocated by other critics, who pointed out that the preferential ballots then in use often gave results detrimental to the most talented candidates. As Condorcet put it, the existing system was so defective that the Academy "winds up electing not the most worthy candidate, but the man whom the majority thinks not unworthy."[41]

37. PV, 14 July 1770. There were precedents for the kind of royal appointment avoided here: the Comte de Brancas in 1758, J.-F.-C. Morandi in 1759.

38. Letters of Mesnard de Chousy to Fouchy, AAS, dossier Sage. Also, P. Dorveaux, "Apothicaires membres de l'Académie royale des sciences. XI. Balthazar-Georges Sage," *Revue d'histoire de la pharmacie,* 23 (1935), 156–159.

39. A balanced assessment is in Henry Guerlac, "Sage," in *Dictionary of Scientific Biography,* ed. C. C. Gillispie (New York, 1975), XII, 63–69.

40. D'Alembert's two memoirs on this subject are in AAS, dossier d'Alembert and dossier Règlements; the latter is published in C. Henry, ed., *Oeuvres et correspondances inédites de d'Alembert* (Geneva: Slatkine Reprints, 1967), pp. 35–50. See Hahn, *Anatomy,* p. 132.

41. Condorcet's observations on a proposal by Borda, in *Hist. Acad.,* 1781 (1784), 31. See Duncan Black, *The Theory of Committees and Elections* (Cambridge, 1958), pp. 156–159, 178–180, 184, and Baker, "Débuts de Condorcet," pp. 254–255. For the ballots then in use, see R. Hahn, "Quelques nouveaux documents sur Jean-Sylvain Bailly," *Revue d'histoire des sciences,* 8 (1955), 343–344.

At the same time, more than one critic suggested that personal or professional prejudices and the disenfranchisement of associates had also had adverse effects in elections. Should associates vote in elections? Should the associates in the same class as the candidates be allowed to vote? Would prejudices within a class result in the names of talented candidates never being presented to the eligible voters?[42]

Although all these questions were debated during meetings of the Academy, not always could the members bring to a vote proposed reforms of the statutes. In any event, most members remained unconvinced that certain changes were necessary or desirable. The broadened franchise in particular was no more acceptable by 1785 than it had been in 1759 or 1769. A revised form of preferential voting was eventually adopted, however, and an effort was made in 1784 to reduce the possibly biased influence of each class in assessing the talent of nonmembers in the same scientific discipline.[43]

Such debates suggest not only that members were concerned with merit, but that merit, however poorly judged, had in fact entered into consideration throughout the century. And yet it is equally clear that other factors also played a decisive role in the minds of voters and ministers. Indeed, it is rare to find any discussion of merit in connection with particular candidates; instead, one is confronted by the common but empty phrase that a given candidate is "worthy" (*un homme de mérite*). For the most part, one may assume that candidates were sufficiently familiar to the scientific community to obviate the necessity for discussion, or that discussion was simply not committed to writing. Nonetheless, one wonders how the *honoraires* and other voters were persuaded of the abilities of candidates in sciences outside their own competence. In the absence of more concrete evidence, the best argument for supposing that merit was a vital criterion in elections is the fact that France's most eminent scientists were all members of the Academy. Admittedly, a good many mediocrities can also be found in the ranks, but this was no doubt inevitable while the available pool of scientific talent remained small; and the number of mediocrities was to decrease in time. By 1784, Lavoisier could rightly claim that "no illustrious savant had died without being a member of the Academy," even if some had not achieved membership as early in life as their abilities warranted.[44]

42. In addition to works cited above, see Bertrand, *L'Academie*, pp. 76–77; Hahn, *Anatomy*, pp. 130–132; and memoirs by Condorcet, Lavoisier, and Fouchy, in AAS, dossier 28 July 1784.

43. PV, 28 July 1784; Black, *Theory of Committees*, p. 180; Hahn, *Anatomy*, pp. 98–101.

44. Lavoisier MS, fol. 3v, in AAS, dossier 28 July 1784. The one exception which comes to mind is crystallographer Romé de l'Isle who was never elected. His reputation is discussed in *Journal inédit du duc de Croÿ 1781–1784*, ed. Grouchy and Cottin, 4 vols. (Paris, 1906–1907), III, 19; *Correspondance inédite de Condorcet et de Turgot 1770–1779*, ed. C. Henry (Paris, 1883), pp. 98, 117–118; Hahn, *Anatomy*, p. 181; and John G. Burke, *Origins of the Science of Crystals* (Berkeley, 1966), pp. 62–63, 80.

Vacancies in the two upper ranks of the Academy require a somewhat different treatment from those at the lowest level, since at no time in the period 1716–1785 did the Academy select an outsider to move directly into the upper ranks. On three occasions, outsiders were indeed confirmed as associates, but all had been the second choices of the voters. On the single occasion when an outsider was thought too distinguished for the lowest rank, the Academy proposed that he be named a "retired" (*vétéran*) associate. Furthermore, only once did the Academy wish to promote an assistant directly to the rank of pensioner.[45] In short, virtually all vacancies in the upper ranks were filled by members from the ranks immediately below.

While one historian has claimed that promotions were based on merit, another maintains that seniority was the normal criterion; still others have wondered why extraordinary merit did not bring rapid promotion.[46] In fact, of some 140 promotions in seven decades, more than three-quarters did occur in order of seniority; the extent to which this was the result of conscious policy remains to be determined. Most exceptions to the seniority pattern date from the Bignon-Maurepas era, when there is evidence that merit may have played some role in promotions. Seniority was also of importance in those early decades, but the adoption of a seniority system seems to date from about mid-century.

When seniority is calculated on the basis of a member's date of election, the records of the Bignon-Maurepas era reveal a considerable number of what came to be called *passedroits*: the passing over, or ignoring the "rights," of a member in order to promote someone with less seniority. A rough count puts the number of *passedroits* at twenty-five, or about 40 percent of all promotions before 1749; and the figure swells if one adds those outsiders named to the associate rank and the single assistant promoted to pensioner. After 1749 there were only six such cases, or about 8 percent. The great majority of *passedroits,* both before and after 1749, stemmed not from ministerial decisions but from the actions of the voters themselves; that is to say, the "irregularly" promoted academician was generally the Academy's first choice for the post.[47] That the number of cases diminished after mid-century may reflect a consistently higher level of talent in the Academy and an increased likelihood that members deserved promotion. At the same time, evidence to be presented below does reveal a shift of

45. The four outsiders named associates were Dortous de Mairan, Bouguer, Bertin, and Jean Darcet (PV, 17 December 1718, 1 September 1731, 9 May 1744, 31 March and 24 April 1784). *Vétéran* status was used only for members (not aspirants) who for some reason could no longer take active part in academic affairs. The unusual promotion was that of Jean Hellot (1739).

46. Above, n. 9; Bertrand, *L'Académie,* pp. 82–83; T. L. Hankins, *Jean d'Alembert: Science and the Enlightenment* (Oxford, 1970), p. 137.

47. These percentages are maxima. I have examined about half of the promotion lists in PV, especially those which other evidence suggested might be irregular; it is thus likely that the unexamined half contains few additional irregularities.

policy among the voters who grew ever more reluctant to assess the merits of their colleagues; instead they moved in the direction of automatic promotion on the basis of seniority. By 1769, d'Alembert could say that for a mediocre member to remain unpromoted was a mere "metaphysical" possibility. And Condorcet later remarked that to deny promotion to the senior associate in his class meant "a kind of censure" of that member rather than a genuine election.[48]

Circumstances surrounding the earlier *passedroits* suggest that the Academy was enforcing two policies: members must be productive, and they must reside in Paris. The two least active members, the luckless Mahieu and the abbé Terrasson, were repeatedly passed over in favor of such men as P.-L. M. de Maupertuis, C.-M. de LaCondamine, P.-C. LeMonnier, and d'Alembert. Chemist G.-F. Boulduc and physicians J.-M.-F. Lassone and J.-B. Senac had to spend much time at Versailles,while the Delisle brothers seemed to be extending indefinitely their mission to Russia; all were passed over in promotions, three of them repeatedly.[49] Although other *passedroits* are not as readily explained, there is no evidence that the membership felt any reluctance to ignore the claims of seniority. After 1749 the six cases of this kind fall into no discernible pattern, but it is worth noting that at least three of the six provoked some debate about the damage done to seniority rights.[50]

The idea that seniority constituted a claim to promotion was certainly present from a very early date, witness Fontenelle's comment in 1705 that to be an *élève* was tantamount to inheritance or survivorship (*une espèce de survivance*).[51] Fontenelle, however, was pointing not to length of service but to sheer membership, and this preference for members over outsiders can be considered a form of seniority. The same sentiment was made more explicit when, in 1725, an anonymous observer summarized Maupertuis' eligibility for promotion: "as a member, [he] deserves to be preferred to an outsider."[52] Once elected, members seemingly bore an indelible mark of merit.

Although seniority defined as length of service can be found in some early documents, emphasis on this criterion becomes apparent only after mid-

48. *Oeuvres . . . de d'Alembert,* p. 42. Condorcet MS, fol. [1r], in AAS, dossier 18 December 1784. Also, Jeaurat to Academy, 17 November 1779, AAS, dossier Jeaurat.

49. Cf. E. Bonnardet, "Jean-Baptiste Terrasson," *Bulletin de l'Oratoire de France,* 36 (October 1939), 328–337, and *Enseignement et diffusion des sciences en France au XVIIIe siècle,* ed. R. Taton (Paris, [1964]), p. 278. For the posts held by Boulduc et al., see the entries in *Index biographique.* For the Delisles, below, n. 66.

50. The five promoted members were d'Alembert (1756), Charles Bossut and Condorcet (1770), Condorcet (1773), Bossut (1779); the sixth, Darcet (1784), was named an associate without ever having been an assistant. The debates of 1756 and 1779 will be treated below; documents of 1779 imply that similar issues arose in connection with Bossut's promotion in 1770. Bossut (MS in AAS, dossier 1 December 1779) says *passedroits* always aroused lively protest; no such evidence has been found before 1756.

51. Fontenelle, *Oeuvres,* I, 127. Also, AOP, B.1.1, fol. 85.

52. AAS, dossier P. LeMonnier.

century. One manifestation of this change dates from 1752 when chemist G.-F. Rouelle made a serious bid to move from the lowest rank directly to the highest. The Academy often did include on lists of nominees for the rank of pensioner the name of one assistant, but reactions to Rouelle's candidacy suggest that the practice was only a matter of form. Rouelle based his campaign on his own merits, and such pretensions induced Jean Hellot—himself the sole example of a promotion of this kind—to advise Macquer, also an assistant, to enter into the competition. Significantly, however, Hellot counseled Macquer not to solicit promotion, but to ask for second place on the list of nominees; such modest behavior would offset the outrageous conduct of Rouelle. Macquer did follow this advice, but neither he nor Rouelle was nominated and the senior associate was promoted.[53]

Since both Rouelle and the second associate, P.-J. Malouin, engaged in this campaign, there was clearly some remaining hope that considerations other than seniority could influence promotions; but such hope, as clearly, was fading. What may well have been the last gasp in behalf of merit dates from 1756, when d'Alembert, the second associate in his class, requested promotion. Perhaps because he foresaw difficulties, d'Alembert in fact proposed that he either be moved to a new *class of physics,* elevated to "retired" pensioner, or made a supernumerary pensioner in his own class. When the Academy and ministry granted the latter request, it was on condition that he receive no pension until his seniors were "placed." There seems to have been no protest from Montigny who had just suffered a *passedroit,* but other academicians are said to have objected to what they deemed a violation of custom. The next year the ministry was to promote Montigny on grounds of his seniority.[54]

Increasing concern about seniority is especially evident in 1758 when, not for the first time, two assistants elected to membership on the very same date contended for a single vacancy at the associate level. Earlier cases of this kind had apparently passed unnoticed, while the Lalande-LeGentil affair of 1758 provoked soul-searching among some of the voters. Neither astronomer could claim to be senior and, furthermore, the votes had been tied when both were elected in 1753. Instead of attempting to evaluate the work of his fellow astronomers, Grandjean de Fouchy resorted to counting up the number of memoirs published by each and trying to decide if all the memoirs had really dealt with astronomy rather than mathematics or

53. Letters of Hellot and others, BN, MS fr 12305, fols. 193, 360, 374, 376–7; MS fr 12306, fol. 212; MS fr 9134, fol. 80; MS fr 23226, fol. 7. An unusually early (1719) reference to seniority rights is in AN, O¹*368, fol. 103.

54. PV, 24 & 31 March 1756; 7 & 14 December 1757. Condorcet, *Oeuvres,* II, 595; III, 74–75. Information and misinformation are in Hankins, *Jean d'Alembert,* p. 73; R. Grimsley, *Jean d'Alembert (1717–1783)* (Oxford, 1963), pp. 158–9; J. Mayer, "D'Alembert et l'Académie des sciences," in *Literature and Science,* Proceedings of the Sixth Triennial Congress, Oxford, 1954 (Oxford, 1955), p. 202.

physics; and he had to conclude, after all, that both had done what was required of them as members. He then went on to suggest a series of rearrangements which would allow both to be promoted. When the Academy nominated only J.-J. Lalande, Fouchy wrote to the Cardinal de Luynes, president for that year, to say that His Eminence's absence from the meeting had brought "malheur" upon the Academy because some members thought it important that Guillaume LeGentil be promoted simultaneously.[55] Whether additional maneuvering followed is unknown. Lalande was promoted.

The subsequent history of the idea of seniority adds little of novelty, although it does much to reveal the temper of academic life and the way in which considerations of seniority hardened within the Academy. To be sure, recognition that there was a seniority system dawned slowly upon some members who insisted that it was still possible to "strive" for promotion.[56] On the whole, however, promotion had become so automatic that even a flagrant case of charlatanism and academic bad taste, which earlier would have provoked "displeasure" and penalties, only aroused discussion but no punishment of the offender who was promoted when his turn came.[57] Nonetheless, there did arise on one occasion a genuine contest for promotion. This was so unusual by 1779 that the several competitors went so far as to present their credentials on the floor of the Academy. This unprecedented display sheds a harsh light on academic politics.

The vacancy of 1779 occurred among the pensioners of the *class of geometry,* and the declared candidates were all associates in the three "mathematical science" divisions: Bézout, Bossut, Jeaurat, and LeGentil.[58] Since Bossut was the only one of the four in geometry, and shifting from class to class had been prohibited since 1768, theoretically there should have been no contest at all. For obscure reasons, however, the Academy decided that Bézout, too, was a legitimate candidate for the post. Bézout, Jeaurat, and LeGentil all claimed seniority if they could be allowed a change of class.

55. Fouchy to [Luynes], n.d., and MS "Etat present de la classe astronomique," in AAS, dossier 1758.

56. Duhamel du Monceau, in Hahn, *Anatomy,* p. 131, and Malesherbes, in *Bullettino di bibliografia,* 18 (1885), 569; both Duhamel and Malesherbes were replying to proposals for reform.

57. See Dorveaux, "Sage," pp. 216–223, for details of Sage's behavior during and after 1778. Lexell, in Birembaut, "L'Académie vue par Lexell," pp. 156, 163–164, offers some explanation for the lack of any punitive action. For an early case of penalized misconduct, see Condorcet, *Oeuvres,* II, 85–87.

58. Relevant documents are in AAS, dossiers Jeaurat, LeGentil, and 1 December 1779 (MS by Bossut which gives some indication of Bézout's point of view). The order of events can be found in PV, 24 November through 7 December 1779, but the arguments in the dossiers were not transcribed into the minutes. Cf. Lexell, in Birembaut, "L'Académie vue par Lexell," pp. 151–152.

Jeaurat went on to hint at his own merit by giving the voters a list of his publications. And Bossut explicitly mentioned merit, only to dismiss the issue. The *règlements,* he remarked, permit the Academy to select freely, to name even an outsider to the upper ranks; but this option had been provided in the expectation that it would rarely be used. If the Academy departed from custom and promoted Bézout, he would no doubt fill the post competently, but competence was a "personal consideration" which should not concern the voters; indeed. to introduce such matters was to open the door to rivalry, bias, and arbitrariness. Seniority, he added, was by far the safer criterion, and he, Bossut, was the senior associate in geometry. The Academy did its duty and nominated Bossut, proposing at the same time that Bézout be made a supernumerary pensioner in his own class of mechanics. Both proposals were confirmed.

When the weighing of merit is said to arouse the worst passions, we have come a long way from the spirit of the *règlements* and from Fontenelle's image of the objective scientific intellect. In 1785, however, Lavoisier showed himself to be well aware of the concerns of his colleagues when he made every effort, in his proposed reform of the Academy's structure, to see that no member would feel his seniority rights to have been prejudiced. That he succeeded to a remarkable degree helps to explain why the reforms were approved.[59]

After mid-century, knowledge of the Academy's seniority system spread among both nonscientists and aspirants to membership. In three instances, unsuccessful candidates for election were even to claim that they had "rights" to a seat in the Academy because they had been candidates for more years than their competitors. A fourth, Jeaurat, was elected after a campaign in which he described himself as "antedating all my competitors as much by my publications as by my solicitations for membership." According to Grandjean de Fouchy, such arguments did carry weight with some members.[60] Among nonscientists d'Alembert played a major role in publicizing the seniority system. As a supernumerary pensioner, he expected to receive automatically the pension vacated by the death of Clairaut in 1765. When St. Florentin proved to be dilatory, d'Alembert informed his friends and the public at large that seniority alone, apart from his abilities,

59. Lavoisier, *Oeuvres,* IV, 574–577, 581–582. Even in 1791, when revolutionary orthodoxy required that talent be rewarded, seniority and other factors continued to weigh more heavily in promotions; cf. AAS, dossier A.-L. de Jussieu, MS dated 7 December 1791 and signed by Jussieu, Jeaurat, and Portal.

60. AOP, B.1.8, fol. 174. Also, Fouchy MS, AAS, dossier 28 July 1784; PV, 11 March 1778 (Demachy); AAS, dossier Descemet, and PV, 1, 8, & 12 February 1783; R. Davy, *L'apothicaire Antoine Baumé (1728–1804)* (Cahors, 1955), p. 64, and P. Dorveaux, "Apothicaires membres de l'Académie Royale des Sciences. XII. Antoine Baumé," *Revue d'histoire de la pharmacie,* 24 (1936), 347.

ought to have secured him the pension. Unaware of bureaucratic complexities, d'Alembert's friends were ready to believe his claim that he was a victim of persecution.[61]

As the seniority system hardened, occasional voices of criticism were to be heard in behalf of the idea of promotion as a reward for talent. Not surprisingly, those who espoused this cause belonged to or sympathized with the small contingent of philosophes in the Academy. Bossut himself, during the debate of 1779, was careful to explain that the seniority system was safest given the Academy's structure, but that a reformed structure would allow the use of more suitable criteria for advancement. Despite earlier emphasis on his own seniority rights, d'Alembert, too, questioned the system when in 1769 he proposed reforms in the Academy's organization. Indeed, both d'Alembert and a leader of the opposition at the time, the abbé Nollet, admitted that merit was the better criterion for promotion. But when d'Alembert, perhaps to mollify his audience, declared that most members were talented and would not, in a reformed Academy, be permanent occupants of a low rank, he confirmed his opponents in their belief that the system was not in need of reform.[62]

Thus far, discussion of promotions has centered around the policies of the Academy rather than the ministry, for the evidence suggests that in this realm the Academy enjoyed considerable autonomy. Although ministerial pressure may have been exerted in this area from time to time, such interference was apparently rare. If we look at the careers of court favorites, men for whom pressure ought to have been applied, it is striking that they did not achieve more rapid promotion than their colleagues; on the contrary, those who spent much time at Versailles sometimes had their promotions delayed or denied. Furthermore, the ministry seldom adopted the expedient of confirming the Academy's second choice in promotions; it is worth noting, however, that most such examples date from the Bignon-Maurepas era.[63]

The reasons behind relative ministerial indifference remain obscure. In 1773, St. Florentin was to explain to an ambitious academician that mem-

61. See any edition of d'Alembert's correspondence of 1765 with Voltaire, Catherine II, and Frederick II. Also, Lagrange, *Oeuvres,* XIII, 38–40, 42, 46, 48. D'Alembert to *Journal encyclopédique,* 15 August, 1 October, 15 December 1765, and in L. Bachaumont, *Mémoires secrets,* 36 vols. in 18 (London, 1784–1789), II, 249, 280. Biographers usually adopt d'Alembert's view of the affair. One reason for St. Florentin's delay was the fact that d'Alembert's claims were being disputed by Arcy and Vaucanson, and the Academy had to debate the issues. PV, 18 May, 3, 7, & 14 August, 16 & 23 November 1765. Also, AN, O[1]*407, fols. 274, 320–321; BN, MS n.a.f. 9544, fols. 2–3, and n.a.f. 21015, fols. 164–165, 167–168. Diderot was one of the few outsiders aware of some of the complexities; cf. Mayer, in *Literature and Science,* p. 204.

62. D'Alembert and Nollet MSS cited above, nn. 22, 40.

63. Promotions of Couplet *fils* (1717), Louville (1719), J.-L. Petit (1725), LaCondamine (1739), and Cassini III (1745). For Jacques Vaucanson, see below, nn. 64, 73. Court favorites will be treated below in the context of the residence requirement.

bership in the Academy was a sufficient distinction for any savant, and it may be that the several ministers believed rank in the Academy to be less important than the fact of membership. More positive evidence reveals that when the Academy began to emphasize the seniority system, the ministry endorsed and upheld this policy. In fact, St. Florentin twice felt obliged to remind the Academy that senior associates had claims to promotion.[64] It seems reasonable to conclude that, by adopting the Academy's policy, the ministry in effect limited its own opportunities to intervene in promotions.

Before attempting to generalize about the Academy's evolution, two more aspects of electoral practice must be discussed: the residence requirement and the appointment of supernumeraries. Both can be used to reveal the extent to which the members and the ministry thought it necessary to enforce the *règlements,* and both have some bearing on the problem of ministerial interference in the Academy's internal affairs.

As noted above, members of the working ranks were supposed to reside in Paris, to attend meetings regularly, and to give evidence of continuing research. Prolonged absence was permissible for such activities as scientific missions, provided that prior consent had been obtained and the absence was of some fixed duration. By 1716 several members had been "excluded" for failure to live up to one or another of these provisions. After 1716, however, exclusions were virtually unheard of, and normal procedure became to retire or "veteranize" absentees, a practice which allowed them to retain some of the privileges of membership. Alternatively, an absentee might be passed over in promotions; increasing reluctance to inflict *passedroits* meant that this expedient was rarely used.[65]

The policies just outlined were, in fact, applied very erratically to all types of absentees. Courtiers L.-C. Bourdelin, Malouin, and L.-G. LeMonnier became pensioners, while G.-F. Boulduc, C.-C. Angiviller, Lassone, Senac, and Joseph Lieutaud all retired before reaching that rank. Among the voyagers, LeGentil was veteranized after an absence of ten years, the Delisles after sixteen, and Joseph de Jussieu after fifteen; at the same time, candidates might be elected to the assistant level while they were away on long missions or when it was known that their departure was imminent.[66] The

64. Cf. promotions of Montigny (PV, 14 December 1757) and Vaucanson (PV, 16 & 23 November 1765, 20 May 1768). Also, letters of St. Florentin in PV, 29 July 1772, and in AN, O[1]*415, fols. 939–940. St. Florentin became Duc de La Vrillière in 1770; his earlier title is retained here to avoid confusion.

65. Hellot MS, pp. 14–16. A rare exception to the residence rule at the very outset was Denis Dodart in 1699, but this was not to set a precedent and did not help Joseph Sauveur's case (PV, 28 February 1699).

66. For absence on mission, see the stipulations in the Delisle brevet, 22 June 1725, AAS, dossier J.-N. Delisle, and Marchand, "Le départ en mission de Delisle." In 1733 a post of pensioner was being "held" for Delisle should he return from Russia; when he delayed, he was veteranized (BN, MS n.a.f. 9186, fols. 10–11). Also, Fouchy MS, AAS, dossier 1758, and A. Chevalier, *La vie et l'oeuvre de René Desfontaines* (Paris, 1939), pp. 32–33, 36. Dombey was

anatomist Bertin was retired shortly after an apparent nervous collapse, while J.-B. Winslow remained on the active roster for some time after his health had failed. Other absentees, among them Buffon, Alexis Fontaine, and d'Alembert, seem never to have been threatened with retirement.[67] In the majority of cases, it is likely that the retirement of absent members was hastened or slowed depending on the availability of other attractive candidates whom the Academy wished to elect or promote. This is evident in 1758–59, following the death of Antoine de Jussieu and the retirement of his brother Joseph; after a dozen years of unchanging composition, the class of botany could be reshuffled to acquire four new young members. Similarly, the retirement of LeGentil in 1770 enabled the astronomers to acquire Cassini IV.[68]

The ministry was quite as erratic as the Academy in enforcing the rules. During the Bignon-Maurepas era, there was some concern about having too many astronomers away on missions at the same time, and the Academy was twice instructed to bring the active roster to full strength by electing supernumeraries. But no objection was raised to the election of Joseph de Jussieu in 1742, although he had already been in South America for seven years. That no one seems to have queried the absence of Buffon and d'Alembert is hardly surprising. Presumably d'Alembert was too eminent, too much in the public eye, and too likely to voice his complaints. Buffon was not only too eminent, but he was a ministerial favorite; the problems created by his performing his duties as treasurer from his estate in Burgundy were quietly resolved in 1772 by the appointment of Mathieu Tillet to serve as his adjunct. Despite such examples, ministerial laxity remains inexplicable, and Condorcet may have been a wise diagnostician in saying that as long as academicians produced *bons ouvrages,* "a milder and wiser administration [than that of Louis XIV] seems to have relaxed the strictness of the rules."[69]

The appointment of supernumeraries has been variously described as a vehicle for royal interference and as a sign of stress in the Academy's structure. Disregarding the unique circumstances of 1716, the first supernumerary was named in 1731 and was to be followed by twenty-three more, the last of them in 1784. Among the earliest such appointments, two fall into a class of their own, the Academy having been told to elect additional mem-

---

advised to seek election while on mission (Laissus, "La succession de LeMonnier," p. 152). Jussieu was retired on condition that he remain eligible for reinstatement on his return, and LeGentil was to be reinstated in 1772.

67. Gabriel Bory proposed in 1788 that the Academy adopt guidelines in allowing members to become *vétérans* (Chapin, "Les associés libres," p. 13), and his proposal resembles rules already in effect in the Academy of Inscriptions.

68. For the astronomers, see Fouchy to [St. Florentin?], n.d., AAS, dossier 27 June 1770. Also Fouchy, MS "Etat present de la classe anatomique," AAS, dossier 1758.

69. Condorcet, *Oeuvres,* II, 87.

bers because too many astronomers were away on missions.[70] A second group of three was appointed because of electoral problems; in one instance, some confusion had attended the counting of ballots, and in the others votes had been tied.[71] Seven received their appointments by request of the Academy,[72] and seven more were the result of ministerial decisions to confirm the Academy's first and second choices to fill the available vacancies.[73] Of the five thus far left unaccounted for, one was Jean Darcet whom the Academy thought too distinguished for election to the lowest rank; the minister thereupon decided that supernumerary associate would be suitable. Two, Tillet and Condorcet, were made supernumerary pensioners when they became permanent officers of the Academy. Finally, L.-G. LeMonnier was promoted at ministerial insistence, and Toussaint Bordenave was a royal appointee not nominated by the Academy. In short, only in the last two cases can the ministry be described as having exercised unusual interference in the normal workings of the Academy; and, as we shall see, these two appointments were to arouse opposition.

While all ministers were willing to create supernumeraries, Maurepas recognized at an early date that the practice ought to be restrained lest the careful structure of 1716 be distorted.[74] In 1778—under circumstances discussed below—Amelot was to promise that he, at least, would no longer initiate such appointments. But the anxiety expressed by Maurepas was not to recur to ministers or members until 1784, after five supernumeraries were created within twelve months. The limited number of seats in the Academy had always meant that rapidity of promotion depended on the mortality or retirement rate among senior members, and some academicians of the 1780's had grown impatient to reach pensioner rank.[75] Earlier in the century, one could hope to rise rapidly by shifting from one class to another, but this practice had been halted by royal order in 1768.[76] Thereafter, the

70. PV, 7 April 1731, 25 June 1735. Fouchy and Cassini III were elected.

71. PV, 28 March 1733, 29 July 1752, 19 & 29 January 1763. Clairaut, C.-F. Geoffroy, and Jeaurat were named.

72. In chronological order, d'Alembert (1756), Arcy (PV, 27 June 1770; AAS, dossier 27 June 1770; Bossut MS, AAS, dossier 1 December 1779; AN, O[1]*412, fols. 325, 356), Bézout (1779), LeGentil (PV, 20 February and 4 March 1782), M.-J. Brisson (PV, 4 December 1782), René Desfontaines (AAS, dossier 1783, *plumitif des séances,* 26 February), Bailly (PV, 11 December 1784).

73. Hellot (1739), Montigny (1757), Lavoisier (1768), Deparcieux (1768), Angiviller (PV, 29 August 1772), Fontanieu (1778), Périer (PV, 19 March 1783). Angiviller and Fontanieu will be treated below. Deparcieux was the Academy's first choice in 1768, but Vaucanson, the second nominee, had seniority and had lost an earlier bid for promotion in 1765.

74. BN, MS fr 22229, fols. 219–220.

75. Most requests for promotion came from the associates; a typical explanation is that by Brisson, in PV, 4 December 1782.

76. The shifting of classes was not as often the road to promotion as was then assumed. There had been thirty-two shifts in the period 1719–1768; only eighteen meant a change in rank. Twelve of the eighteen passed over members with greater seniority, but the latter were almost invariably men bypassed more than once. For continuing debate on the possibility of

only way to rapid advancement became to request supernumerary status. Such requests might of course be denied, but denials were rare and a sufficient number was granted after 1780 to make plain to the interested parties that the practice was becoming an abuse. The Baron de Breteuil, perhaps weary of special pleading, suggested in 1784 that future requests of this kind should not be forwarded to him unless approved by two-thirds of the voters. A more complex solution was then offered by Condorcet who proposed that an associate be forbidden to ask for the rank of supernumerary pensioner unless he had been an associate for twenty years. In the end, intricacies were abandoned, and a committee of academicians asked the minister to forbid the Academy to ask for supernumeraries. This prohibition became part of the new *règlements* adopted in 1785.[77]

Surveying the history of the Academy's electoral practices, a few clear patterns emerge. In the election of new members, the rules required—and everyone agreed in principle—that candidates demonstrate their abilities. In fact, however, a variety of pressures often did influence the voters. That such pressures tended to be of a personal kind—that is, rivalries, loyalties, the use of influence—is hardly surprising when it is recalled that the few surviving accounts stress that personalities rather than scientific issues dominated meetings of the Academy. Further research would probably reveal that philosophical issues and professional differences also played their part in the outcome of elections. Although there is evidence that merit was not ignored, the relative importance of this criterion cannot be determined with any precision. At the same time, in filling the two upper ranks, the Academy did at first show some willingness to weigh merit or at least the proper performance of academic duties; after mid-century, however, members were emphasizing seniority and avoiding debate on the abilities of their colleagues. Bossut was doubtless correct in thinking that the seniority system reduced strife within the Academy, but the system also effectively stifled the spirit of the *règlements*.

The Academy's relationship to its protector also underwent some change during these decades. Broadly speaking, it is likely that ministerial interference was more frequent than we know, and it is apparent that such interference was more characteristic of the Bignon-Maurepas era than afterwards. Denying the Academy's first choice was an expedient used rarely throughout the century, but most instances antedate 1749. Preelection campaigning probably occurred regularly under every ministry, but the surviv-

---

shifting classes after 1768, see AN, O[1]*412, fols. 325, 356; Bossut MS, AAS, dossier 1 December 1779; PV, 7 December 1779.

77. Breteuil in PV, 24 April 1784, and quoted in Bertrand, *L'Académie,* pp. 71–72. Condorcet's proposal included other conditions, one of them a two-thirds vote (AAS, dossier 18 December 1784), A Breteuil proposal of late 1784 is in Lavoisier, *Oeuvres,* IV, 592–3. The minister's attitude was provoked by the effort of Charles to become a supernumerary assistant; for the refusal of requests by Charles and Antoine Baumé, see AN, O[1]*495, fols. 43–44; 489, fol. 91. The committee's solution is in Lavoisier, *Oeuvres,* IV, 570.

ing evidence suggests that here, too, the practice was more common before 1749.

To judge from the content and the tone of early documents, the Academy enjoyed a close working relationship with Maurepas and especially with Bignon; put another way, ministerial surveillance was constant and was seemingly accepted without resentment by the savants. The later decades, by contrast, might be characterized as a period of relative ministerial compliance. Not only did the ministry adopt the seniority system developed by the Academy, but the number of times that the Academy's first and second nominees were both confirmed may indicate reluctance to dispute the Academy's decisions or to exercise independent judgment. Similarly, while Amelot promised to create no more supernumeraries, such requests originating in the Academy generally received approval. Indeed, this decline in active surveillance was to suggest to the aging Grandjean de Fouchy that the Academy had long since entered into an agreement with the ministry by which the voters' expressed preferences would always be granted; Fouchy's memory was unreliable, but his statement seems an accurate reflection of the atmosphere in which the Academy functioned.[78] Compared with Bignon and Maurepas, only Malesherbes among the later ministers seems to have combined an informed judgment in scientific matters with an intimate knowledge of academic affairs. If Argenson was a cultivated patron of letters, St. Florentin, Amelot, and probably Breteuil were uninspired courtiers. It is thus understandable that the initiative should have passed to an Academy willing to regulate itself.

When ministerial interference became rarer, and often cruder, such interference was to provoke crises and protests. The first disturbance of this kind dates from 1758 and was an isolated affair, while the decade of the 1770's was marked by a series of conflicts between the Academy and the ministers of the Maison du roi. Were the ministers acting as "despotically" as critics alleged? On what grounds did the Academy defend itself? Did these conflicts alter the relationship between Academy and crown?

The crisis of 1758 began when St. Florentin urged that L.-G. LeMonnier be promoted to supernumerary pensioner. Aware of LeMonnier's absenteeism, St. Florentin privately admitted that the Academy's rules were thus "not favorable" to him, while Fouchy, after rehearsing possible precedents, declared to a colleague: "It would perhaps be unfair always to enforce the rules rigidly. But it would be a mockery of the law to go so far as to think that it does not apply to an academician who for six years has sent [the Academy] no work, especially when his competitor is as faithful and hardworking as M. Guettard."[79] Fouchy's nice distinction between bending and breaking rules was not ignored by St. Florentin, who promised the

78. AAS, dossier 28 July 1784.

79. Fouchy to [Morand?], n.d., AAS, dossier 1758. St. Florentin to LeMonnier, 2 August 1758, AN, O¹*400, fols. 454–455. Also, PV, 22 & 29 July, 5 August 1758. A draft of the

Academy that the LeMonnier promotion would set no precedent. If LeMonnier and St. Florentin got their way, it can hardly be a coincidence that the next year was marked by the retirement, on grounds of absenteeism, of two courtier-academicians.[80] Eventually, however, all parties returned to the usual erratic enforcement of residence rules, and more than a decade of calm was to follow before the Academy faced another objectionable example of ministerial intervention.

Conflicting precedents, complicated by other issues of principle and personality, marked the crises of the 1770's. The first of these is well known since it involved the career of Condorcet. Fouchy proposed to St. Florentin in 1773 that Condorcet be made his assistant, to succeed fully to the post of secretary when Fouchy should retire or die. St. Florentin approved and directed the Academy to discuss the suitability of this appointment. Votes were divided on the issue, with a vocal minority, supporters of the candidacy of astronomer J.-S. Bailly, protesting that for the Academy to approve an appointment which already had royal consent was a travesty of an election. Condorcet did obtain the post, but in 1776, upon Fouchy's retirement, asked that a new vote be taken in order to test freely the Academy's wishes. Although Bailly and his friends were present at the new election, no rival candidate was proposed and Condorcet's appointment was unanimously endorsed.[81]

Three features of this conflict are especially pertinent here. In the first place, the Academy itself was divided by personal and philosophical differences, with the majority supporting Condorcet and St. Florentin. Secondly, the minority invoked what were deemed to be the relevant precedents, and discovered that Fouchy and his predecessor, Dortous de Mairan, had each been the sole nominee presented for the crown's approval; they pointed out that Condorcet, by contrast, was the crown's nominee presented for the Academy's approval. St. Florentin, however, realized that a new precedent had been established in 1772 by the appointment of Mathieu Tillet as assistant to the treasurer. Tillet's appointment had been suggested to the minister by Buffon, and the Academy had then given its approval without murmur.[82] Academicians were thus not on the safest of grounds in citing precedent, a matter which may help to explain the third important

---

Academy's protest, probably by Morand, is in AAS, dossier 29 July 1758. LeMonnier and Guettard had equal seniority.

80. For the retirement of Lassone, St. Florentin to Fouchy, 3 February 1759, AAS, dossier Lassone; for Lieutaud, Condorcet, *Oeuvres,* II, 401, 404, and AAS, dossier Joseph Lieutaud.

81. Baker, "Débuts de Condorcet." The Condorcet letter sought by Baker (p. 254n) exists in a draft in BN, MS n.a.f. 5151, fol. 92. For Macquer's role see Ahlers, "Un chimiste," pp. 61–63.

82. PV, 12 & 19 December 1772; AAS, dossier 1772. The precedent of Tillet, invoked by St. Florentin, is not mentioned by Baker in his otherwise admirable analysis. For Tillet's reputation, see Lexell, in Birembaut, "L'Académie vue par Lexell," pp. 161–162.

feature of this affair: the fact that, apparently for the first time, members asserted their right to freedom of choice in elections. This claim, however, was so much the position of a frustrated minority that it should not be taken to represent the principles of most members in 1773. Indeed, it was Buffon, responsible for Tillet's appointment, who led the protest in behalf of liberty, while, ironically, the nucleus of philosophes in the Academy was using that kind of political pressure which, in principle, it deplored.[83]

Once launched, the claim to freedom of choice in elections had a brief but dramatic career. When surgeon Toussaint Bordenave, who had not been nominated for an existing vacancy, was appointed supernumerary assistant in 1774, the membership united in its protest.

> The Academy has seen with sorrow the appointment of Bordenave to a post of assistant in Anatomy without his having been nominated by the Company. By unanimous decision, the Academy asks [St. Florentin], in his capacity as academician, president [of the Academy], and Minister, to make known to His Majesty that this appointment is wholly unprecedented in the Academy's history and entirely contrary to the rules which the Monarch himself has given the Company and which have always been observed; and that, besides, such an appointment henceforth makes voting pointless and, in interfering with the liberty of elections, heralds the destruction of the Academy.[84]

This time the precedents were unambiguous, and Bordenave was shifted to "veteran" status two weeks later. It is to be noted, however, that another court favorite, Angiviller, had recently (1772) been made a supernumerary associate without ever having been an assistant; but Angiviller had been the Academy's second nominee for the post, and the appointment aroused no comment. Worth noting, too, is the fact that Angiviller's nomination had been engineered, in the expectation that he would be confirmed. Despite the Academy's assertion of its liberties, one may reasonably suspect that had pressures succeeded in placing Bordenave's name on the list of nominees, there might have been no crisis in 1774.[85]

The third rebellion of the 1770's once more involved the position of L.-G. LeMonnier. A supernumerary pensioner, he could normally have expected to move to regular pensioner status upon the death of Bernard de Jussieu in 1777. The absenteeism of LeMonnier, the terms of the Academy's protest, the reply of Amelot—all were virtually the same as in 1758. Again the ministry was temporarily victorious, although later in the year another court

83. That Condorcet was not himself entirely devoted to the Academy's rights and liberties is made clear by Baker, "Débuts de Condorcet," esp. pp. 253–254, 259. See also the remarks on Arcy in Condorcet, *Oeuvres,* II, 387, and quoted in Hahn, *Anatomy,* p. 131, also p. 134.

84. Fouchy MS, n.d., AAS, dossier Bordenave. Also, PV, 16, 19, & 26 March 1774, and BN, MS fr 9134, fol. 99. For Macquer's role see Ahlers, "Un chimiste," pp. 65–69. Condorcet, *Oeuvres,* II, 542, later misrepresented the issues in his eulogy of Bordenave.

85. LeRoy to X, n.d. [1772], AAS, dossier LeRoy. Macquer (B.N., MS fr 9134, fols. 100–101, 133–134) implies that there was so much pressure before the election that the voters feared to nominate Bordenave lest the same happen in future.

favorite and absentee was to retire from active membership and LeMonnier was to do the same in 1779. The brief flirtation with academic liberties in 1773 and 1774 had given way to the Academy's more customary citation of rules and precedents.[86]

The last flurry of the decade took place in 1778 when Cornette and Fontanieu were nominated to a vacancy in the *class of chemistry* and both were confirmed, Fontanieu as supernumerary. Although others had been named in similar fashion, the Academy decided to protest against the appointment of a supernumerary. Fontanieu was veteranized, and the minister promised to refrain from such appointments in future.[87]

Reviewing the several crises of the decade, it should now be apparent that only the Bordenave affair is a clear case of unprecedented ministerial interference. And it is the one case in which the Academy was unanimous in claiming that its liberties were in jeopardy. Uniquely true of the Bordenave affair, in addition, is the rapidity with which the minister retreated. By contrast, the debates about Fontanieu continued for many months, while the appointment of Condorcet and the promotion of LeMonnier can be described as temporary ministerial victories followed by solutions acceptable to the Academy. Indeed, a glance at the years after 1778 reveals that relations between the ministry and the Academy were little affected by the crises just discussed. The year 1779 saw the confirmation of Lamarck, the Academy's second choice, and this was accepted with the normal lack of protest. Six supernumeraries were confirmed, five of them by request of the Academy.[88] And no one seems to hae mentioned the residence rule when botanist René Desfontaines was elected, although he was known to be planning a long foreign expedition. In short, the Academy had reverted to normalcy, and the ministry, except in the case of Lamarck, gave evidence of return to compliance with the Academy's wishes.

In view of the Academy's behavior before 1773 and after 1778, one must inevitably ask why there was so concentrated a period of rebellion and what meaning is to be attached to the idea of academic liberties. The fact that these crises arose in the 1770's suggests a parallel with public sentiment about "liberty" and "despotism" in the years after the Maupeou coup of 1771. Indeed, one observer specifically applied these political terms to the

86. For the retirement of Angiviller, PV, 6 & 10 December 1777; AAS, dossier Angiviller; BN, MS fr 12305, fol. 15. For LeMonnier's efforts not to retire in 1777, and the sequel: PV, 15 November 1777, 19 December 1778; AAS, dossier 1777, entries for 29 November in the *plumitifs* by Condorcet and Lavoisier; AAS, dossier L.-G. LeMonnier, letter to his brother, 28 November 1777; BN, MS n.a.f. 5152, fols. 289–290.

87. PV, 14 & 18 March 1778; 20, 27, & 30 January 1779.

88. See the lists given above, nn. 72, 73. After 1778 only J.-C. Périer may not have been a case of academic request, but no evidence on this subject has been found. His inclusion in n. 73 (rather than n. 72) was an educated guess.

Bordenave affair.[89] Such language, however, seems not to have entered into academic vocabularies—a circumstance true even in the Académie française, dominated by the philosophes, during bitterly contested elections in 1772.[90] Unlike the literary institution, the Academy of Sciences included few known philosophes and many opponents of its own leading philosophe, d'Alembert. Furthermore, had there been an important heightening of political awareness among the savants after 1771, such sentiments ought to have left some trace in connection with the appointments of Tillet and Angiviller the next year. No such trace has been found. However provocative the political parallel, there is no evidence that these large issues affected thinking about the Academy's internal affairs during this decade. In fact, a disenchanted visitor to the meeting rooms in the Louvre was to remark in 1773: "It is a pity . . . that personal rivalries are more important there than matters of science, and that people hardly listen to one another, so much is each concerned with his own aims, with his own business, and with attending meetings only out of fashion or self-interest."[91] Narrowly personal concerns may indeed help to explain the acceptance of the Tillet and Angiviller appointments (both men were well liked), just as they help to explain some of the antagonism toward Condorcet.

If academicians were often self-centered, they also were able on occasion to enlarge their vision in defense of their institution. Except in 1774, the specific issue in each crisis was, in fact, quite insignificant: the residence of pensioners, royal appointment of the secretary, the creation of a supernumerary. When seen against a background of past ministerial behavior and academic willingness to bend the rules and to submit to pressures, the selection of such matters as grounds for protest makes little sense. These crises do become more comprehensible if viewed in the light of growing ministerial neglect after 1749. The academicians of the 1770's were protesting less against specific ministerial actions than against the fact that the ministry was once more becoming very active and exerting pressures to which the members were no longer accustomed. What is more, the relatively delicate maneuvers of Bignon had been replaced by the bluntness of St. Florentin.

When texts and contexts are examined, it is clear that the Academy did not demand that complete independence which *la liberté des suffrages* im-

89. Bachaumont, *Mémoires secrets,* XXVII, 209–210. Details are garbled in this account. A greater variety in public sentiment after the coup than is usually presented by historians is indicated by D. Hudson, "In Defense of Reform: French Government Propaganda during the Maupeou Crisis," *French Historical Studies,* 8 (1973), 51–76.

90. An excellent account of the royal veto of the elections of Jacques Delille and J.-B.-A. Suard, and the sequel, is in Lucien Brunel, *Les philosophes et l'Académie française au dix-huitième siècle* (Paris, 1884), esp. pp. 244–260.

91. *Journal de Croÿ,* III, 54–55.

plies. The only occasion on which "liberty" was emphasized by more than a few members was in connection with Bordenave, and the word was then intended to mean both freedom from undue pressure and the necessity to abide by the rules and precedents. The latter emphasis is more characteristic of other crises, when academicians were asserting their right to be the interpreters of the *règlements* and of precedents. Within these guidelines, whose existence guaranteed a degree of academic autonomy, members were well aware that pressure would still be exerted by their colleagues, by aspirants to membership, and by the ministry. Indeed, when three academicians—two of them with reputations for radicalism—tried to find remedies for the exercise of "authority" by ministers, they could only suggest that the Academy should be careful to nominate men of talent so that no minister would be tempted to use his powers or judgment.[92] Clearly, then, the Academy's objections were not to pressure, but to excessive pressure; and the only effective recourse was to insist on the rules and precedents. Clearly, too, the binding force of the *règlements* was meant to apply more to the ministry than to the Academy itself; the best illustration of this attitude is to be found in the issue of supernumeraries, with the academicians requesting such appointments and the ministry denied the initiative in these matters.

The royal bureaucracy probably deserves some of the credit or blame for inadvertently teaching the academicians about their rights. It will be recalled that *la liberté des suffrages* was being defended by Bignon as early as 1726. At times, to be sure, such pronouncements were merely polite refusals to exert the influence the ministers knew they possessed. On the other hand, if scientific competence were to be judged, Bignon and Maurepas knew the academicians to possess more expertise than they, and they therefore realized that a certain freedom of judgment was essential in the Academy. Above all, ministers knew it to be their duty to enforce the provisions of the *règlements,* and they were careful to inform the Academy whenever particular irregularities were intended not to constitute precedents. In a sense, then, ministerial notions of academic liberties were remarkably similar to those expressed by the academicians. To exert pressure was considered normal, while undue pressure or departure from precedent was a matter requiring special explanation. These lessons may well have been reinforced in the

92. D'Alembert, Arcy, and Montigny, 1 April 1778, in Bertrand, *L'Académie,* pp. 76–77. This document is unique in also attacking the problem of electioneering within the Academy itself. The original text cannot now be located, but PV, 1 April 1778, reveals that the debate centered not on elections to the working ranks but to the categories of *associé libre* and *associé étranger.* When the Revolution was some months old, Cassini IV suggested another solution to the problem of authority: the Academy should present only one nominee, to be confirmed or rejected by the king. This was the system in use in the Académie française where rejections had occasionally aroused both academic protest and public agitation. See R. Hahn, "L'Académie royales des Sciences et la réforme de ses statuts en 1789," *Revue d'histoire des sciences,* 18 (1965), 15–28.

1750's by the reading to the Academy of Jean Hellot's history of the application of the *règlements* from 1699 through about 1753. Certainly, the first LeMonnier affair, in 1758, revealed that some savants had acquired a good working knowledge of rules and precedents.

By the 1780's, both the Academy and the ministry might be diagnosed as victims of arteriosclerosis. The Academy's repeated refusals to countenance a variety of reform proposals were usually couched in terms of predictions that to change the rules would impair the quality of the institution.[93] On a lower level, members did not hesitate to remind each other that reform might endanger seniority rights. Simultaneously, in using precedents to defend its growing autonomy, the Academy appealed to the letter of the law and the inviolability of custom. Self-interest, the weight of custom, a narrow legalism, and a sense of corporate and individual rights pervaded the membership during the decade before the Revolution. It may seem ironic that some earlier concern for merit had become a minority cause, and a radical one, even while the intellectual climate of the Old Regime was moving in the opposite direction. The Academy, however, had gone through a process not unknown in other facets of life under the Old Regime: a young institution, capable of some flexibility and some spirit of innovation, had become an established corporation. That the ministry, too, was suffering from the same complaint is evident from St. Florentin's remark to an importunate academician: "one must introduce nothing new."[94] No minister—and certainly no one of the caliber of St. Florentin—could follow the example of Bignon in 1699 and arrive with a new set of rules in hand. Academicians instead had to be wooed into accepting certain changes, and in 1785 Breteuil and Lavoisier appealed to the members' self-interest and emphasized that to reject a new *règlement* would be to spurn a token of royal generosity toward a valued corporation.

93. For a reasoned defense of certain electoral practices, see Lavoisier, AAS, dossier 28 July 1784.

94. St. Florentin to Arcy, 20 May 1770, AN, O[1]*412, fol. 356.

IV

STRUCTURE AND FUNCTION IN EARLY MODERN SCIENCE

EARLY CONCEPTS OF THE MICROVASCULAR SYSTEM: WILLIAM HARVEY TO MARSHALL HALL, 1628–1831

GENEVIEVE MILLER

KNOWLEDGE about the structure and function of small blood vessels has evolved during the three and a half centuries since William Harvey's discovery of the circulation of the blood initiated the search for the means by which blood passes from arteries to veins. Studies may be grouped into four periods, the first of which established the existence of capillary vessels, perfected methods for observing them, and incorporated them into current medical theories. The second period, extending from the early eighteenth century to 1831 when Marshall Hall published his *Critical and Experimental Essay on the Circulation of the Blood,* was concerned with experimental studies of vessels in living animals. After the enunciation of the cell theory and the development of histology, the third period was devoted primarily to the study of the small vessels in excised organs, until around 1920 when the modern study of living organs *in situ* began. This fourth period has witnessed continuously evolving methods for the study of living tissue, as well as the *ex vivo* study of blood vessels by electron microscopy.

WILLIAM HARVEY

William Harvey's discovery of the circulation of the blood had enormous consequences for medical thought. It clearly controverted earlier theories of the structure and function of the vascular system, and it destroyed ancient ideas of disease causation. As Hermann Boerhaave, a Dutch clinician and teacher of the early eighteenth century, told his students, "Before the Discovery of the Circulation of the Blood it was impossible that any one should

The author is greatly endebted to Dr. Edward H. Bloch, Professor of Anatomy at Case Western Reserve University, for his constructive criticism. This study was supported by National Institutes of Health grant GM 11050-02.

explain those Functions which depend only on the Circulation of the Blood."[1] Harvey's discovery immediately stimulated anatomical investigations of the cardiovascular system in order to demonstrate how the blood crosses from the arterial to the venous system and to ascertain the cause of the blood's motion in different parts of the body.

Harvey himself never observed the pathway between arteries and veins.[2] He believed that the blood flowed from the arteries "into the veins and the porosities of the flesh" (*in venas et porositates carnis*) from whence it was again collected into veins and returned from the circumference to the center.[3] He saw that arteries became more like veins in structure as they receded from the heart, that the pulse diminished proportionally, and that the "ultimate hairlike arterial branches (*divisiones capillares*) seem to be veins not only in structure but also in function" since they either have no perceptible pulse or only have one when the heart beats very violently.[4] He thought that the motive forces responsible for driving the blood from the periphery back to the heart were the movements of the limbs and the compression of muscles which caused the blood to move "from capillary veins into venules and thence into comparatively large veins" (*e venis capillaribus in parvas ramificationes, et inde in majores*).[5] The only cause of the pulse was the impulse of the blood. No motion was derived from the contraction of the arterial walls.[6]

Thus it is clear that Harvey did not postulate closed pathways between arteries and veins. The blood flowed from the arteries into "porosities of the flesh" which differed in size in different parts of the body: "Hence it is more reasonably supposed that the blood in its circuit crosses more slowly through the kidneys than through the substance of the heart; more rapidly through the liver than through the kidneys; through the spleen than through the liver; through the lungs than through the muscles or through any other viscera according to the comparative rareness of their compositions."[7]

Knowing that Galen and his seventeenth-century followers such as Jean Riolan believed that there were direct anastomoses between arteries and veins through which the blood could flow back and forth in either direction, Harvey tried to find them. He boiled various viscera and needled out the parenchyma in order to isolate the vascular fibers, but only in the cerebral

1. Hermann Boerhaave, *A Method of Studying Physick* (London, 1719), p. 265.

2. For a detailed refutation of the common assumption that Harvey postulated the existence of capillaries, see Yehuda Elkana and June Goodfield, "Harvey and the Problem of the 'Capillaries'," *Isis,* 59 (1968), 61–73.

3. William Harvey, *Movement of the Heart and Blood in Animals,* trans. Kenneth J. Franklin (Oxford, 1957), pp. 87, 189.

4. Ibid., pp. 100, 207.

5. Ibid., pp. 90–91, 192.

6. William Harvey, *The Circulation of the Blood,* trans. Kenneth J. Franklin (Springfield, Ill., 1958), pp. 59–60, 148–149.

7. Ibid., pp. 52, 142; see also pp. 46, 137.

plexus, in the spermatic preparative vessels, and in the umbilical vessels did he find "openings of veins and of arteries mutually inosculating."[8] In each of these three instances Harvey observed that the arterioles disappeared obliquely within the coats of the larger veins, similar to the insertion of the ureters into the bladder and of the bile duct into the duodenum, with the result that the walls of the veins formed a sort of valve to prevent a reverse flow of blood. This was further proof to Harvey of the unidirectional flow of the blood.[9]

After Harvey several views prevailed about how the blood was transported from arteries to veins.[10] Some thought that both arteries and veins had open ends whereby the blood was carried by arteries into the tissues where it was dispersed and then collected again in veins. Others asserted that there were direct anastomoses between arteries and the veins which carried the thicker blood back to the heart, while the smallest arteries conveyed thinner blood used to nourish the parts. Others postulated that the body contained innumerable glands that transferred blood from artery to vein, while separating part of it out as a secretion. Disease was attributed to stagnation of blood, either in blood vessels or in the glands, producing the same symptoms as those observed by ancient authorities such as Hippocrates and Galen. The methods of treatment for illness used by physicians holding these theories were still identical with those of physicians who had no knowledge of the circulation.

MARCELLO MALPIGHI

A few years after Harvey's death the first recorded observation of the transfer of blood from arteries to veins was made by Marcello Malpighi, professor of medicine at Bologna. A skilled anatomist, Malpighi was investigating various animal viscera in order to obtain a more detailed picture of their structure. Employing the recently invented microscope as an aid,[11] he turned his attention first to the lungs. Almost immediately he observed significant new details which he published in 1661 in the form of two letters to his friend Giovanni Borelli.[12]

8. Ibid., pp. 74, 158–159; also pp. 25–28, 119–120.
9. Ibid., pp. 76–78, 160–162.
10. Archibald Pitcairn, *The Whole Works,* 2d ed. (London, 1727), p. 37.
11. According to Boerhaave, Malpighi preferred to use microscopes made by the English optician J. Marshall. Hermann Boerhaave, *Academical Lectures on the Theory of Physic* (London, 1743), II, 261.
12. *De pulmonibus observationes anatomicae* (Bologna, 1661). Trans. James Young, *Proceedings of the Royal Society of Medicine,* 23 (1930), pt. I, 1–11; Latin text with Italian translation in Marcello Malpighi, *De pulmonibus,* trans. Luigi Belloni (Milano, 1958). For a detailed description of events leading to Malpighi's publication and an analysis of its content, see Howard B. Adelmann, *Marcello Malpighi and the Evolution of Embryology* (Ithaca, N.Y.: Cornell University Press, 1966), I, 172–198.

The current opinion, originally derived from Galen, held that the lungs were an undifferentiated mass of flesh (*parenchyma*) formed from the blood. Malpighi saw, to the contrary, that they consisted of very light, thin membranes which formed "an almost infinite number of orbicular vesicles and cavities."[13] In order to observe them more clearly, he devised a method of washing out the blood from severed lungs of animals, of drying, and finally of inflating them.

Malpighi tried to delineate the course of the blood vessels in a dog's lung by inflating the pulmonary artery and introducing mercury. At first he had inflated the lungs and injected colored liquids in order to determine whether or not there were anastomoses between pulmonary arteries and veins. The results were uncertain. When he injected mercury he frequently saw it break out of the vessels into the interstitia. For a more intensive study of this problem he chose frogs' lungs because of "the simplicity of the structure, and the almost complete transparency of the vessels which admits the eye into the interior."[14]

When he first examined the living frog's lungs without a microscope. Malpighi thought that the blood broke out of the arterial vessels into spaces between them, from whence it was again collected by open vessels which led to veins. But when he observed the dried lung of a frog by means of a microscope, he saw "things even more wonderful," a network of tiny vessels, and concluded that the blood "is not poured into spaces but always works through tubules, and is dispersed by the multiplex winding of the vessels."[15] He found the same to be true of a tortoise's lung.

Malpighi developed several methods for observing these tubules. A dried preparation was made by ligating the pulmonary vein of a living frog, so that when the frog was killed and the lungs were dried, they would be filled with blood. The tiny vessels could then be seen either with a simple microscope when the preparation was held in the direction of the setting sun, or with a compound microscope if the lung were placed on a crystal plate and illuminated from below by a lighted candle. A living lung preparation could also be seen in either of these ways, so that one could observe the blood in motion.[16]

From these observations on frogs and tortoises, Malpighi inferred that similar subvisible vascular networks must be present in the lungs of "perfect animals," and that anastomoses may join vessels together in other parts of the body, similar to those he saw in the urinary bladders of frogs. He also observed the circulation in the mesentery of frogs.[17]

13. Young's translation, p. 2.
14. Ibid., p. 7.
15. Ibid., p. 8.
16. Ibid., p. 9.
17. Malpighi, *Opera posthuma* (Venice, 1698), p. 91.

Malpighi attempted to deduce the function of the lungs from his anatomical observations. He decided that because of the extensive proliferation and ramification of the blood vessels, the lungs must serve as an organ to mix the various components of the blood and to give them the proper shape to be assimilated by the body. In his first letter to Borelli he stated: "Therefore, in order that this mixing may succeed best, and that the smallest part of the white (serous) may fall between and touch the smallest part of the red, and the mass of the blood be renewed and made by steady mixing, Nature has made the lungs."[18] He concluded his second letter by reasserting that the blood is "arranged and prepared into the nature of particles of flesh, bone, nerve, etc., while it enters the myriad vessels of the lungs. It is conducted into divers very small threads [*stamina*]. Thus a new form, situation, and motion is prepared for the particles of the blood, from which flesh, bone, and spirits may be formed."[19]

ANTONI VAN LEEUWENHOEK

Although Malpighi made the first direct observations, the first intensive investigator of the capillary circulation was Antoni van Leeuwenhoek, the pioneer microscopist of Delft. Leeuwenhoek's observations were accepted and repeated by later researchers such as Stephen Hales and Albrecht von Haller. They formed factual bases for Hermann Boerhaave's system of medicine, which was built on the premise that most physiological functions were performed by various-sized particles circulating in elastic tubes.

Studies of the blood and small vessels interested Leeuwenhoek for nearly sixty years, beginning with his earliest reports of the red globules of his own blood made in 1674.[20] In his early observations he sometimes confused capillary blood vessels with nerve fibers and lymphatics, but in 1683 he published illustrations of the direct anastomosis of arteries and veins in the intestinal villi of a hog and the following year he observed capillaries in the brains of a turkey cock, a sheep, an ox, and sparrows.[21]

On 7 September 1688 he directed a letter to the Royal Society of London

18. Young's translation, p. 5.

19. Ibid., p. 10.

20. For a detailed account of Leeuwenhoek's research on the microvascular system, see G. Miller, "Leeuwenhoek's Observations on the Blood and Capillary Vessels," in *Medicine, Science, and Culture, Historical Essays in Honor of Owsei Temkin,* ed. Lloyd G. Stevenson and Robert P. Multhauf (Baltimore: The Johns Hopkins Press, 1968), pp. 115–122. An invaluable analytical index to Leeuwenhoek's published works appeared in F. J. Cole, "Leeuwenhoek's Zoological Researches," *Annals of Science,* 2 (1937), 1–46, 185–235; a modern critical edition of all of Leeuwenhoek's letters is being published in Holland: *Alle de brieven van Antoni van Leeuwenhoek,* vols. 1–10 (Amsterdam, 1939–1979) referred to hereinafter as *Collected Letters.*

21. Letter of 28 December 1683, *Philosophical Transactions of the Royal Society,* 14 (1684), 587–589; Letter of 25 July 1684, *Phil. Trans.,* 15 (1685), 883–889.

which contained the first detailed description of the capillary circulation.[22] Studying the tadpole of a frog, Leeuwenhoek saw the capillaries in the tail and external gills, noting that the blood passed from arteries to veins through vessels which admitted only one red blood corpuscle at a time. In the tail the main vessels were axial, with the capillaries forming loops. He also saw capillary circulation in the mature frog and in the fins of stickleback, minnows, and bream.

From his evidence in lower animals that the blood passed from the arterial to the venous system only through these minute vessels which allowed only one red corpuscle at a time to pass through, Leeuwenhoek concluded that the same must be true for higher animals and man. This is the first time that anyone made a general statement about the capillary circulation.

A few months later, on 12 January 1689, he reported additional observations on capillaries of fish and Crustacea, and gave the illustration of an "aalkijker," a special apparatus which he constructed to see the circulation in the eel.[23]

As early as 1675, Leeuwenhoek stated that circulation in the small veins was necessary for the nourishment of the body,[24] and this belief was repeated in the 1688 letter. He believed that artery, capillary, and vein were a continuous vessel of identical substance and proportional thickness. The function of the capillaries was to transfer the most tenuous substances from the blood to the tissues through their delicate walls. He emphasized several times that the capillary vessels have no ends.[25] At first he thought that some diffusion of blood occurred also in arteries, but in 1689 he noted that arteries have thick solid coats which do not permit even the thinnest liquids to escape.[26] Like his contemporaries who were influenced by the mechanical philosophy of René Descartes, Leeuwenhoek thought in mechanical terms of a fluid of different-sized particles flowing in various-sized vessels and distributing particles according to the size of the apertures in the walls.

Studies of the small vessels interested Leeuwenhoek until the end of his life. He observed capillaries in various fish such as perch, pike, minnows, carp, flatfish, shrimp, and flounder. He noted them in the hind feet of the crab, the feet and tail of lizards, and the foot of an Indian scorpion, as well

22. Original Dutch with English translation in *Opuscula Selecta Neerlandicorum de Arte Medica,* 1 (1907), 38–81. A facsimile edition of the Latin text from Leeuwenhoek's *Arcana naturae detecta* (Delft, 1695), pp. 165–186, with an introduction by A. Schierbeek appeared as Vol. II of *Dutch Classics in the History of Science, Mathematics and Exact Sciences* (Nieuwkoop, 1962).

23. *Arcana naturae* (Delft, 1695), pp. 187–206. This instrument with five interchangeable lenses is preserved at the History of Science Museum in Leyden.

24. Letter of 26 March 1675, *Collected Letters,* I, 289, 291.

25. Letter of 13 June 1722, *Phil. Trans.,* 32 (1722), 151.

26. Letter of 1 April 1689, *Arcana naturae* (Delft, 1695), pp. 218–219; letter of 26 May 1717, *Epistolae physiologicae super compluribus naturae arcanis* (Delft, 1719), p. 350.

as in the wing membrane of the bat, the gills and comb of the cock, the ear of a rabbit, the leg of the flea and louse, and in a newly hatched duck.[27]

INJECTION OF CAPILLARIES

At the same time as Leeuwenhoek, other scientists were also beginning to observe the capillaries, either by means of the microscope or by injecting the vessels with colored liquids. In Dublin, William Molyneux, an Irish devoté of the new learning who had organized the Dublin Philosophical Society on the model of the Royal Society of London, demonstrated the circulation in the water-newt at meetings of the Dublin Philosophical Society in May and June 1684.[28] In Holland, the art of injecting vessels which had been initiated by Jan Swammerdam before 1666 and brought to perfection by Frederick Ruysch[29] was used, as Boerhaave wrote, "to view the whole arterial System from the Heart, even to where their smallest Twigs join with the incipient Veins."[30] Swammerdam injected mercury into the vessels of a frog's lung and colored wax in the vessels of various fish in order to demonstrate their course.[31] In 1676 their fellow countryman Stephen Blankaart proved by injecting a dark liquid that pathways connect arteries and veins in a muscle excised from a dog, although he mistakenly asserted that the muscle fibers themselves were the passages and that they contained valves similar to those in the veins.[32] This notion that the blood circulated through the muscle fibers was denied by Leeuwenhoek.[33]

WILLIAM COWPER

Because all these observations had been made on lower animals, some still questioned whether capillary circulation also occurred in man and higher animals. In reply, William Cowper, the English surgeon, made an abdominal incision in a living ten- or twelve-day-old kitten and examined the omentum with a compound microscope. He saw the globules of blood

27. See Cole, "Leeuwenhoek's Zoological Researches."

28. *Phil. Trans.,* 15 (1685), 1236–1237; Thomas Birch, *History of the Royal Society* (London, 1756–1757), IV, 303 ff. See also K. Theodore Hoppen, "The Royal Society and Ireland. William Molyneux, F.R.S. (1656–1698)," *Notes and Records of the Royal Society of London,* 18 (1963), 125–135.

29. See F. J. Cole, "The History of Anatomical Injections," in Charles Singer, ed., *Studies in the History and Method of Science* (Oxford, 1921), II, 301–309.

30. Boerhaave, *Academical Lectures on the Theory of Physic* (London, 1742), I, 18.

31. Jan Swammerdam, *Bibel der Natur* (Leipzig, 1752), pp. 327–328, 349.

32. Stephen Blankaart, *Tractatus novus de circulatione sanguine per tubulos,* reprinted in *Anatomica practica rationalis, sive rariorum cadaverum morbis denatorum anatomica inspectio* (Amsterdam, 1688), pp. 306–307.

33. Letter of 3 March 1682 to Robert Hooke, *Collected Letters,* III (1948), 391.

moving from arteries to veins. The experiment was repeated on the omentum and mesentery of a young dog.[34]

Cowper also noted the change in capillary circulation in the extremities induced by fatigue. After frequently seizing the hind foot of a live frog in order to put it under the microscope, he noticed that the blood globules had receded from the membrane between the toes but returned gradually and restored the circulation.[35] Similarly he noted in lower animals that the size of the capillaries is not uniform in the same individual.[36] When he injected arteries with melted wax mixed with oil of turpentine and colored with vermilion, he found that in man and the quadrupeds he succeeded in getting the wax to pass from arteries to veins only in the lungs, spleen, and penis. Furthermore he found it impossible to observe with the naked eye the capillary circulation anywhere in either lower or higher animals.[37]

POPULAR OBSERVATIONS

During these years and later, observation of the capillary circulation became a favorite activity of microscopists. In 1702 a description of James Wilson's newly invented pocket microscope contained the method of observing the circulation of the blood in "Living Creatures" which included fish, eels, and the foot of a frog.[38] Among the plates of John Harris' *Lexicon technicum, or an Universal English Dictionary of Arts and Sciences* (second edition, 1708–1710) was a drawing of Marshall's double microscope for viewing the circulation of the blood. Forty years later, Henry Baker's book *The Microscope Made Easy*[39] repeated many of Leeuwenhoek's observations using eels, flounders, newts, crabs, shrimp, grasshoppers, and spiders. The transparent membrane between the toes of a frog's hind foot was used most commonly. In Berlin, J. N. Lieberkühn devised a spectacular new method, the solar microscope, which projected the greatly enlarged image on a screen. Examining the mesentery of a living frog by this method, Baker reported, "Several of the Vessels were magnified to above an Inch in Diameter, and the Globules of the Blood rolling through them seem'd near as large as Pepper-Corns."[40] However, an astute researcher such as Albrecht von

34. *Phil. Trans.,* no. 280 (July-Aug. 1702), 1181–1183.
35. Ibid., pp. 1184–1185.
36. Ibid., p. 1183.
37. *Phil. Trans.,* no. 285 (May-June 1703), 1389–1390.
38. *Phil. Trans.,* no. 281 (Sept.-Oct. 1702), 1241–1247.
39. London, 1742, pp. 121 ff.
40. Ibid. p. 135; Henry Miles, "Some Remarks Concerning the Circulation of the Blood, as Seen in the Tail of a Water-eft, Through a Solar Microscope," *Phil. Trans.,* no. 460 (April-July 1741), 725–729.

Haller did not find the solar microscope useful for his work because of the poor definition and chromatic aberration.[41]

IATROMATHEMATICS

In addition to microscopic investigations, research on the vascular system during the seventeenth century proceeded along another line. Because of the great success of mathematical studies in physics and astronomy, confidence was high that a similar application could be made to living creatures. Throughout Europe investigators calculated the force of the heart, the relative size and strength of vessels, the specific gravity and velocity of blood in order to apply laws of mechanics and hydraulics.[42] Iatromathematicians such as Archibald Pitcairn and Lorenzo Bellini regarded the arterial system as a cone with the apex at the heart and the minute vessels of the periphery as the trunk. Viewing the heart as the sole motor force, they thought the velocity of the blood decreased as it entered the increasingly narrow vessels, which also produced friction and tended to form obstructions. Bellini supposed that the obstructions produced inflammation and inflammatory diseases, a view adopted by Boerhaave and widely held for over a century.[43]

On the other hand, another iatromathematician, Borelli, on the basis of erroneous calculations of the force of the heart, disagreed with Harvey's view that it was solely responsible for the circulation, and asserted that the veins drew up the blood by capillary attraction, comparing the veins with thin glass tubes which draw up fluids because they adhere more strongly to the sides of the tube than to themselves.[44]

HERMANN BOERHAAVE

The system of medicine of Hermann Boerhaave, which incorporated much of this iatromathematical reasoning with the observations of Leeuwenhoek, illustrates in detail ideas about the structure and function of the vascular system which were common during the eighteenth century. In lectures at the University of Leyden and the numerous editions of his *Insti-*

41. *Deux mémoires sur le mouvement du sang* (Lausanne, 1756), p. 179.

42. A good summary may be found in Kurt Sprengel, *Histoire de la médecine,* trans. A. J. L. Jourdan (Paris, 1815), V, 131–194.

43. The geometric "proof" is given in L. Bellini, *Opuscula aliquot ad Archibaldum Pitcarnium* (Pistoria, 1696), p. 140 (De motu bilis, prop. 26), and pp. 190–192 (De fermentis, prop. 37 and 38).

44. For a full treatment see the dissertation by Samuel Hambacherus, *Dissertatio physico-medica de theoriae physicae tubulorum capillarium ad corpores humanum applicatione* (Halle, 1743).

tutes of Medicine, Boerhaave disseminated a medical version of the new corpuscular philosophy throughout Europe.[45]

For Boerhaave, studies of the vascular system were of crucial importance because the motion of the heart and blood was regarded as the key to life, which was defined as a "constant Progress of the Fluids by certain Pipes or Vessels, assisted by the Impulse of the Solids, preserving a corruptible Body from Decay."[46] The German clinician Friedrich Hoffmann summed up this basic assumption with the statement "*Vita nihil aliud est quam motus sanguinis,*" life is nothing else than the motion of the blood.[47] Boerhaave, an exponent of the mechanical philosophy, thought in terms of discrete particles of matter in motion as the ultimate cause of physiological activities: "The human Body is an Assemblage of small elastic Solids, by whose conjunct and regular Actions, Life and Health are produced. The Head or first Spring of Motion, in these elastic Solids is the Heart."[48]

The vascular system was considered a system of conical tubes decreasing in size as they withdrew further from the heart. They transported blood particles of various sizes which separated out in branching vessels of appropriate diameter. The heart originated the motion of the blood by pumping it into the arterial system. The aorta and arterial walls helped to propel the blood forward by means of their elasticity, being dilated by the thrust of blood from the heart during systole. Their contraction then pushed the blood onward toward the extremities where the minutely branched vessels either joined with the venous system or ended in various ducts or pores.[49]

The veins had a similar shape, increasing in size as they approached the heart. Their number and capacity were greater than that of arteries. Their walls were much thinner, less elastic, and without a pulse or the ability to contract which arteries had. Valves served to prevent the blood from regressing while the heart contracted.[50]

In addition to blood vessels there were smaller vessels of decreasing size to accommodate even smaller particles. Boerhaave accepted Leeuwenhoek's conjecture that red blood corpuscles were spherical in shape and were composed of six smaller particles which were yellow in color and constituted the serous part of blood. These in turn broke down to six even smaller colorless particles corresponding to lymph, and it was not known how much further

45. Lester S. King, *The Medical World of the Eighteenth Century* (Chicago, 1958), pp. 65–75, gives an excellent summary of Boerhaave's medical system.

46. Introductory Preface to Boerhaave's *Institutions in Physick,* 2d ed., trans. J. Browne (London, 1715), p. xvii.

47. Cited in Robin Fahraeus, "Die Grundlagen der neueren Humoralpathologie; die frühe Geschichte der Mikrozirkulation," *Virchows Archiv für Pathologische Anatomie und Physiologie und für Klinische Medizin,* 333 (1960), 181.

48. Note 1 to paragraph 4, Boerhaave's *Academical Lectures,* I, 7.

49. Ibid., II, 1–12 (Paragraph 132).

50. Ibid., pp. 12–17 (Paragraph 133).

the subdivision could be carried. The whole blood contained the "Materials of all the other Juices formed in every Part of the Body" including the nutriment. The size of the red blood corpuscles was determined by the diameter of the capillaries of the lungs; the smaller particles entered orifices in the vascular system accommodated to their size.[51] The globules not only subdivided, but reunited again as the need arose. Although the red globules were naturally spherical, they were also flexible and, as Leeuwenhoek had observed, assumed an oval shape in the capillaries.[52]

In Boerhaave's system the heart exercised other functions besides providing the motive power for the circulation of blood. In the right ventricle it received the venous blood which it attenuated by a shaking action in order to mix in the chyle. The blood was then pushed into the lungs where, as Malpighi had also believed, "a very intimate Commixture of all the new and old Juices" occurred.[53] It was necessary for the crude particles of chyle to pass through the lungs so that they would be broken up, polished, rounded, and formed into tiny globules of different sizes suitable for their ultimate nutritive functions.[54] They were molded into their spherical shapes by the shape of the walls and the pressure of the air in the lungs. The air was denied any other function.[55]

Boerhaave followed the doctrine of Bellini that the heat of the blood arose from the friction of the blood globules rubbing on each other and on the walls of the vessels, especially the capillaries. The amount of heat was proportional to the blood's velocity and the number of red globules it contained.[56]

STEPHEN HALES

Because of inadequate methods and measuring devices, the work of most of the early advocates of quantification was grossly inaccurate and of little value for the further development of knowledge of the vascular system. Stephen Hales, the English clergyman-scientist, is the exception. Convinced that "the animal Fluids move by Hydraulick and Hydrostatical Laws,"[57] an experimentalist par excellence, he devised numerous methods of studying vegetable and animal fluids moving within their vessels and he reported his

51. Ibid., pp. 172–177 (Paragraph 226). "Hence the Arteries and Veins are not only sanguiferous but also *serous,* lacteal, lymphatic, carrying Oil, Water, Spirits, &c. nor can we tell where the Progression, or Division of the Vessels into lesser Series, terminates" (p. 219).

52. Ibid., p. 362.

53. Ibid., p. 92.

54. Ibid., pp. 90 ff. Particles which easily become spherical in shape were considered alimental, those which become spherical with difficulty as medicinal, and those which cannot become spheres were poisons (p. 99).

55. Ibid., p. 126.

56. Ibid., pp. 154, 157.

57. Stephen Hales, *Statical Essays: Containing Haemastaticks* (London, 1733), II, xvii.

results to the Royal Society of London which sponsored their publication in two volumes of *Statical Essays.*[58] Hales subscribed to the current view that health of an animal "consists principally in a due Equilibrium between its Fluids and Solids,"[59] and he undertook to discover "the Force and Velocity with which these Fluids are impelled; as a likely means to give a considerable Insight into the animal Oeconomy."[60] He hoped to correct the conflicting and inaccurate work of predecessors and to improve the current method of studying the small vessels.

Hales' pioneering work on the measurement of blood pressure, for which he is best known, also included the capillary circulation. In a series of well-planned experiments on dogs he proved that various liquids flowed through the capillary vessels of the intestines at varying rates, depending upon the substance of the liquid and its temperature. After bleeding dogs to death he slit open the intestines longitudinally, poured measured quantities of various liquids through the aorta, and measured the time required for all the liquid to pass through the capillary openings—the longer the time the more constricted the vessels. He discovered that warm water relaxed the vessels, cold water constricted them, as did also brandy and decoctions of Peruvian bark (quinine), camomile flowers, and cinnamon. Hales summarized as follows:

> We see in this and the three foregoing Experiments, how the Vessels of the Body are manifestly contracted or relaxed, by different Degrees of Warmth, Heat or Cold, or from the different Qualities of the Fluids which pass thro' them, as to Restringency or relaxing: And such Qualities of the Fluids must have very considerable Effects on the finer capillary Vessels, whose Coats bear a much greater Proportion to the contained small cylindrical Fluid, than in larger Vessels. Tho' it is not to be imagined that the Effects are so sudden and great in a live Animal, as in these Experiments; because in a live Animal, the several Fluids which are taken in, are more gradually and in smaller Proportion blended with the Blood.[61]

He added that substances which constricted the vessels probably also raised the blood pressure.

Hales also injected the intestine with vermilion and saw the successively smaller vessels down to "nearly 1/3240th part of an Inch, that is so fine, that only single Blood globules can pass them into the Veins."[62] He noted the difference in blood pressure of arteries and veins (10 or 12 to 1) and attributed it to the resistance which blood met in the capillaries.[63] He also estimated the blood pressure in the capillaries. In the lungs there was evidently a

58. London, 1727 and 1733.
59. Ibid., II, 126.
60. Ibid., II, Introduction, p. [1].
61. Ibid., II, 136.
62. Ibid., p. 52.
63. Ibid., p. 55.

freer passage in the capillary vessels, for he calculated that the velocity of blood in lung capillaries was forty-three times greater than in the capillaries of muscles.[64] By viewing with a strong light the circulation in a living frog's lung, he observed the various paths that the blood followed on the surface of the vesicles. Following Boerhaave, he thought that the blood acquired a greater heat in the lungs because of friction of the corpuscles on the walls of the vessels, but that it was simultaneously cooled by the refrigerant action of the air. The florid color was caused by "the strong Agitation, Friction and Comminution" caused by passing through the lungs, although unlike Boerhaave, he admitted that the blood in the lungs might "receive some other important Influence" from the air.[65]

He attempted to improve injection methods because he felt that those currently practiced introduced an abnormal force that might burst the fine vessels. For this purpose he invented a method of inserting melted wax at the same pressure as normal blood pressure.[66] Although he did not succeed in fixing the colored liquid in many communications between arteries and veins of his experimental dog, he achieved this in the stomach, intestines, urinary bladder, and gall bladder. In the latter, by means of the microscope, he saw the blood flow directly from artery to vein without passing through any intermediate glands, as some contemporaries believed.[67] He was aware that there was not a uniform system in the body for the passage of blood from arteries to veins: in some parts there were direct anastomoses, in others capillary networks.[68] He also attempted to estimate the number of capillary arteries in a man, concluding that there were nearly one million in the body exclusive of the lungs, which contained over three and a half million.[69]

EMPIRICISM IN MEDICAL RESEARCH

Hales' efforts to quantify biological phenomena terminated a period, for even while he was doing his most important work a new trend had started: empiricism in medicine and biological research. This development stemmed from the philosophical directives of Francis Bacon and John Locke that only empirical observations were valid avenues to new knowledge. Therefore one should not construct theories based on analogies or deductions from other knowledge; only direct observations and personal experience were sound. Furthermore, contrary to the habits of followers of Descartes who not only applied mathematics to all possible situations but also tried to construct

64. Ibid., pp. 63–68.
65. Ibid., p. 105.
66. See ibid., pp. 145 ff.
67. Ibid., p. 150.
68. Ibid., p. 152.
69. Ibid., pp. 70–72.

systems, that is, all-inclusive rationalizations to explain phenomena, the British-derived empirical philosophy taught that each problem should be treated individually, avoiding the formation of systems of thought to explain everything.

In the history of research on the vascular system Albrecht von Haller and Abbé Lazaro Spallanzani both made important contributions stemming from their empirical methods. Both rejected the attempt to reason by analogy from mechanical laws to those of living creatures, and insisted that one should not establish any theories without prior observations and experimentation.

ALBRECHT VON HALLER

Haller, a disciple of Boerhaave, was concerned with the general problem of the cause of motion in the various parts of the body. Over twenty years of experimentation by him and his students during his professorial years at the University of Göttingen and later as a private researcher at Berne resulted not only in his classical work on the contractility and irritability of the various parts of the animal body,[70] but also a book on the motion of the heart and blood. His *Deux mémoires sur le mouvement du sang, et sur les effets de la saignée; fondés sur des expériences faites sur des animaux* (Lausanne, 1756) was the most important work on the subject during the eighteenth century.[71] A skilled anatomist, he described in detail the vascular system of both warm-blooded and cold-blooded animals, reviewing carefully the work of others. His book is thus a commentary on other authors, agreeing or disagreeing according to the results of his own observations.

Haller studied Leeuwenhoek's work very carefully and repeated his experiments. He admitted that he had little to add to Leeuwenhoek's description of the capillary vessels.[72] He disagreed with Boerhaave, however, who had postulated that tiny vessels too small to transport red blood corpuscles branched off from arteries carrying a liquid which was different from blood. Haller, "an observer less prompt to believe what he hoped," regarded these tiny vessels carrying a yellow fluid as blood capillaries so narrow that only a single file of red corpuscles could be carried, and he noted that the distinguished French investigator Senac had already corrected Boerhaave's error.[73] However Haller did not attempt to destroy Boerhaave's system by denying that smaller channels actually might exist for the circulation of

70. "De partibus corporis humani sensibilibus et irritabilibus," *Commentarii Societatis Regiae Scientiarum Gottingensis,* II, 114–158. Reprint of the English translation is in *Bulletin of the Institute of the History of Medicine,* 4 (1936), 651–699.

71. See the detailed study of P. Wobmann, "Albrecht von Haller, der Begründer der modernen Hämodynamik," *Archiv für Kreislaufforschung,* 52 (1967), 96–128.

72. *Deux mémoires,* pp. 12–13, 207–210.

73. Ibid., p. 15.

fluids finer than blood, because he had observed a silver colored fluid in the muscles of eels!

Reviewing the current information about the composition of blood and the size and shape of red corpuscles, unlike some other investigators who described them as oval, Haller observed only circular disks, although he agreed that he had seen them change shape especially in tracing the curves of small vessels.[74] Like Senac he disagreed strongly, however, with Leeuwenhoek's and Boerhaave's assertion that the red corpuscles disintegrated into smaller globules which then reunited again,[75] an assumption basic for Boerhaave's system of physiology. Also he thought there was very little difference between the arterial blood of the pulmonary vein and venous blood of the vena cava.[76]

The origin of the blood's motion was a problem of central concern to eighteenth-century investigators. Was it caused solely by the contraction of the heart, as Harvey had thought, or did the blood vessels also contribute to it? Haller undertook to answer the question by a number of experiments on both warm and cold-blooded animals. He measured the distance the blood was propelled when different arteries of goats, dogs, rabbits, and sheep were cut;[77] he tied various arteries and observed how the pulse disappeared below the ligature.[78] He controverted the current theory that inflammation was caused by an obstruction in the blood vessels, by demonstrating in an artery of the frog's mesentery that an obstructed vessel did not become engorged with blood and swell, but instead the blood detoured through other neighbouring vessels past the obstruction.[79] The sole cause of the pulse was the impulse of the blood against the elastic walls of the arteries, although there were some arteries with such rigid walls that no pulse could be felt.[80]

Haller's most distinctive work, which was to influence medical thought for several generations, was his investigation of the walls of various parts of the vascular system in order to determine whether or not they have any contractile force. In the previous century Jean Pecquet and Thomas Bartholin were among those who asserted that the blood was propelled by arterial contractions.[81] Haller however denied this. Using a Culpeper microscope,

74. Ibid., pp. 19–20.
75. Ibid., pp. 26–27.
76. Ibid., p. 30.
77. Experiments 41, 42, 43, 46, 49.
78. Experiments 47, 48, 51, 52, 53, 54, 57, 58.
79. Pp. 43, 244, and Experiment 54, p. 204. See also Cowper's observation of the establishment of collateral circulation, *Phil. Trans.*, no. 285 (1703), p. 1391.
80. *Deux mémoires*, p. 40.
81. See the literature citation in Georg Wedemeyer, *Untersuchungen über den Kreislauf des Bluts und insbesondere über die Bewegung desselben in den Arterien und Capillargefässen, mit erklärenden Hindeutungen auf pathologische Erscheinungen* (Hannover, 1828), pp. 1–3, and Sprengel, *Histoire de la médecine*, IV, 169 ff.

he could see no contraction in the arterial vessels of frogs, fish, and other cold-blooded animals, and he stated that blood flowed through these vessels as through immobile glass tubes.[82] He denied also that the muscles helped to maintain the circulation by expelling the blood *within* them during contraction, but he granted that the movement of muscles compressed the veins between them and contributed to maintain the venous circulation.[83] Only heat and cold, as in the case of the flow of nutritive fluids in vegetables, had any effect on the amount of blood in the vessels: heat appeared to dilate them so that they yielded more readily to the blood's action, and cold produced the opposite effect. Also gravity contributed to the motion of the blood, as one saw when the mesentery of frogs was tilted different ways or with the everyday experiences of different positions of the human body, and the varicose legs of workers who stood up constantly.[84]

Haller also dismissed Robert Whytt's theory that the circulation was partly maintained by oscillatory contractions of the walls of the capillary vessels stimulated into action by the blood moving through them.[85] He insisted that since he could not observe such contractions with his microscope they did not exist, although at least one of his contemporaries had published experimental evidence to the contrary, and Jean de Gorter, another Boerhaave disciple, had based a new theory of local inflammation upon it.[86]

LAZARO SPALLANZANI

Knowledge of the vascular system in the mid-eighteenth century was thus inconclusive. The conflicting evidence, controversies, imprecise definitions, and crude techniques did not bring any final answers to the causes of the blood's circulation or the nature of the capillary vessels, and the subject still remained open for investigation. In 1771 the Italian experimentalist Lazaro Spallanzani (1729–1799), who three years before had published a report of his observations on the circulation of the salamander,[87] began to work on

82. Haller cited fourteen experiments to prove this, cf. p. 236, note g. See also his "Dissertation on the Sensible and Irritable Parts of Animals," *Bull. Hist. Med.*, 4 (1936), 680–681.

83. *Deux mémoires*, pp. 142–144.

84. Ibid., pp. 148–151.

85. *An Essay on the Vital and Other Involuntary Motions of Animals* (first published Edinburgh, 1751) in: *The Works of Robert Whytt* (Edinburgh, 1768), pp. 52–54. More complete discussion is found in "An inquiry into the causes which promote the circulation of the fluids in the very small vessels of animals," ibid., pp. 211–253. Haller, *Deux mémoires*, p. 93.

86. Josias Weitbrecht, *Commentarii Academia Scientiarum Imperialis Petropolitana*, VI, 276; VII, 320; VIII, 339–340, cited in Sprengel, *Histoire de la médecine*, IV, 169. Jean de Gorter, "Exercitatio medica de motu vitali," published in *Exercitationes medicae quatuor* (Amsterdam, 1737).

87. *Dell'azione del cruore ne' vasi sanguigni, nuove osservazioni* (Modena, 1768).

new aspects of the subject. Using a different microscope and new techniques he not only confirmed much of Haller's work on cold-blooded animals, but he also added new data, particularly about warm-blooded animals.

Spallanzani's attention was drawn to the vascular system of the developing chick because one of his friends was repeating Haller's experiments on the formation of a chick. His friend was in the midst of showing him the gradual evolution of the embryonic organs of an incubated egg, when Spallanzani's attention was arrested by the "beautiful net-work of vessels" of the umbilical cord. Placing the egg on the object-stand of a Lyonet microscope, he was "transported with joy" to see so clearly the circulation of blood.[88] The Lyonet microscope consisted of interchangeable simple lenses like those of Leeuwenhoek, but they were mounted on a five-jointed brass arm above a horizontal table illuminated by a mirror from below.[89] This freed the hands of the observer and provided flexibility in lenses and light sources.

Spallanzani described his method as follows:

> I exposed that part of the chick, which I wished to examine, to a beam of light, admitted into a darkened chamber through a hole in the window-shutter, by which means the eye, not being dazzled by any superfluous rays, could readily distinguish the hidden springs of the vascular organization. This method gave me, besides, the advantage of examining the chick at leisure; because the heat of the solar light partly supplying that which is necessary to the growth and life of the animal, the exercise of its functions continued, for a long time, unimpaired. I conducted, in the same manner, a great number of experiments upon the eggs of European and Indian fowls; and so discernible was the circulation wherever vessels could be seen, that is, upon the umbilical cord, the membranes, and even the body of the chick, that I succeeded in repeating the principal experiments that I had made upon salamanders, frogs, &c. So complete was the identity of the phenomena, that I no longer hesitated to apply the results of the facts which I had observed, in cold-blooded animals, to all those with warm blood, and consequently to the human species.[90]

Like Haller, Spallanzani demonstrated that the mechanical laws of hydraulics cannot be directly applied to living bodies. He found by experiment that there was no relationship between the retardation of the blood in the capillaries to the obstacles encountered, nor was there any ratio between the velocity of the circulation in arteries near the heart to its speed in the capillaries.[91] While acknowledging that the general laws of hydraulics had

88. Abbé Spallanzani, *Experiments upon the Circulation of the Blood, throughout the Vascular System: On Languid Circulation: On the Motion of the Blood, Independent of the Action of the Heart: and on the Pulsations of the Arteries* (London, 1801), p. 131. This book was first published in Italian in Modena, 1773.

89. See the description in Pierre Lyonet, *Traité anatomique de la chenille* (The Hague, 1762), appendix pp. 1–10.

90. Spallanzani, *Experiments*, pp. 131–132.

91. Ibid., p. 256 ff.

some influence on the circulation, he contended that "their power is counterbalanced by opposite causes, inherent in the sanguiferous system."[92]

Repeating Haller's pulse experiments on both the cold-blooded salamander and the warm-blooded chick embryos, Spallanzani confirmed that there is a pulse in the smallest arteries of both cold and warm-blooded animals.[93] He also confirmed Haller's view that the heart alone provides the motive force for the circulation, since he saw only "a complete *inertia* in the coats of the capillary branches, and a retardation rather than an augmentation of velocity in the circulation."[94]

Ironically, the very phenomenon which Haller had done so much to elucidate, the irritability of muscle fibers, became a continuing source of conflict in views about the circulation. For while Haller and Spallanzani both could not see any evidence of irritability or contractility in the capillary vessels, there were numerous contrary observations by others,[95] and the application of the doctrine of irritability to the vascular system led to new theories of inflammation which dominated medical thought and practice in the latter part of the century. The unsolved problems about the cause of the blood's circulation led to works such as Georg Prochaska's *Controversae quaestiones physiologicae* of 1778.[96] Xavier Bichat in his *Anatomie Générale* of 1801 summarized the current confusion as follows: "We see by the different views that I have presented, that almost all authors have described in an inaccurate manner the motion of the blood, and what loose ideas they have had of it. Experiments have only served to confuse them; it is a work that requires to be entirely done again, either with the materials that many respectable authors have already amassed, especially Haller, Spallanzani, Weitbrecht, Lamure, Jadelot, &c. or with new facts."[97]

JAMES CARSON

Meanwhile there was a growing tendency to consider bodily processes in vitalistic terms, assuming that the presence of life contributed an element in physiology which transcended ordinary physical law. John Hunter, for

92. Ibid., p. 260.

93. Ibid., pp. 261–262.

94. Ibid., pp. 272–273.

95. They included Gauthier Verschuir (1766), P.-A. Fabre (1770), C. L. Hoffmann (1779), C. Kramp (1786), and H. van der Bosch (1786). See literature citation in Sprengel, *Histoire de la médecine,* V, 351–354.

96. *Controversae quaestiones physiologicae, quae vires cordis et motum sanguinis per vasa animalia concernunt* (Vienna, 1778).

97. Xavier Bichat, *General Anatomy Applied to Physiology and Medicine,* trans. George Hayward (Boston, 1822), I, 351. F.-B. de B. de Lamure had published *Recherches sur la cause de la pulsation des artères* (Montpellier, 1769) and Jean-Nicolas Jadelot, *Mémoire sur la cause de la pulsation des artères* (Nancy, 1771).

example, insisted that the blood was alive, and as stated by one author "the language of the schools on the subject may be said to be, that it is in vain to apply the laws of hydrostatics to the motion of the blood, a living fluid flowing in living tubes."[98] James Carson of Liverpool undertook to refute this opinion. When he was a medical student at Edinburgh, Carson had heard a lecture at the Royal Medical Society which discussed the uselessness and even danger of studies in mathematics and the physical sciences for medical students. In order to disprove this he selected the circulation of the blood as a suitable subject. Discovering that the heart was not the sole motor force and that additional arterial contraction was not sufficient either, Carson in 1815 was among the first to discuss the role played by atmospheric pressure.[99]

Following the suggestions of Erasmus Darwin who had claimed that the veins were absorbent vessels which drew out blood from the capillaries,[100] Carson reasoned that the major cause of the blood flow from the capillaries back to the heart was the pressure of the atmosphere upon the expanded elastic lungs during inspiration. Thus he gave one of the earliest statements that the intra-thoracic pressure was subatmospheric and that it was a necessary mechanism for maintaining the circulation.[101] The major credit for this discovery has been assigned to David Barry who later received more publicity for his studies.[102]

A. P. WILSON PHILIP AND JAMES BLACK

Another British investigator, A. P. Wilson Philip, explored the relationship between the nervous system and the circulation, as a by-product of studies on the manner in which poisons destroy life. By means of observations made on the capillaries of the membrane between the toes of a frog's

98. James Carson, *An Inquiry into the Causes of Respiration; of the Motion of the Blood; Animal Heat; Absorption; Muscular Motion; with Practical Inferences*, 2d ed. (London, 1833), p. 258.

99. For details see Lord Cohen of Birkenhead, "James Carson, M.D., F.R.S., of Liverpool," *Medical History*, 7 (1963), 1–12. His idea first appeared in his Edinburgh dissertation (*De viribus quibus sanguis circumvehitur*, 1799) and *An Inquiry into the Causes of the Motion of the Blood; with an Appendix, in Which the Process of Respiration and Its Connexion with the Circulation of the Blood Are Attempted to Be Elucidated* (Liverpool, 1815). Other statements on the influence of the air pressure on the circulation occur in P. G. Schacher, *De anatomica praecipuarum partium administratione* (Leipzig, 1710), reprinted in Haller's *Selectus dissertationum anatomicarum*, VI, tab. I, quoted in S. Th. von Sömmerring, *Gefässlehre, oder, von Herzen, von den Arterien, Venen, und Saugadern* (Frankfurt am Main, 1792), p. 97, and in John Huxham, *Observations on the Air and Epidemic Diseases* (London, 1759), p. ix.

100. Erasmus Darwin, *Zoonomia, or the Laws of Organic Life*, Vol. I (London, 1794).

101. Detailed account in Cohen, "James Carson," pp. 2–8.

102. David Barry, *Experimental Researches on the Influence Exercised by Atmospheric Pressure upon the Progression of the Blood in the Veins* (London, 1826); see C. S. Breathnach, "Sir David Barry's Experiments on Venous Return," *Medical History*, 9 (1965), 133–141.

hindfoot, Wilson Philip observed in 1815 that the capillary circulation continued briskly after the spinal marrow had been destroyed or after the heart had been extirpated. Thus he assumed that the capillary vessels possess the power to propel blood.[103] In other experiments he showed that the velocity of blood flow in the capillaries could be accelerated by stimulating them by concentrated sunlight, or as Hales had done earlier, by the application of brandy or by friction, and conversely that the speed could be decreased by applying an infusion of opium or tobacco.[104]

With the growing recognition that different laws governed capillary action than the action of the large vessels, the question whether the capillaries possess independent contractility became a central one. In 1825, James Black published *A Short Inquiry into the Capillary Circulation of the Blood, With a Comparative View of the More Intimate Nature of Inflammation,* based upon experiments on the webs of frogs' and young ducks' feet and on the mesenteries of the frog. After applying ammonia and other solutions, he concluded that the blood's motion in capillaries depended on "some automatic power in the part itself."[105] He described the phenomenon now called the "red reaction" by physiologists, while observing the independent contractility of capillaries.[106] He stated "that the capillary vessels, under certain irritations, also enlarge and diminish in their diameters, with corresponding alterations in the velocity and in the volume of the blood circulating through them, without either the heart, brain, spinal marrow, influencing, or sympathizing in, the local affection, are facts, which experiment has fully confirmed."[107] He also was among the first to apply the notion of "tonus" to blood vessels, although it was in a speculative context. Not being able to refrain from speculating about the ultimate cause of capillary contractility, Black ascribed it to a "principle of vitality," shared also by the blood and many other parts of the body, which caused the small vessels to have a "tensive or an erectile state."[108]

103. A. P. Wilson Philip, "Some Additional Experiments and Observations on the Relation Which Subsists between the Nervous and Sanguiferous Systems," *Phil. Trans.*, pt. 2 (1815), 438–446. For biographical details see William H. McMenemey, "Alexander Philips Wilson Philip (1770–1847), Physiologist and Physician," *Journal of the History of Medicine,* 13 (1958), 289–348.

104. A. P. Wilson Philip, *An Experimental Inquiry into the Laws of the Vital Functions,* 3d ed. (London, 1826), pp. 285–286.

105. James Black, *A Short Inquiry into the Capillary Circulation of the Blood; with a Comparative View of the More Intimate Nature of Inflammation: and an Introductory Essay* (London, 1825), p. 70.

106. See the letter by J. F. Fulton, "James Black's Description of Contractility of Capillaries (1825)," *The Lancet,* 217 (1929), 1010–1011.

107. Black, *Inquiry,* pp. 83–84.

108. Ibid., p. 91. A footnote by Black on p. 106 states, "The essential nature of the living principle may be safely left to future discovery, if it be within the grasp of human science; but we may as fairly argue on its properties and laws, as the chemists do on the subject of caloric; though the most eminent of them are not agreed, whether it is a quality of matter, or a material substance."

MARSHALL HALL

The work of Wilson Philip and Black brought the first, inconclusive period of investigation to a close. Marshall Hall's *Critical and Experimental Essay on the Circulation of the Blood, Especially as Observed in the Minute and Capillary Vessels of the Batrachia and of Fishes,* published in London in 1831, set a new pattern for more precise research. His method made use of the new improved achromatic microscope; his aim was greater accuracy. "The microscope will," he said, "be the means of the next improvements in anatomy."[109] With the goal of "perfect accuracy," Hall first defined precisely the subject which he was investigating:

> The minute vessels may be considered as arterial, as long as they continue to divide and subdivide into smaller and smaller branches. The minute veins are those vessels which gradually enlarge from the successive addition of smaller roots. The true capillary vessels are obviously distinct from each of these. They do not become smaller by subdivision, nor larger by conjunction; but they are characterized by continual and successive union and division, or anastomoses, whilst they retain a nearly uniform diameter. . . . The term capillary must be reserved and appropriated to designate vessels of a distinct character and order, and of an intermediate station, carrying red globules, and perfectly visible by means of the microscope.[110]

Hall's book contains ten plates, elegantly showing the distribution of vessels in the tail of the stickle-back, the web of a frog, the mesentery of the toad, the lungs of the salamander, frog, and toad. In addition to his observations, Hall also repeated various experiments which had been designed to explain the causes of the blood's circulation. Among his conclusions was the assertion that the arteries possess irritability and hence have the power to contract, while the capillaries do not.

ASSESSMENT

Although the capillary vessels were first observed over three hundred years ago, we have seen that more than half of the period was occupied with inconclusive investigations which led to little positive knowledge. This slow progress was not for lack of interest in the subject, since current medical theories postulated that many diseases arose from obstructions or abnormal circulation in the small vessels.

Real progress could not occur until a number of requirements for accurate biological experimentation had been satisfied. For a study of the capillaries the primary requirement was an instrument which would permit clear, direct observation of the vessels. The achromatic microscope was not de-

109. Edition of Philadelphia, 1835, p. 28.
110. Ibid., pp. 3–4.

veloped until the 1820's. The second requirement was adequate technique in experimentation and the handling of experimental animals, combined with the realization that many observed phenomena were caused by the experiment itself. Marshall Hall was aware of this problem, and criticized much earlier work on this basis. A third requirement was precise definition of the problem and confining one's attention strictly to it. A final requirement was an understanding of what current scientific tools and methods were applicable to this particular problem.

During the first 170 years of research on the microvascular system contemporary scientific and philosophical ideas were a strong influence. In the first half of the period the attempt to quantify all biological phenomena resulted in naive calculations of blood pressure and volume, and the strength of the vessels in man. More reliable data came from the empirical approach of researchers such as Leeuwenhoek, Haller, and Spallanzani, who limited themselves to observations of the small vessels of lower animals under different experimental conditions. At the same time, however, vitalistic ideas began to influence biological investigations. Toward the end of the period increasing attention was paid to the interaction of the vascular system with the nervous system.

The early nineteenth century opened with confusion, so far as knowledge of the microvascular system was concerned. Thus Marshall Hall's insistence upon returning to essentials, in the form of precise definitions, and his emphasis on accuracy, were necessary prerequisites for subsequent stages of microvascular research.

ASPECTS OF FORM AND STRUCTURE: THE RENAISSANCE BACKGROUND TO CRYSTALLOGRAPHY

CECIL J. SCHNEER

THE beautiful plane polyhedral surfaces of natural crystals, their structural and their geometrical relationships, their classification and their genesis are the concern of modern crystallography. Renaissance neo-Platonists intoxicated with Pythagorean form and number and harmony determined new directions for natural philosophy and set standards for future scientific concern. Platonic qualities and preoccupations colored the new thought. The high Renaissance search for harmony in nature never turned directly on crystals or crystal forms. Nevertheless in their insistence on an underlying Pythagorean basis for nature, in their continual preoccupation with Pythagorean images and models, the artist-naturalists of the quattrocento and cinquecento laid the framework for a new science. "Which are the harmonic numbers and which are not and whence the difference?" Plato asked in the Republic.[1] On the one hand realistic empiricists determined to reproduce with scientific accuracy the images before them, on the other hand completely given over to the Pythagorean aesthetic of number and geometric form, they set the standards and procedures for a universal architecture, an architecture justly famous for having inspired the Copernican revolution but not as clearly acknowledged as the force behind other developments. The inspiration for the painting of pictures and the making of images was Platonic and Hermetic, but by also demanding ideas which could be painted or modeled in clay or brass and harmonies which could be

This essay was developed from a memorial address for C. Doris Hellman read in part at the Twentieth Anniversary Dinner of the Metropolitan New York Section of the History of Science Society, 1 May 1973.

I shall always be grateful to Henry Guerlac who first introduced me to the history of science and brought me to share his delight in the achievement of the mind of the past.

1. Plato's *Republic, VII,* 541c.

played on actual strings, the Renaissance opened the door for modern science. Not so much the unnatural alliance between Plato and Democritus, as Alexandre Koyre wrote,[2] but that between the idea and the literal image, between the philosopher and the artisan, gave rise to crystallography.

In emphasizing Platonic influences, my aim is not to show the mystical and occultist underpinnings of science, but rather to demonstrate the drives toward order and rationality, that is to say toward science, within the neo-Platonic-Pythagorean activities of the fifteenth and sixteenth centuries, a period which might well be characterized as the Age of Harmony. Nowhere is this better illustrated than in the work of Johann Kepler, a man whose lifework appears a sustained drive toward a single goal. Kepler did not so much fail to achieve a mathematical physics or a modern celestial mechanics as he succeeded in achieving a world harmonics, the accomplishment of a Renaissance goal which most of us now remember only in so far as we see in Kepler the precursor of a science and a world of which he never dreamed.

As one of his many interests, Kepler made observations on the form and internal structure of crystals, a subject hardly approached again until the twentieth century. This paper is intended to sketch some aspects of the fifteenth- and sixteenth-century background out of which Kepler was to build the views of the relationships between crystals and solid geometry expressed in his works from the *Mysterium cosmographicum* of 1595 to the *Harmonice mundi* of 1619,[3] and particularly in his essay on the snowflake.[4] Kepler was not alone in his Pythagorean preoccupation with the harmony of the spheres. On the contrary to a significant extent the whole of the scientific academy, particularly Copernicus before and Galileo after, were similarly motivated and influenced. As late as the English Restoration period, Robert Hooke was to refer to himself and his friends as "Pythagoreans" and Isaac Newton in 1704 wrote of seven colors, which he related numerically to the seven chords of the musical octave in just intonation.[5]

Harmony in the Renaissance was a topic that included much more than simply music.[6] It was a particular branch of arithmetic identified with the

2. Alexandre Koyré, "Gassendi: Le Savant," in *Pierre Gassendi 1592–1655* (Paris: Albin Michel, 1955), p. 62.

3. Johannes Kepler, I, *Mysterium Cosmographicum* (1938), VI, *Harmonice Mundi* (1940), in *Gesammelte Werke* ed. Walther von Dyck, Max Caspar, and Franz Hammer (Munich: C. H. Beck).

4. Cecil Schneer, "Kepler's New Year's Gift of a Snowflake," *Isis*, 51 (1960), 531–545.

5. "Lines divided after the manner of a musical chord . . . as the Numbers 1, 8/9, 5/6, 3/4, 2/3, 3/5, 9/16, 1/2, and so to represent the chords of the Key, and of a Tone, a third Minor, a fourth, a fifth, a sixth Major, a seventh and an eighth . . . will be the Spaces which the several colours [red, orange, yellow, green, blue, indigo, violet] take up." Isaac Newton, *Opticks* (Based on 1730 ed.; New York: Dover, 1952), p. 128.

6. Angelo Policiano, "Opusculum, quod Panepistemon inscribitus," *Isagogue harmonica,* sub Cleonidae nomine, trans. Georgii Vallae (Venice: Bevilaqua, 1497). Two kinds of music are

good, the true, and the beautiful. It was the fundamental attribute of the Pythagorean One, who had become by virtue of history, God the Creator. Geometric form and structure as well as the length of the chords of the lyre define harmonious proportions. The essence of Renaissance Pythagoreanism was in the relatively specific and literal definition of harmony as simple integral ratio, or proportion. "God in the creation of the world employed arithmetic, geometry, music, and astronomy; we also avail ourselves of these arts in seeking the proportions of things, elements and motions."[7]

The interest of Renaissance art and science in mathematization was an outgrowth of the search for harmonious relationships in nature. Science in the sense of reasoned knowledge meant ratio and proportion. The arts representing nature were the human analogue of the divinely placed reflections or images of the Archetype, "traces of the intrinsic secrets of Nature," Luca Pacioli called them.[8] Renaissance man turned to harmony and proportion to find the order and dignity denied him in the morass of religious conflicts. What task is more pious, Marin Mersenne wrote in 1636, than to reveal the perfect and harmonious workings of the universe and thereby shed glory on the Creator?[9]

The geometric form of the universe was regarded as that of the sphere, the ultimate perfect solid. But the form was also required to be harmonic; the spheres generated music. To the literal Renaissance mind, the music of the spheres meant that the configurations and motions of the heavens are governed numerically by specific key integer relationships 2:1, 3:2, 4:3, the perfect harmonic ratios of the octave, the fifth, and the fourth. It was not sufficient to save the appearances, as Simplicius averred, in terms of perfect circles and uniform speeds.[10] The phenomena were also to be saved as perfect harmonies. "The eternal mind acts like a musician rendering all to harmonic proportions."[11] The planetary motions had been expressed as harmonies also by Ptolemy, but in his *Harmonia* rather than his *Almagest*. The *Harmonia* was immensely useful to the development of music, but

---

distinguished, natural and artificial. Natural music is either human or mundane. The structure of mundane music is that of the celestial motions; the aspects of the stars correspond to the perfect harmonies, motions to the east or west determine pitch; the altitudes prescribe the modes.

7. Nicolo Cusano, *De docta igorantia,* libro 2, cap. 13, in Piero della Francesca, *De prospectiva pingendi,* ed. G. Nicco Fasola (Florence: G. C. Sansoni, 1942), p. 17.

8. Luca Pacioli, *De divina proportione,* (MSS Bibl. Ambrosiana & 170 Sup.); facs. ed. Franco Riva, *Fontes Ambrosiani, 31* (Verona: Mardersteig, 1956); The quotation is from the Spanish translation of the 1509 Venice publication, *La divina proporción,* by Ricardo Resta (Buenos Aires: Losada, 1946), p. 151.

9. "God has put us in this world to be the spectators of his work and to consider the actions and the movements in order to admire the wisdom and the power of the artisan." Marin Mersenne, *Harmonie universelle* (Paris: Sebastien Cramoisy, 1636), fol. 103.

10. Simplicius, *Source Book in Greek Science,* ed. M. R. Cohen and I. E. Drabkin (Cambridge: Harvard University Press, 1966), p. 97.

11. Cusano, *Idiota, De Mente,* libro 3, cap. 6, in Piero, *De prospectiva pingendi,* p. 15.

offered no such assistance (until Kepler's revolution) to the calculating and predicting astronomer as did the *Almagest.*

Music and art as much as science were attempts, then, to penetrate to the innermost recesses of the world-order, attempts, as Pierre Duhem described ancient astronomy, based on a priori metaphysical and Pythagorean assumptions rather than on an empirical physics of machines which could be modeled in clay or brass.[12]

Among the ancients, Aristoxenus represented the empirical musician, basing his musical diapason on sense perception rather than arithmetic. In the Renaissance, it is the father of Galileo, Vincenzo Galilei, who challenged his master Gioseffo Zarlino of Chioggia, opposing empiricism to the new concept of the tempered scale. Was music then ever considered literally representational of nature? In 1589, Zarlino wrote (attributing to Diodorus Siculus) that there was formerly a time when men had the aspect of beasts without reason [ratio] and lived apart, hiding in caves; however, they were peaceful and ate only health foods. Then by some chance they discovered fire and cooking, banded together and made laws. In imitation of the nests of birds, they built houses and took clothing from animals. Thus they discovered architecture which consists of order and disposition, adopting the compass and rule, and so they progressed to the arts. And music, they made in imitation of the *ucelli,* the little birds, as they had developed architecture in imitation of their nests.[13] What a charming fable, three hundred years before anthropology! This is music as representation, as an empirical art.

But Zarlino also gives us the Pythagorean account of the origin of music in Egypt.[14] It was Thoth, the ibis-headed (Hermes Trismegistus), who invented number and writing and dice, who made the seven-chorded lyre, one chord for each of the seven planets to bring the celestial music to human ears.[15] The angular disposition of the planets was also related to the harmonies by the Renaissance music theorists following Ptolemy; this time, through the geometric harmonies of the regular polygons. For example the

12. Pierre Duhem, *The Aim and Structure of Physical Theory* (New York: Athenaeum, 1962).

13. Aristoxéne, *Eléments harmoniques,* French trans. Charles-Émile Ruelle (Paris: 1870), pp. 50, 51. Vincenzo Galilei, *Dialogo . . . della musica antica et della moderna* (Florence: G. Marescotti, 1581), pp. 33, 53. Gioseffo Zarlino, "Sopplimenti musicali," libro 8, cap. 4, *De tutte l'opere* (Venice: Francesco dei Francheschi, Senese, 1589), p. 12.

14. His source is Macrobius, *In somnio Scipio,* lib. 2, cap. 1, Zarlino, *Opere,* "Istitutioni harmoniche," p. 72.

15. Hermes made the first lyre from the shell of a tortoise. It had seven strings. He taught Orpheus to play it and Orpheus taught Thamyris and Linus. Trying to teach Hercules, Linus reproved him and was killed for it. The women of Thrace killed Orpheus and threw the original lyre of Hermes into the sea where it drifted until it came ashore on the Isle of Lesbos. Fishermen brought it to Terpander who showed it to the priests of Egypt as his own invention. There on the altars of the Gods, Pythagoras found it and brought it to Greece. Nicomaque de Gerase, *Manuel d'harmonique et autres textes relatifs a la musique, fragments,* trans. and ed. C.-E. Ruelle (Paris: Baur, 1881) p. 41.

opposition, when planets are diametrically opposite each other on the celestial sphere, the trine, the quartile, the sextile (planets at 120, 90, and 60 degrees) represented the octave, the fifth, the fourth, and the minor third (5:6) respectively.[16]

In the 1595 astronomy of Kepler, the polygons triangle, square, hexagon and so on were replaced by the sequential inscription of regular solids of Plato and later in 1619 by the five regular solids and one of two additional regular polyhedra which he had discovered. In his attempts to save the planetary appearances both as harmonic ratios and somehow as perfect solids, Kepler considered the ratios of the numbers of sides, edges, and corners of the solids and the ratios of the radii of their circumscribing to inscribed spheres.[17] In 1602 as an outgrowth of his consideration of a magnetic-like force emanating from the sun, he related the area of the spreading force to the velocities of the planets. The volumes of regular solids as functions of the radii of the inscribed and circumscribed spheres were a principal part of the neo-Platonic mystique. When Kepler compared the ratios of the volumes of the circumscribing spheres, everything fell into numerical order. The areas of the second law were related to the periods which in 1595 he tried desperately to relate to harmonic ratios of the orbital radii. The volumes of the spheres circumscribing the five perfect solids were proportional to their radii cubed. Kepler as physicist provided Kepler as numerologist with the tables of ratios for comparison. The idea of which he was so proud in 1595, replacing polygons with polyhedra, led him in 1618 to the cubes of the radii and his third or harmonic law. In the first and second laws, Kepler had saved the appearances in the sense of reconciliation with geometric form and a mathematical regularity of motion (rather than uniform speed). With the third law he was to save the phenomena by demonstrating the true harmony of the disposition and periods of the planets; and the disposition in turn was harmonic not only in the sense of the ratios of the seventeenth-century musical space, but also in the use of the perfect solids.[18] Kepler was successful in representing the universe as a gigantic harp, the ancient goal of the Pythagoreans, because he was a Copernican.

16. Ingemar Düring, "Ptolemaios und Porphyrios über die Musik" (German translation "Harmonielehre des Klaudios Ptolemaios"), *Goteborgs Högskolas Arsskrift,* 40 (1934), 126–135. The study of aspects goes back to Babylonian astrology (p. 275).

17. Kepler, *Mysterium,* pp. 28, 46, 48.

18. Ibid., p. 46 compares the ratios of the inscribed to the circumscribing spheres times 1000 (column 1), with the Copernican ratios of the distances (column 2); ratios rather than dimensions are given to correspond with music.

		1000	Saturn
Cubo	577	572	Jupiter
Pyram.	333	290	Mars
Dodec	795	658	Earth
Icos	795	719	Venus
Oct	577	500	Mercury

All previous attempts were doomed to failure or at least to "inelegant" solutions because in a geocentric universe, not even an approximation to an orderly set of chord lengths can be derived from the planetary configurations. There is no consistent rule which will give even the order of the planetary distances from the earth—for example those of Mars, Venus, and Mercury.[19] In spite of this difficulty, however, Kepler's predecessors from the Pythagorean Philolaus on had constructed such harmonic universes.[20] I will cite only one example, that of the Minorite friar Francesco di Giorgio da Veneto, 1460–1540 (not to be confused with the architect Francesco di Giorgio). In 1525, a century before Kepler's *Harmonices mundi,* Giorgio published *De harmonia mundi* in which he rigorously derived the architecture of the universe from the principles of harmony, successfully inasmuch as he felt not the slightest concern for the astronomical appearances, reserving his empiricism for music! It was the musical or aural appearances which were saved.

I have found no direct references by Kepler to this earlier *Harmonies of the World* but the system of human proportions of Giorgio's *Harmonia* influenced Albrecht Dürer (according to Erwin Panofsky), and through Dürer passed on to Kepler.[21] Giorgio's *De harmonia mundi* is itself an architectural construct in the form of a sonata in three canticles, each with eight tones, numerous *modula,* and so forth. He represented the harmonic ratios of human physiology and psychology, astronomy, and architecture, as well as theology, philosophy, and history. The influence of Marsilio Ficino and his Hermeticism permeates the work, as well as the influence of Pico della Mirandola. It brought Mersenne's scathing criticism down upon it but not until after it had undergone a second Latin edition in Paris in 1545, and a French translation in 1579.[22] Giorgio, or Zorzi as he is sometimes identified, was steeped in the Ficinian and Pichinian neo-Platonism which saw in the *Pimander* of Hermes Trismegistus an authority of biblical

19. "Then the primary planets, by radii drawn to the earth, describe areas in no wise proportional to the times; but the areas which they describe by radii drawn to the sun are proportional [to the times] of description." Isaac Newton, "The System of the World" book 3, *Mathematical Principles* (Berkeley: University of California Press, 1962), Phenomenon 5, p. 495.

20. "The nature of the cosmos is harmony.... The cosmos and the things that are in the cosmos are harmonically composed." Philolaus, ed. Antonio Maddalena, *I Pitagorici* (Bari: Gius. Laterza and Figli, 1954), p. 50.

21. Erwin Panofsky, *The Life and Art of Albrecht Dürer* (Princeton: Princeton University Press, 1971), pp. 268, 270.

22. a. Francisci Georgi Veneti Minoritanae Familiae, *De Harmonia Mundi totius cantica tria,* Biblioteca Archigymnasio, Bologna, MS A235–237; b. (Venice: Bernardinum de Vitalibus, 1525); c. (Paris: Antonium Burthelium, 1545); d. ed. Réné Benoit (Paris, 1546); e. Fr. trans. Le Fevre de la Boderie, *L'Harmonie du mond* (Paris: Jean Macé, 1579). Another Paris edition under the title *Promptuarium rerum theologicarum et philosophicarum* appeared in 1566.

dimensions, and which attributed a spurious antiquity as well as a theologically dubious sanctity to the Hermetic corpus.[23] At the summit of the system was the Archetype, the plan in the mind of God. Sets of image-models such as the planet-pantheon replicated on hierarchical levels the gradations between man and God.

The Renaissance artist was primarily a maker of images. Duhem quotes Le Fevre de l'Etaples who wrote in 1503 that the astronomer in composing a correct representation of the heavens and their movements, imitates "the good and wise Artisan of all things" who produced the true heavens and their motions.[24] Man is the model of the world, Leonardo da Vinci wrote, called by the ancients the microcosm (*mundo minore*), duplicating in his person, not only the fire, earth, air, and water of the innermost spheres, but in his bones, the rocky mountains, in his pulse, the oceanic tides.[25] In the *Divine Proportion,* Pacioli wrote, "all measures and their denominations are derived from the human body and all the species of proportions and proportionalities entering the intrinsic secrets of nature are signed in it by the [hand] of the Highest."[26] This Hermetic-Pythagorean approach was diametrically opposed to the traditional Aristotelian-Averroist view of the harmonies of the spheres which treated them literally and, finding them physically impossible, dismissed them out of hand. First, the planets were above the sphere of air, but also they must move with their spheres "like the nails in a ship," and finally, the motion of the spheres is necessarily frictionless. It was impossible that the motions of the spheres could generate music.[27]

Piero della Francesca was a representative artist of the quattrocento. Aristotelian and therefore physicist enough to attempt to represent nature (rather than seeing himself as a constructor of ikons), he devised the science of projective geometry to do so in the most exact and precise manner possible. Piero's *De prospettiva pingendi* was the painter's *Almagest* for it set out to do for the artist what Ptolemy did for the astronomer, to provide the mathematical apparatus for the recreation of the image of the world; to make a mathematical science of painting, a task which Dürer also was to

23. Francesco Giorgio Veneto, "Testi scelti," ed. Cesare Vasoli, *Archivio di filosofia,* 1 (1966), 79–104. Frances Yates, *Giordano Bruno and the Hermetic Tradition* (New York: Random House, 1969).

24. Pierre Duhem, *To Save the Phenomena* (Chicago: University of Chicago Press, 1969), p. 57.

25. Giuseppe De Lorenzo, *Leonardo da Vinci e la geologia* (Bologna: Zanichelli, 1920), p. 8. Cf. Cecil Schneer, "Leonardo da Vinci, Geology" *Dictionary of Scientific Biography,* vol. 8, ed. Charles C. Gillispie (New York: Scribner's, 1973), 241–243.

26. Pacioli, *La divina proporción,* Resta trans., p. 151.

27. Nicole Oresme, *Le livre du ciel et du monde,* trans. A. D. Menut, ed. A. D. Menut and A. J. Denomy (Madison: University of Wisconsin Press, 1968). chaps. 17, 18, pp. 409–487.

undertake.[28] To this end Piero also introduced natural landscape and realistic portraiture into his painting.

Geometric perspective for the artist Piero was empirical physics, the science of the literal representation of the world. This is Piero as a naturalist. But Piero was also a Pythagorean with all the a priori assumptions of ideal beauty and form and harmony. The proportions of the human figure, the microcosm, were the visible analogues of the ratios of string lengths which made for harmony, and the forms of the five regular solids, the Platonic solids. "We will never admit that there are more beautiful visible bodies than these, each in its kind," Plato said.[29]

Because in the *Timaeus* the god imposes harmony and proportion and the geometric form of the regular solids on the cosmos, Piero della Francesca as Pythagorean turned to mathematics to impose his assumptions of ratio and geometric form on the nature he portrayed as physicist-naturalist. He had studied mathematics in his youth. He early was associated with Leon Battista Alberti, a painter, also a master of architecture and perspective, and a pioneer scientist of mechanics. In the provincial Tuscan town of Borgo San Sepulchro and at the court of Urbino, he composed his work on perspective which he dedicated to the duke, Federigo da Montefeltre, and also after 1484, a book in three parts titled *De quinque corporibus regularibus,* which he dedicated to Guidobaldo, the ducal heir and successor.[30] It was an arithmetic geometry of the Platonic solids based on Euclid's Books 13, 14, and 15 (the latter two attributed to Hypsicles), but also original in that it calculated numerical cases so that it was a practical geometry for the working artist and architect. Piero began with the regular polygons, went on to the five regular solids, the ratios of their sides, axes, areas, volumes, and the diameters of the circumscribing spheres and finished with the proportions of the solids inscribed one within the other. It is here then that the computational geometry employed by Kepler first entered the Renaissance.

In Borgo San Sepulchro, Piero had a pupil named Luca Pacioli, a Franciscan friar and leader of the Italian mathematical renaissance. He was the principal source of Leonardo's mathematics. Piero included Pacioli in at least one painting and recommended him to Leon Battista Alberti, with whom Pacioli lived for some months in Rome.[31] The *Summa de arithmetica, geometria, proportioni, et proportionalita* which Pacioli published in Venice in 1494 was a mathematical encyclopedia bringing together the works of Euclid, Boethius, Jordanus Nemorarius, Sacrobosco, and Regiomontanus, a

28. Piero della Francesca, *Petrus Pictor Burgensis De prospectiva pingendi Codice Parmense,* 1482 (?). Albrecht Dürer, *Unterweisung der messung mit dem Zirkel und Richtscheit,* facs. of 1525 edition, ed. Alvin Jaeggli (Zurich: Josef Stocker-Schmid, 1966).

29. Plato, *Timaeus,* trans. H. D. P. Lee (Baltimore: Penguin, 1965), pp. 53, 72, 73.

30. Piero dei Francheschi da Borgo San Sepulcro, *De quinque corporibus regularibus et de corporibus irregularibus* (Codice Urbinate-Vaticano 632, Vatican Library, + 1492).

31. Aldo Mieli, "Prólogo," in Pacioli, trans. Resta p. 17.

work which the historian of mathematics Gino Loria said marked the beginning of Italian algebra.[32] In 1496, Ludovico el Moro, the duke of Milan, brought Pacioli to the court to join the group of scholar-engineer-artists that included Leonardo himself and the musician-theorist Franchino Gafurio.[33] History remembers this group as the Accademia Leonardo da Vinci, a worthy Milanese counterpart of the Platonic Academy of Florence or the court of the Montefeltre at Urbino.[34] Pacioli was on close terms with Leonardo. They escaped the city together in 1499 when it was overrun by the French, and lived together for a time in Florence.

In 1498 before the fall of Milan, Pacioli composed a *Compendium of the Divine Proportion,* which he dedicated to Ludovico. It began with a mathematical treatment of the divine proportion (the ratio which Euclid treated as the division of a line into mean and extreme segments, and which Kepler called the Divine Section, later known as the Golden Section). It is the limit of the Fibonacci series, 1, 2/1, 3/2, 5/3, 8/5, . . . = 1.618 . . . and it enters into the geometry of the regular solids. Pacioli's book continued with the geometry of the regular solids largely derived from Euclid. Two manuscripts are preserved, one of which came to the Ambrosiana Library in Milan as a gift of Count Arconati along with the Leonardo material now known as the *Atlantic Codex.* The other with inferior illustrations is in the library of Geneva. The illustrations in the Arconati manuscript might well be from the hand of Leonardo, the "ineffable left hand" which, Pacioli wrote, had illustrated the work later published in Venice with crude woodcuts. Some of the illustrations appear independently in the *Atlantic Codex* itself.[35] All Leonardo's virtuosity is engaged in representing the complex solids in perspective, especially as he undertook to illustrate them as *elevatus,* that is, starred by the superposition of other solids on the faces, and also truncated (*abscisum*) and skeletal (*vacuo*). The addition of regular pyramids to the faces produces a quasi-regular polyhedron. In some cases such as that of the *octahedron elevatum,* known as the *stella octangula,* it

32. Quoted by Giuseppina Masotti-Biggiogero, "Della vita et delle opere di Luca Pacioli," in Pacioli, ed. Riva, p. 221.

33. Franchino Gafurio, b. 1451, was concert master of the cathedral of Milan and the court of Ludovico Sforza; he is believed to be the *Musician* of Leonardo's portrait (Pinacoteca Ambrosiana, Milan). Gafury, *De harmonia instrumentali libri tres,* Bibliothéque Nationale, Paris, MSS Latin 7208, 1500, on the theory and practice of music is a detailed specification of the harmonies through mathematics, astronomy, physiology, and psychology. The tetrachord is linked to both the four seasons and the four elements; the planets are also the modes, presided over by the muses, the seven virtues are congruent with the seven varieties of the diapason, and so on.

34. "Not an academy in reality, but a society of fraternal spirits, an 'invisible college,'" Giuseppina Fumagalli, "Leonardo: ieri e oggi," in *Leonardo Saggi e Richerche,* ed. Achille Marazza (Rome: Istituto Poligrafico dello Stato, 1954), p. 412. The title "Accademia Leonardo da Vinci" appears in decorative engravings made at Milan by Leonardo.

35. Luca Pacioli, *Divina proporzione opera a tutti gli ingenii perspiaci . . .* (Venice: Paganus Paganinus du Briscia, 1509). Leonardo, *Codex Atlanticus.* fol. 263r.

requires a mathematical lawyer to establish that it is *not* a regular solid in Euclid's definition. Leonardo's *icosahedron elevatum* is very close to the great stellate dodecahedron which the mathematical lawyers have ruled is a regular solid, each of its twenty sides a regular pentagonal star. This icosahedron is one of the two additions to the Platonic solids reputed to have been discovered by Kepler who was also credited with the *stella octangula.*[36] The other of the two Keplerian solids was also attempted by Pacioli and Leonardo but less successfully. Pacioli did not understand the stellation procedure. He placed a pyramid of five equilateral triangles on each of the pentagons of the quintessence, the dodecahedron.[37] Kepler simply extended the sides to intersection leading to a pyramid of five isosceles triangles on each pentagon.

Kepler nowhere mentions Pacioli or Leonardo, he also nowhere claimed the stellate polyhedra as his own invention; but by the seventeenth century, this kind of geometry was already part of the neo-Platonic apparatus considered as necessary for the education of the architect-engineer, along with music theory, perspective, and mechanics.[38] The solids were represented as inscribed one within the other (compare Kepler's polyhedral cosmology a century later) in a hanging glass model, part of the background to Jacopo da Barbari's portrait of Pacioli explaining Euclid to a handsome pupil, possibly Guidobaldo, duke of Urbino. In 1489, Pacioli was engaged in constructing models of the solids for Guidobaldo, and in 1494 he dedicated his *Summa* to Guidobaldo.[39] At some point, Pacioli translated into Italian Piero's book of the *Five Regular Solids* which had also been dedicated to Guidobaldo, and in 1509, he published the *Divina proportione* in Venice with his translation as the third part, changing the dedication, however, and omitting his master Piero's name as author.[40] He also included Piero's measurements and drawings for the system of human proportions which he expounded in the second part of the book.

Albrecht Dürer came to Venice in 1494 while Pacioli was teaching mathematics there, and again in 1505, drawn over the Alps like Copernicus, who was also in Padua from 1501 to 1503 and who had also studied painting at one time. In this Venetian environment (which included Padua), Giorgio Valla, the humanist scholar, was translating and editing Euclid's Books 14

36. Schneer, "Kepler's Snowflake," pp. 531, 532n.

37. Pacioli, *Divina proporzione,* (1509), tavola 31.

38. Cf. Giacomo Barozzi da Vignola, *La prospettiva . . . con commento del P. Egnazio Danti* (Bologna: Lelio dalla Volpe, 1744), first edition was 1583; also Cecil Schneer, "Steno: On Crystals and the Corpuscular Hypothesis," in *Steno as Geologist,* ed. Gustav Scherz (Odense: Odense University Press, 1971), p. 304 n.; also Ferdinando Galli Bibiena, 1657–1743.

39. The portrait is in the Pinacoteca di Napoli. Luca Pacioli, *Summa de arithmetica geometria proportioni et proportionalita* (Venice: Paganus Paganinus, 1494).

40. Girolamo Mancini, "L'opera 'De corporibus regularibus' di Pietro Francheschi detto Della Francesca, usurpata da fra Luca Pacioli, Memoria," *Atti de' lincei, memorie, classe di scienze morali,* Ser. 5, vol. 14, 453 (1913).

and 15 (Hypsicles) on the regular solids. He also translated Aristarchus, Ptolemy, Cicero, Proclus Diadochus, Eusebius, and Pliny on astronomy and natural history; Cleomedes, Nicephoros, and Psellus on harmony; also Galen, Hippocrates, Rhazes, Aristotle Aphrodisias, Averroes, Apollonius, Philo, and Athenagoras. Valla, a cousin of Lorenzo Valla, had been tutor to the sons of El Moro when Leonardo arrived at the court of Milan in 1482.[41] Pacioli was his successor not only in Milan, but later in Venice.

Both at Venice and his native Nuremburg, Dürer was on close terms with Jacopo da Barbari, who "showed him two figures male and female . . . constructed by geometrical methods." Jacopo also combined in his painting the Pythagorean quest for an ultimate perfection (the image of which would be the human form) with the perspective science of the representation of nature.[42] Dürer's *Introduction to Measurement* of 1525 is one of four books which he wrote on art and architecture as applied mathematics. He intended to carry out in Germany that program of converting the sense impressions of the artist as naturalist into the science of Pythagorean-hermetic imagemaking which in Italy had begun with Filippo Brunelleschi and the discovery of geometric perspective. His *Introduction* is modelled on the *Perspective* of Piero and the *Divine Proportion* of Pacioli. Like the latter it treats and illustrates the five solids but goes beyond Pacioli in the representation of Archimedean and other solids. Like the Pacioli, the Dürer work constructs an alphabet, that is, a type face, using perfect circles and harmonic proportions. Like the Pacioli and the Piero works, the Dürer book treated the ideal proportions of the human face and figure, and related these to the canons of architecture. But Dürer, who in Nuremberg was involved in astronomy and every facet of humanist culture, also included in his *Introduction* practical means for the construction of sundials and other instruments, methods which he must have learned as an apprentice to the goldsmiths. The construction out of sheet brass of these sundials and of globes may have inspired Dürer to mark off templates for the construction of first, tesselations filling a plane, and then, of tesselations on a plane which could be cut and folded into polyhedra. This construction for the cube only had appeared in Eutocius, one of the mathematicians translated by Valla and studied by Dürer. Dürer did not simply print templates of six squares to be folded into the six sides of the cube, but worked out constructions as complex as the cube modified by the octahedron and the trisoctahedron (thirty-eight sides).[43]

By 1578 when Christopher Clavius was publishing his first *Euclid,* the

41. Giorgius Valla, 1447–1500. Collections of his translations were published (Venice: Bevilaqua, 1497, 1498); and posthumously by his adopted son, Johannes Valla. Cf. J. L. Heiberg, "Beitrage zür geschichte Georg Valla's und seiner bibliothek," *Beihefte zum centralblatt für bibliothekswesen,* 16 (1896).

42. Panofsky, *Dürer,* p. 35.

43. Ibid., p. 260; Cf. Giorgius Valla, *De expetendis et fugiendibus rebus* (Venice: Aldi Romani, 1501); Cf. Dürer, *Unterweisung,* fol. N5r,v, and fig. 42, fol. N4v.

method of Dürer was already in use, as for example in P. Egnatio Danti's 1573 *Vignola.* Although Clavius mentions Dürer as the originator of the device, Kepler, who refers to Dürer elsewhere, finds it unnecessary. In Clavius' *Euclid,* there is a sixteenth book extending the geometry of the mutual inscription of the Platonic solids. It was written by François de Foix, comte de Candalla, and it was the source of the proportions for Kepler's polyhedral-cosmological model. Kepler refers frequently to Clavius' mathematical and astronomical works, and in these Clavius uses Pythagorean drawings of lattices of points to illustrate squared and cubic numbers, as well as precise drawings to illustrate the mutual inscription and circumscription of polyhedra.[44]

It is a convenient point at which to place the origin of crystallography. Kepler may have been the first to associate the regular plane polyhedral forms of crystals with an internal structure (the association of crystal forms with the Platonic solids is much earlier and might explain the association with the elements). In his little pamphlet on the snowflake, Kepler describes the relationship between the hexagonal form of the snowflake and "material necessity" that is, a regular packing of droplets of water. But in the snowflake pamphlet it is Kepler at work as a Pythagorean, fitting the appearances to a priori assumptions.[45] Kepler the observer and naturalist who created a new astronomy out of a discrepancy of eight minutes of arc in the orbit of Mars would have known that, had the *formative urge* intended to construct a dodecahedron in the silver ore of the inlaid panel which he saw in the Royal Palace at Dresden, it would have formed one with equilateral pentagonal sides and every angle exactly 108 degrees, rather than with isosceles pentagons. He would have noticed that the triangles of the "front part of the icosahedron" exhibited at the baths at Boll were once again, isosceles but not equilateral or equiangular. But then, he wrote, excusing himself from possible error, "I have only knocked on the door of chemistry and I see how much remains to be said."[46]

By this time alchemists, pharmacologists, chemists, mining mineralogists, and metallurgists had developed an empirically based practical science which occasionally employed a crude morphology as a means of identification of inorganic species.[47] Crystallography was in the same position of

44. Christophoro Clavio, *Euclidis posteriores lib. IX accessit XVI De solidorum, . . .* a Francisco Flussate Candalla adiectum . . . (Rome: Bartholomaeum Grassium, 1589). In his collected works, Kepler cites Clavius nearly one hundred times; he also acknowledges Dürer. In the *Mysterium,* chap. 13, "Computation of the spheres which are inscribed and which circumscribe the polyhedra"; Kepler cites Candalla, p. 46 *passim.*

45. Johannes Kepler, *Strena seu de nive sexangula* (Francofurti ad Moenum: Godefridum Tampach, 1611); the pertinent passages are in Schneer, "Kepler's Snowflake," p. 544.

46. Kepler, *Strena* (1611), p. 24.

47. Conrad Gesner, *De omni rerum fossilium genere, Gemmis, Lapidibus, Metallis . . .* (Tiguri: Iacobus Gesnerus, 1565) contains qualitative geometric descriptions of crystals as; "Pyrites, octahedrons, dodecahedrons from which fire is drawn." (p. 78).

sophistication relative to these crude observations of rectangular or hexagonal or pointed specimens as the mathematical astronomy of Ptolemy had been, compared to the physics of Aristotle. In the seventeenth century, Descartes, Bartholinus, Boyle, Hooke, and Huygens developed the early suggestions of Kepler of a relationship of an internal packing of structural units to the external morphology of crystals. It was Robert Hooke who in 1665 in the *Micrographia* showed the regularity of angle independent of infinite variety in the sizes of the crystal faces; Huyghens who in 1695 anticipated Bergmann and Romé de l'Isle in relating the external shape of the calcite rhombohedron to an ellipsoidal shape of the packing particles, and most important, did this quantitatively.

There are reasonable grounds to suppose that Nicolas Steno, who became bishop of Titopolis, was familiar with Hooke's *Micrographia* when he took the dramatic step of abandoning the Pythagorean assumption of relating the five Platonic solids to natural forms. As Kepler had once seen the necessity of abandoning circles and uniform velocities for the ellipse and the law of areas, Steno in the 1669 Prodromus to a never-to-be-written *Dissertation on Solids contained within solids,* abandoned the requirement of the 90 degree angle, the hexagon, pentagon, and the equilateral triangle of the perfect solids. Steno took the method of Dürer for the construction of models of mathematical polyhedra from templates or tesselations to construct templates for models of crystals of hematite from the island of Elba. His drawings of quartz crystals show that he fully understood the principle of Hooke, namely the independence of the angle from the accidents of sizes and truncations of the faces of crystals. For Hooke, this was a necessary consequence of the regular stacking of spherical corpuscles, atomism in short. Steno, whose training and early work had been in anatomy as an empiricist and observer, and who in Italy had become a convert to Catholicism and was shortly to abandon the world for the religious life, was hostile to the idea of atomism. Forced to concentrate on the surface form, his rejection of the internal periodicity of packed spheres meant a rejection of the perfect angles, 90 degrees, 60 degrees, and so on, and opened the way for his measurements of actual interfacial angles. These are not given in the Prodromus or his notes, but we know that he had to make them because the templates for the models are given and are accurate.[48]

In a real crystal, individual faces of a symmetric form are never exactly congruent but are always similar (Hooke's principle). To represent them on a model, as Steno has done, not only for one set of symmetric forms but for several sets on two different crystals of the same mineral, Elba hematite, Steno had to measure the angles accurately and to grasp the idea that corresponding angles were the same and were characteristic not only from

48. Schneer, "Steno."

face to face within the same set, but from specimen to specimen of the same species. It is the first law of crystallography. To complete the model, he made all the faces of each symmetric form not only similar according to the principle of Hooke, but even congruent. These observations in their departure from 30 and 90 degrees would be clearly at odds with the primitive Pythagorean atomism of the seventeenth century. After the example of Kepler, Steno had saved the appearances but at the expense of the assumptions. Before Newton, he was combining the theoretical astronomer beloved of Plato with the observing physicist, or naturalist of Aristotle. The idea of an internal periodicity which he had discarded was actually premature. It would require a century of observations to establish what the twentieth-century French crystallographer Georges Friedel called "the morphological basis of the reticular hypothesis," that hypothesis which had been expressed by René Just Haüy in 1802, and hailed by Georges Cuvier.

A mathematical science of crystals became possible when Kepler saw Platonic solids in crystals of ore at the silver court of Dresden and when Kepler, Descartes, and Hooke, saw a model for this in the number-lattices of Euclid's Book 13, but the physical science began only when Steno, constructing models of hematite from Elba using the method of Dürer, went beyond the Pythagorean ideal in the manner of Piero and Leonardo to make an exact science of the representation of nature. As Kepler before him had abandoned the circle for the ellipse, Steno abandoned Platonic solids for actual measurements of mineral species. As Kepler's laws were to wait nearly seventy-five years for Newton, Steno's discovery would not be understood and finally formulated in contemporary terms for over a century.

V

THE REVOLUTIONARY ASPECTS OF MODERN PHYSICS

HOW TO GET FROM HAMILTON TO SCHRÖDINGER WITH THE LEAST POSSIBLE ACTION: COMMENTS ON THE OPTICAL-MECHANICAL ANALOGY

THOMAS L. HANKINS

In the second part of his famous paper of 1926 on wave mechanics, Erwin Schrödinger wrote at length about the optical theory that Sir William Rowan Hamilton had published almost a century earlier. Hamilton had found that he could write the laws of geometrical optics and the laws governing the motion of particles in the same mathematical form. Of course the physical quantities used were different, but the equations themselves were formally the same in both cases. This formal analogy between his theory of optics and his theory of dynamics was extremely satisfying to one with Hamilton's metaphysical predilections, but in the nineteenth century it was nothing more than an analogy, and Hamilton had no experimental evidence that would indicate a more fundamental unity between the properties of light and the properties of material particles.

By 1926 the situation was quite different. Classical mechanics had failed in the realm of the very small and had been replaced by quantum mechanics. The work of Louis de Broglie suggested that in the world of the atom, material particles might have wave properties like those of light. The optical-mechanical analogy was becoming not only an analogy of mathematical formalism, but also an analogy relating the essential properties of matter and light.

In his paper of 1926, Schrödinger exploited the Hamiltonian analogy to show how his famous wave equation was related to the principles of classical mechanics and optics. He pointed out that just as ray optics failed when one attempted to apply it to the realm of the very small, so had Hamilton's classical mechanics of particles failed at approximately the same point. The

This essay appears, in a slightly different form, as a chapter in my book *Sir William Rowan Hamilton,* published by The Johns Hopkins University Press in 1980.

diffraction and interference of light passing through very small apertures could not be explained by the geometry of rays; only the wave theory could account for the observations satisfactorily.

In the transition from the world of our common experience into the world of the atom, classical mechanics had to give way to quantum mechanics just as ray optics had to give way to wave optics. Schrödinger speculated that these failures might be related; that there was some fundamental connection between the physics of optics and the physics of particles that caused their classical formulations to fail at approximately the same place in the scale of magnitude. If such a fundamental connection existed, it would explain in part why Hamilton was able to find a common formulation for optics and mechanics in his characteristic function. These are not two different sciences at all, but different aspects of the same science. Schrödinger wrote:

> *We know today, in fact, that our classical mechanics fails for very small dimensions of the path and for very great curvatures.* Perhaps this failure is in strict analogy with the failure of geometrical optics, . . . that becomes evident as soon as the obstacles or apertures are no longer great compared with the real, finite, wave length. Perhaps our classical mechanics is the *complete* analogy of geometrical optics and as such is wrong and not in agreement with reality; it fails wherever the radii of curvature and dimensions of the path are no longer great compared with a certain wavelength, to which in q-space a real meaning is attached. Then it becomes a question of searching for an undulatory mechanics, and the most obvious way is by an elaboration of the Hamiltonian analogy on the lines of undulatory optics.[1]

If the Hamiltonian analogy were a complete analogy, then the wave properties which allow one to pass satisfactorily from the macroscopic to the microscopic world in the case of light could be applied equally well to particles and provide a similar satisfactory transition for mechanics. A "wave mechanics" might be expected to apply equally well to the macroscopic and microscopic worlds. This is exactly what Schrödinger found.

A closer look at the optical-mechanical analogy reveals that there are really two analogies involved, or perhaps it would be better to say that the analogy has two aspects. These two aspects are more pronounced in Hamilton's writing than in Schrödinger's and therefore it is worthwhile keeping them distinct. (1) The first part of the analogy concerned the debate over the physical theory of light that was raging fiercely at precisely the time Hamilton was writing his optical papers (1825–1833). The debate was whether light was a wave or a particle and Hamilton could not ignore the question. To his great pleasure and surprise, he found that his mathematical optics

1. "Quantisierung als Eigenwertproblem (Zweite Mitteilung)," *Annalen der Physik,* (4) Bd. 79. 1926. The translation is from the *Collected Papers on Wave Mechanics,* trans. from the 2d German ed. of the *Abhandlungen* J. F. Shearer and W. M. Deans (London: Blackie and Son, 1928), p. 18. I have altered an occasional word in the translation in order to make it more faithful to the German.

applied equally well to waves *and* to particles.[2] His theory stood in a sense entirely outside the debate, because it could not decide between the conflicting theories. It also stood outside the debate because Hamilton considered only geometrical optics and did not treat the phenomena of diffraction and interference which require the wave theory. Therefore his theory of optics never really came to grips with the problem of determining the physical nature of light. Hamilton's system described both waves and particles, because it associated with any bundle of rays a series of surfaces orthogonal to them. The mathematics described the rays and surfaces at the same time. If you preferred the particle theory of light, you would use the ray aspect of the equations. If you preferred waves, the orthogonal surfaces and not the rays would have physical meaning, since they represented the successive positions of an expanding wave front.[3] Hamilton believed in waves of light largely for metaphysical reasons, but as he recognized, his own optical theory could never resolve the conflict.

The very fact that Hamilton's theory could not decide between waves and particles was significant for Schrödinger. What Hamilton could never imagine in the 1820's was apparent a century later. It is not a matter of "either-or," but a matter of "both-and." Light quanta and subatomic particles exhibit both particle and wave properties simultaneously. At the close of his Nobel Prize Address of 1933, Schrödinger asks us to understand this anomaly by appealing to the model that appears in Hamilton's very first optical paper. A subatomic "object," light quantum or particle, has "longitudinal continuity" along a ray and "transverse continuity" on a surface orthogonal to the ray.[4] It is these two forms of continuity that best conform to the dual nature of elementary particles.

(2) The second part of the optical-mechanical analogy as derived by

2. William Rowan Hamilton, *Mathematical Papers,* vol. I, ed. A. W. Conway and J. L. Synge (Cambridge: Cambridge University Press, 1931), 10, 14, 314. In a letter to Samuel Taylor Coleridge, 3 October 1832, Hamilton writes: "my chief desire and direct aim [has been] to introduce harmony and unity into the contemplations and reasonings of Optics, considered as a portion of pure Science. It has not even been necessary, for the formation of my general method, that I should adopt any particular opinion respecting the nature of light." Quoted from Robert P. Graves, *Life of Sir William Rowan Hamilton,* I (Dublin: Hodges, Figgis, 1882), p. 592.

3. The orthogonal surfaces give the successive position of wave surfaces of constant phase, but there is no direct analogy between the velocity of the light wave and the velocity of an imagined light particle moving along the ray. In fact the mathematical relationship is an inverse one. Hamilton, *Mathematical Papers,* I, 277–279. See also Cornelius Lanczos, *The Variational Principles of Mechanics,* 4th ed. (Toronto: University of Toronto Press, 1970), pp. 274–275.

4. "The Fundamental Idea of Wave Mechanics" published in *Science and the Human Temperament,* trans. James Murphy and W. H. Johnston (New York: Norton, 1935), p. 192. The wave-particle duality of quantum mechanics leads one naturally to ask if there is any real difference between electrons and light, or is it just a historical accident that electrons were first observed as particles and light as waves? An examination of the theory indicates that it was not an accident and that there are good reasons physics developed as it did. See R. E. Peierls et al., "A Survey of Field Theory," *Reports on Progress in Physics,* 18 (1955), 471–473.

Hamilton did not necessarily have anything to do with waves at all. He showed that he could describe the geometry of light rays and the motion of particles by the same mathematical formula. The analogy is a mathematical one and is revealed by comparing Hamilton's optical papers of 1827–1833 with his papers on mechanics of 1834–1835. In this part of the analogy we remain entirely in the classical world of geometrical light rays and billiard-ball particles.

The analogy is one of mathematical formalism. Two completely different sciences—geometrical optics and classical mechanics—are shown to have a mathematical, but not physical, equivalence. It is this second aspect, the formal mathematical equivalence of optics and mechanics, that is usually given the name of the optical-mechanical analogy. For an appreciation of Hamilton, it is necessary to keep in mind the difference between the mathematical theories and the physical theories of optics. In one sense it could be said that Isaac Newton had an optical-mechanical analogy, because he explained light by the mechanical motion and interaction of particles. He sought for each optical phenomenon an explanation in terms of the mechanical action of particles. But Hamilton's optical-mechanical analogy was quite different. By the application of mathematics, he was able to show that the entire science of geometrical optics and the entire science of rational mechanics could be given the same formulation.

In 1926, Schrödinger complained that Hamilton's theory had been misunderstood and that his optics had been unjustly ignored. He lamented the fact that other contemporary descriptions of Hamilton's theory had failed to emphasize its connection with wave propagation: "Unfortunately this powerful and momentous conception of Hamilton is deprived, in most modern reproductions, of its beautiful intuitive raiment, as if this were a superfluous accessory, in favor of a more colourless representation of the analytical connections."[5] This criticism immediately raises the question of the actual fate of Hamilton's ideas, and how Schrödinger came to revive them in 1926? If the optical-mechanical analogy was "lost" in the years between Hamilton's papers and those of Schrödinger, it was because the optical papers slipped from notice, while the mechanical theory with the additions of C. G. J. Jacobi received a great deal of attention. In 1895, H. Bruns published his theory of the "Eikonal," a new theory of geometrical optics which Bruns applied to the analysis of optical instruments. The Eikonal is basically a restatement of Hamilton's Characteristic Function, the fundamental equation of Hamilton's theory of optics. For Bruns it was a new discovery. He had apparently been entirely unaware of Hamilton's optical theory, had taken the well-known Hamilton-Jacobi equation of mechanics, and applied it to optical problems, repeating what Hamilton had done sixty years ear-

5. Schrödinger, *Collected Papers,* p. 13.

lier.[6] This would certainly indicate that Hamilton's optical papers had not reached a wide audience.

Yet the analogy did survive in British books on mechanics. The famous *Treatise on Natural Philosophy* by William Thomson (Lord Kelvin) and Peter Guthrie Tait (1867) contained a lengthy discussion of Hamilton's Principle of Varying Action and the differential equations derived from it. The authors explained: "Irrespectively of methods for finding the 'characteristic function' in kinetic problems, the fact that any case of motion whatever can be represented by means of a single function . . . is most remarkable, and, when geometrically interpreted, leads to highly important and interesting properties of motion, which have valuable applications in various branches of Natural Philosophy. *One of the many applications of the general principle made by Hamilton led to a general theory of optical instruments,* comprehending the whole in one expression."[7] And there follows a reference to Hamilton's optical papers and an example of his principles applied to "common optics." The analogy is there, but it is treated as a curiosity—not as a matter of significance. The authors continue: "The now abandoned, but still interesting, corpuscular theory of light furnishes a good and exceedingly simple illustration [of Hamilton's theory]." Obviously Thomson and Tait believed that the victory of the wave theory of light deprived the optical-mechanical analogy of any real physical significance.

An equally famous book on mechanics. E. T. Whittaker's *Treatise on the Analytical Dynamics of Particles and Rigid Bodies* (1904), contains a lengthy development of Hamilton's principle from the theory of contact transformations. In the preface Whittaker states: "I may mention that the new explanation of the transformation theory of Dynamics . . . sprang from a desire to do justice to the earliest great work of Hamilton's genius: . . . The origin of the method is to be found in a celebrated memoir on optics, which was presented to the Royal Irish Academy by Hamilton in 1824: the principles there introduced were afterwards transferred by their discoverer to the field of dynamics."[8] Whittaker goes on to say: "In order to follow Hamilton's thought, we must refer to the connexion between dynamics and optics—a connection which is perhaps less obvious in our day than in his when the corpuscular theory of light was widely held."[9]

6. J. L. Synge, "Hamilton's Method in Geometrical Optics," *Journal of the Optical Society of America,* 27 (1937), 75–77, and "Hamilton's Characteristic Function and Bruns' Eiconal," ibid., 27 (1937), 138–144. Hamilton's optical papers were published in the *Transactions of the Royal Irish Academy,* a journal that was practically unknown on the Continent. His papers on mechanics were published in the *Philosophical Transactions of the Royal Society* and had wide circulation.

7. *Treatise on Natural Philosophy,* 1962 ed. retitled *Principles of Mechanics and Dynamics* (New York: Dover, 1962), I, 354 (italics mine).

8. Second ed. (Cambridge: Cambridge University Press, 1917), pp. v and 288.

9. Ibid., p. 288.

There is great irony in these lines by Whittaker. He recognized the connection between Hamilton's dynamics and optics, but, like Thomson and Tait, he could not grant to it any physical meaning. Light was composed of waves in an electromagnetic field, matter was composed of particles, and never the twain could meet. He published a year too early. In 1905, Albert Einstein showed that the photoelectric effect could only be explained by granting a particle nature to light.

While the British writers emphasized the dynamical aspect of Hamilton's theory, they never lost sight of the optical-mechanical analogy, or of the fact that Hamilton had first derived his theory from a study of optics. In Germany, however, because of the influence of Jacobi, the optical aspect of Hamilton's work received very little attention and the emphasis was almost entirely on mechanics. Jacobi had accused Hamilton of not fully understanding the implications of his own work, and he showed that the problem of integrating Hamilton's partial differential equations of motion can be made much more general and much simpler.[10] The criticism is valid if the integration of the differential equations is the major point at issue. But this does not seem to have been Hamilton's aim. The British authors, even in the earliest expositions of dynamics, defended Hamilton against the criticisms of Jacobi. Arthur Cayley, in a lengthy "Report on the Recent Progress of Theoretical Dynamics" for the British Association in 1857 wrote:

> The method given for the determination of [the characteristic function], or rather of each of the several functions which answer to the purpose, presupposes the knowledge of the integral equations; it is therefore not a *method of integration,* but a theory of the representation of the integral equations assumed to be known. I venture to dissent from what appears to have been Jacobi's opinion, that the author missed the true application of his discovery; it seems to me, that Jacobi's investigations were rather a theory collateral to, and historically arising out of the Hamiltonian theory, than the course of development which was of necessity to be given to such theory.[11]

Jacobi's interpretation of Hamilton's work did tend to obscure the optical-mechanical analogy, and yet the analogy was scarcely "lost" even in Germany in spite of his supposedly nefarious influence and in spite of the general ignorance of Hamilton's optical papers. We can more properly say that it was ignored, but remained available to anyone who might wish to use

10. C. G. J. Jacobi, "Über die Reduction der Integration der partiellen Differentialgleichungen erster Ordnung zwischen irgend einer Zahl Variabeln auf die Integration eines einzigen Systems gewöhnlicher Differentialgleichungen," *Crelle's Journal,* 18 (1837); 97–162. See also Hamilton, *Mathematical Papers,* II, 613–621, and Lanczos, *Variational Principles,* pp. 254–262. Upon reading Jacobi's comments Hamilton wrote: "When I can find leisure to take up the subject again, I do not despair of showing him that in the way of generalisation the tables may be turned." Graves, *Life,* II, 247. The generalization to which Hamilton here refers is probably his Calculus of principal relations which he developed in 1836 prior to Jacobi's paper.

11. *British Association Reports* (1857), p. 10.

it. Thomson and Tait's *Principles of Natural Philosophy* was translated into German in 1871 at the suggestion of Hermann von Helmholtz and became an important source for German physicists. Whittaker was known, too; Schrödinger cites him in his 1926 papers.[12]

A major effort to "recover" the optical-mechanical analogy for physics was made during the 1890's by Felix Klein. In his *Vorlesungen über die Entwicklung der Mathematik im 19. Jahrhundert,* he recounts the history of his attempts. Klein tells us that Hamilton's optical foundations were ignored and his results were "snatched away" by Jacobi whose name replaced that of Hamilton, leaving Hamilton as only an insignificant precursor of Jacobi.[13] Klein is quite indignant about the whole matter. He claims that he learned of the true state of affairs by his travels, but was unsuccessful in his attempt to make Hamilton's optical and mechanical works known in Germany. Since Schrödinger acknowledges these efforts in 1926, we know that Klein's labor was not all in vain. In the summer of 1891, Klein worked out the Hamilton method making all of mechanics a method of optics in n-dimensional space, including Jacobi's later developments. And in the same year he lectured on the subject to the Naturforscherversammlung in Halle. He complained that his notes lay around for twenty years in the reading room at Göttingen, but attracted no enthusiasts.[14] Only two mathematicians joined in the effort to exhume Hamilton's optics. The first was Eduard Study, who in 1905 wrote a biographical sketch of Hamilton in honor of his hundredth birthday, followed by an article in which he developed Hamilton's geometrical optics using the theory of infinitesimal contact transformations.[15] The second attempt to revive Hamilton was by Georg Prange, who entitled his "*Habilitationsrede*" "W. R. Hamilton's Significance for Geometrical Optics."[16] Prange echoed the criticism of Cayley, that the integration of the Hamilton-Jacobi equation had not been Hamilton's major interest. He claimed that the great oversight of those who followed Jacobi was to miss Hamilton's work in geometrical optics, both for its theoretical and for its practical importance.

Neither of these articles by Klein's protégés had any noticeable effect. Klein attributed this neglect in part to the debate in Germany over the teleological implications of the Principle of Least Action. On this subject he

12. Schrödinger, *Collected Papers,* p. 13.

13. Klein, *Vorlesungen . . . ,* (Berlin: Verlag von Julius Springer, 1926), p. 198.

14. Ibid., p. 198.

15. Eduard Study, "Sir William Rowan Hamilton," *Jahresbericht der deutschen Mathematiker-Vereinigung,* 14 (1905), 421–424, and "Über Hamilton's geometrische Optik und deren Beziehung zur Theorie der Beruhrungstransformationen," ibid., pp. 424–438.

16. Georg Prange, "W. R. Hamilton's Bedeutung für die geometrische Optik. Habilitationsrede, gehalten am 26. Februar 1921," (*Jahresbericht der deutschen Mathematiker-Vereinigung* Halle), 30 (1921), 69–82. In 1933, Prange gave one of the most faithful accounts of Hamilton's theory: "Die allgemeinen Integrationsmethoden der analytischen Mechanik," *Encyklopädie der mathematischen Wissenschaften,* band IV, heft 4, pp. 505–804.

blamed Helmholtz and Max Planck.[17] Helmholtz spent a great part of his later years studying action principles because he believed it highly probable that least action was the one universal law pertaining to all processes in nature.[18] His pupil, Heinrich Hertz, took an intermediate position and argued that Hamilton's method is "not based on any physical foundation of mechanics, but that it is fundamentally a purely geometrical method, which can be established and developed quite independently of mechanics, and which has no closer connection with mechanics than any other of the geometrical methods employed in it."[19] And yet Hertz also worried about the teleological nature of variational principles, and recognized that the problem would not go away just by claiming that teleology has no place in physics.[20]

Planck, on the other hand, was willing to go much further into metaphysics, which elicited a few predictable snorts from Klein. In his most extreme statement, Planck claimed that the Principle of Least Action "possesses an explicitly teleological character. . . . In fact the least action principle introduces a completely new idea into the concept of causality: the *causa efficiens* . . . is accompanied by the *causa finalis* for which . . . the future—namely a definite goal, serves as the premise from which there can be deduced the development of the processes which lead to this goal."[21] Although Planck was not consistent in his position (at other times he rejected all teleological principles in mechanics)[22] his occasionally misguided enthusiasm brought action principles under suspicion, or so Klein argued.

The great enthusiasm for action principles that reached its height in the 1880's can also be observed in British physics. Sir Joseph Larmor published an article in 1884 entitled "On Least Action as the Fundamental Formulation in Dynamics and Physics," in which he argued that all phenomena can be unified to the extent that they can be described by a common action

17. Klein, *Vorlesungen . . .* , p. 199. Hamilton himself had rejected the notion that his principle had any teleological implications. He wrote: "Although the law of least action has thus attained a rank among the highest theorems of physics, yet its pretensions to a cosmological necessity, on the ground of economy in the universe, are now generally rejected. And the rejection appears just." *Mathematical Papers,* I, 317.

18. Leo Koenigsberger, *Hermann von Helmholtz,* trans. Frances A. Welby (New York: Dover, 1906), p. 350. See also Wolfgang Yourgrau and Stanley Mandelstam, *Variational Principles in Dynamics and Quantum Theory* (Philadelphia: Saunders, 1968), p. 163.

19. Heinrich Hertz, *The Principles of Mechanics Presented in a New Form* (New York: Dover, 1956), p. 32.

20. Ibid., p. 23.

21. Max Planck, *Scientific Autobiography and Other Papers,* trans. Frank Gaynor (New York: Philosophical Library, 1949), pp. 179 and 181. Planck's reply to Schrödinger's letter describing the new discoveries in 1926 is characteristic: "I find it extremely congenial that such a prominent role is played by the action function. . . . I have always been convinced that its significance was far from exhausted." Planck to Schrödinger 2 April 1926 in Erwin Schrödinger, *Letters on Wave Mechanics,* ed. K. Przibram, trans. Martin J. Klein (New York: Philosophical Library, 1967), p. 3.

22. Yourgrau and Mandelstam, *Variational Principles,* p. 165.

principle.[23] He then proceeded to show how very disparate parts of physics, including optics, can be brought under this single umbrella. (The Second Law of Thermodynamics was the one troublesome exception.) And yet Larmor, as Thomson and Tait before him and Whittaker after, did not see any particular significance in the optical-mechanical analogy. Larmor was an Irishman with an interest in the history of physics and a large reservoir of pride in his homeland. He heaped praise on Hamilton, but apparently never read his optical papers. In 1927, after Schrödinger had revived the optical-mechanical analogy with his wave mechanics, Larmor dug out the optical papers and described them in a "Historical Note on Hamiltonian Action." He found the papers "very dishevelled in form, doubtless from the distraction of [Hamilton's] wide range of philosophical and poetic interests." The later optical papers, "enforc[ed] the fundamental ideas, but perhaps also still further confus[ed] their application by excess of detail only partially relevant." These criticisms were all perfectly valid, but they sound suspiciously like an excuse for Larmor having overlooked the papers in the 1880's.[24]

Actually Larmor needed no excuse. In the 1880's quantum mechanics did not exist and there was no experimental evidence to indicate that the wave theory of light might not be the complete explanation of optical phenomena. Moreover Hamilton's analogy was limited to geometrical optics and was not adequate to deal with the phenomena that had confirmed the wave theory. It is Klein's position that needs explaining, and the only explanation that I can give is that the optical-mechanical analogy seemed more important to a mathematician like Klein, who recognized its beauty and unifying power, than to a physicist who could not yet think of photons and matter waves.

The question now presents itself, how did Schrödinger become aware of the optical-mechanical analogy, and how did it lead him to wave mechanics, if, indeed, it had any influence at all? Schrödinger derived his famous wave equation in the first paper from the Hamilton-Jacobi equation without any reference at all to Hamilton's optical-mechanical analogy. It appears to depend far more on the work of Louis de Broglie than on that of Hamilton. The analogy appears only in the second paper where it is elaborated at length. In a footnote Schrödinger thanks Arnold Sommerfeld for calling his attention to Felix Klein's campaign to foster Hamilton's ideas.[25] It looks as if Schrödinger may have first derived his wave equation on his own, showed it to Sommerfeld who mentioned Klein's work to him, and then developed the connection between wave mechanics and classical mechanics through

23. Sir Joseph Larmor, *Mathematical and Physical Papers,* I (Cambridge: Cambridge University Press, 1929), 31 and 70.

24. Ibid., p. 640.

25. Schrödinger, *Collected Papers,* p. 14.

the optical-mechanical analogy. In other words the second paper may have been merely an attempt to justify a new theory that had in fact been obtained in a very different way.

Sommerfeld was the obvious person to put Schrödinger onto Klein's ideas. He had been Klein's assistant and clerk of the famous Göttingen reading room, where Klein claimed that his papers on the optical-mechanical analogy collected dust for twenty years.[26] It is hardly possible that Sommerfeld did not read them. Moreover Sommerfeld's own work in 1916 had shown the close link between the action integral and the quantum conditions for the hydrogen atom. In his famous book *Atomic Structure and Spectral Lines,* from which most physicists learned their quantum mechanics in the 1920's, he wrote:

> Up to a few years ago it was possible to consider that the methods of mechanics of Hamilton and Jacobi could be dispensed with for physics and to regard it as serving only the requirements of the calculus of astronomic perturbations and mathematics . . . [but] since the appearance of the papers [of Epstein, Schwarzschild, and Sommerfeld] in the year 1916, which . . . link up the quantum conditions with the partial differential equations of mechanics, it seems almost as if Hamilton's method were expressly created for treating the most important problems of physical [quantum] mechanics.[27]

The importance of Hamilton's method for the older quantum theory had indicated that it was likely to be the most successful approach for any future theory.

Schrödinger also cites another important paper by Sommerfeld and J. Runge on geometrical optics that follows very closely Hamilton's arguments. The most interesting section (and the section that Schrödinger cites) is entitled "The Eikonal and the Limits of Geometrical Optics."[28] Sommerfeld recognizes that the Eikonal and the Hamiltonian theory of the Characteristic Function are equivalent in geometrical optics; and following a suggestion by Pieter Debye, he works out the limits at which wave optics may be correctly approximated by ray optics. Schrödinger goes out of his way to emphasize Debye's role in this paper, because he was working closely with Debye at Zurich at the time. Sommerfeld's suggestions were obviously important for Schrödinger, but probably not crucial, and we still do not know if Schrödinger's pursuit of Hamilton's optical-mechanical analogy came before or after his discovery of the wave equation.

As is usually the case with historical investigations of this kind, the truth

26. Constance Reid, *Hilbert, with an Appreciation of Hilbert's Mathematical Works by Hermann Weyl* (New York: Springer Verlag, 1970), pp. 47 and 103.

27. Trans. from the 3d German ed. by Henry L. Brose (London: Methuen, 1923), pp. 555–556.

28. A. Sommerfeld and J. Runge, "Anwendung der Vektorrechnung auf die Grundlagen der geometrischen Optik," *Annalen der Physik,* 4, bd. 35 (1911), pp. 289–293. Cited in Schrödinger, *Collected Papers,* p. 18.

does not seem to lie at either extreme, but somewhere in the middle. I wish to show that the immediate search for the wave equation was a response to the work of de Broglie, but this does not mean that Hamilton's analogy was not also a contributing factor. Schrödinger was thoroughly acquainted with the optical-mechanical analogy long before Sommerfeld suggested that he look at Klein's papers. More than that, his training and previous research in large part determined the direction that his thoughts took in 1926.

Several authors have given careful attention to the events that led up to Schrödinger's discovery.[29] In 1925, Schrödinger was working on the quantum statistics of gases. While reading a paper by Einstein on the same subject, he was "suddenly confronted with the importance of de Broglie's ideas" which were mentioned in the paper. In his dissertation for the Sorbonne, de Broglie had treated the concept of "matter waves" ascribing wave properties to electrons. Schrödinger first followed Einstein's lead in trying to apply de Broglie's new concept to the quantum statistics of gases. By early November 1925, he wrote that he now regarded the particles of a gas as merely wave crests on a background of waves. Soon afterwards he was working to apply the same theory to the electron orbits of the hydrogen atom, and discussed his efforts with Pieter Debye, also in Zurich, Alfred Landé, and Wilhelm Wien. But finding the correct wave equation for his theory was a big problem. His first relativistic equation was a failure (because it did not take into account the then-unknown phenomenon of electron spin), and he confessed to Wien: "I must learn more mathematics in order to fully master the vibration problem—a linear differential equation, similar to Bessel's, but less well known, and with remarkable boundary conditions that the equation 'carries within itself' and that are not externally determined."[30] He then turned to search for a nonrelativistic wave equation and was successful in January 1926.[31]

Schrödinger readily confessed that his inspiration had come from de Broglie and that his wave mechanics should be regarded as an extension of de Broglie's theory. In his thesis de Broglie had returned again and again to emphasize the importance of the optical-mechanical analogy, although he attributed it to Pierre de Fermat and Pierre-Louis Moreau de Maupertuis rather than to Hamilton.[32] Anyone reading the thesis could not miss a point

29. Max Jammer, *The Conceptual Development of Quantum Mechanics* (New York: McGraw-Hill, 1966), pp. 255–258; Armin Hermann, "Erwin Schrödinger," *Dictionary of Scientific Biography,* 12 (1975) pp. 217–233; V. V. Raman and Paul Forman, "Why Was It Schrödinger Who Developed de Broglie's Ideas," *Historical Studies in the Physical Sciences,* 1 (1969), 291–314; and J. Gerber, "Geschichte der Wellenmechanik," *Archive for History of Exact Science,* 5 (1968–1969), 349–416.

30. Schrödinger to Wilhelm Wien, 27 December 1925, Hermann, "Schrödinger," p. 219.

31. Jammer, *Conceptual Development,* pp. 257–258.

32. Schrödinger, *Collected Papers,* p. 20, and Louis de Broglie, "Recherche sur la théorie des quanta," *Annales de physique,* 3 (1925): "Fermat's principle applied to the wave, becomes

made so emphatically. Most German physicists, however, were not enthusiastic about de Broglie's ideas. De Broglie had been brashly critical of Niels Bohr and Sommerfeld and had earned the reputation in German circles of being an ill-tempered crank. But Schrödinger's interest in quantum statistics and his objections to the probabilistic interpretation of quantum theory put him in the same camp with Einstein against the Copenhagen school. He found de Broglie's use of the optical-mechanical analogy congenial.[33]

These were the immediate events leading up to Schrödinger's discovery. It was a straightforward attempt to find a wave equation for the electron orbits of the hydrogen atom satisfying the quantum conditions. Because of the great importance given to the optical-mechanical analogy by de Broglie, we can scarcely regard it as an afterthought that entered Schrödinger's mind only after the first paper was finished. Less than a month separates the first and the second papers. It is difficult to believe that the ideas in the second paper were worked out entirely in such a short interval. Moreover, Schrödinger mentions that he originally intended to give a more physical or "intuitive" representation of the wave equation, but decided instead to put it in a more "neutral" mathematical form, indicating that he probably did have a vibrational model in mind even when he wrote the first paper.[34] It is likely that the "intuitive" model was something close to the presentation that Schrödinger actually made in his second paper.

In the short autobiographical sketch attached to his Nobel Prize address of 1933, Schrödinger paid special tribute to his first teacher at Vienna, Friedrich Hasenöhrl, who had been killed at a relatively young age in the first world war. If he had lived, Schrödinger believed that Hasenöhrl might well have been receiving the prize for the discovery of wave mechanics in his place.[35] Cornelius Lanczos has recently judged the importance of Hasenöhrl: "It is no accident that Schrödinger repeatedly and emphatically referred to his outstanding teacher at the University of Vienna, Fritz Hasenöhrl. Hasenöhrl was one of the few theoretical physicists who fully recognized the importance of Hamilton's work and gave full account of it in his lectures. Thus it was completely in line with Schrödinger's theoretical

---

identical to the principle of least action applied to the particle. The rays of the wave are identical to the trajectories of the particle" (p. 22). "Guided by the idea of a profound identity between the principle of least action and Fermat's principle, I have been led, from the beginnings of my research on this subject, to admit that . . . the possible dynamic trajectories of the [particles] coincide with the possible rays of the [waves]." (p. 45).

33. Raman and Forman, "Why Was It Schrödinger?" pp. 293–303. These authors see a preparation for, if not an anticipation of, de Broglie's ideas in an earlier paper by Schrödinger, "Über eine bemerkenswerte Eigenschaft der Quantenbahnen eines einzelnen Elektrons" (1922), (Raman and Forman, pp. 303–310).

34. Jammer, *Conceptual Development,* p. 260, and Schrödinger, *Abhandlungen zur Wellenmechanik* (Leipzig: J. A. Barth, 1928), p. 12.

35. *Les Prix Nobel en 1933* (Stockholm: Imprimerie Royale, 1935), p. 87.

background to take de Broglie's geometrical optics and change it into a physical form of optics, with the result that he arrived with necessity at his famous equation."[36] Beginning in 1906, Schrödinger sat through a four-year cycle of daily lectures by Hasenöhrl on theoretical physics. The emphasis was on Hamiltonian mechanics and the theory of eigenvalue problems in the physics of continuous media.[37] The direction of Hasenöhrl's lectures is indicated by one of his articles in the Festschrift for Ludwig Boltzmann (1904) entitled "On the Application of Hamilton's Partial Differential Equation to the Dynamics of Continuously Distributed Masses."[38] In this paper Hasenöhrl claimed that he was making the first attempt to employ Hamilton's partial differential equation in the solution of problems of motion in a continuous medium. He selected the simplest example—that of the vibrating string—and obtained a solution using Hamiltonian methods. While he did not derive a wave equation (in fact the beauty of the method is that it permits a solution without deriving a wave equation) Hasenöhrl was obviously headed in the right direction for Schrödinger.

Among Schrödinger's manuscripts are three university notebooks entitled "Tensor analytische Mechanik." It is possible that some of the notes were taken at Hasenöhrl's lectures, although they are more likely from the years 1918 to 1920 when he returned to Vienna after World War I.[39] Schrödinger had the ambition to continue the tradition of Hasenöhrl's lectures in a new

36. Cornelius Lanczos, "William Rowan Hamilton—An Appreciation," *University Review*, National University of Ireland, 4 (1967), 155. Lanczos' statement carries weight because he was for many years a colleague of Schrödinger's at the Dublin Institute for Advanced Study.

37. *Les Prix Nobel en 1933*, p. 87; William T. Scott, *Erwin Schrödinger: An Introduction to His Writings* (Amherst: University of Massachusetts Press, 1967), p. 2; and Stefan Meyer, "Friedrich Hasenöhrl" a eulogy in *Physikalische Zeitschrift*, 16 no. 23 (1915), 432. According to Armin Hermann, Schrödinger began to attend Hasenöhrl's lectures in 1907 during his third semester at the University of Vienna. He was greatly impressed by Hasenöhrl's inaugural lecture on the work of his predecessor, Ludwig Boltzmann ("Erwin Schrödinger," pp. 217–218).

38. Friedrich Hasenöhrl, "Über die Anwendbarkeit der Hamiltonschen partiellen Differentialgleichung in der Dynamik kontinuierlich verbreiteter Massen," *Festschrift Ludwig Boltzmann* (Leipzig: Verlag von Johann Ambrosius Barth, 1904), pp. 642–646.

39. Thomas Kuhn, John L. Heilbron, Paul Forman, and Lini Allen, *Sources for History of Quantum Physics: An Inventory and Report* (Philadelphia: The American Philosophical Society, 1967), p. 124. Microfilm 39, E. Schrödinger Papers, to c. 1920. I am extremely grateful to Professor John L. Heilbron and to the Bancroft Library, Berkeley, California, for making copies of these manuscripts available to me. The manuscripts appear to be lecture notes. In some cases it is possible to find brief jottings which are later written out in full, as if Schrödinger had later expanded on notes taken hurriedly during lecture. This would tend to indicate that the notes are from his student days. But a reference in the notes to Schwarzchild's papers of 1917 proves that at least a portion of the notes came from the postwar years, 1918–1920, when Schrödinger returned to Vienna. The blank notebooks in which Schrödinger wrote these notes were printed in Vienna, another indication that the notes were probably taken in Vienna. I wish to thank Professor Rudolf Peierls and Professor John A. Wheeler for their assistance in reading these manuscripts and for their helpful comments on this paper.

teaching career at Czernowitz, but when that opportunity collapsed with the Russian occupation of the city after the war, he returned to teach at Vienna.[40] These notes were probably prepared for lectures there. The third book contains a section entitled "Analogy to Optics, Huygens' Principle and Hamilton's Partial Differential Equation." It is a beautiful development of the optical-mechanical analogy, with the mechanical version of the equations on the left-hand side of the page, and the optical version on the right-hand side. He develops the optical theory in detail, derives Hamilton's "Principle of Varying Action," derives from it the optical equivalent of the Hamilton-Jacobi equation, and proves the existence of "wave" surfaces orthogonal to the optical rays. There is no attempt to derive a wave equation, of course, but that would have been premature in 1918. The passage concludes with a section on the "Direct Transfer [of the Optical Theory] to Mechanics."

When Schrödinger came to search for the wave equation in 1925, his familiarity with problems of wave propagation and his exposure through Hasenöhrl to the methods of Hamiltonian mechanics applied to motion in a continous medium, must have stood him in good stead. With de Broglie pointing the way, exploiting Hamilton's analogy was the obvious path to follow.

In summary we can conclude that the optical-mechanical analogy was not as "lost" or "misunderstood" as Schrödinger and Klein before him had claimed. Schrödinger may have borrowed Klein's complaint (in almost the same strong words) to dramatize the novelty of his discovery. But it is unlikely that he learned anything from Klein directly. Schrödinger was already well versed in the analogy long before he began his search for the wave equation. There is no evidence that he ever read Hamilton's optical papers, but he did not have to. Hamilton's optical-mechanical analogy remained "alive" through the century following his formulation of it in the 1820's, first in the writings of British physicists, and then in the campaign by Klein. Hasenöhrl elaborated the theory in lectures that Schrödinger followed closely. As experimental evidence mounted for a dual wave-particle description of light and electrons, the optical-mechanical analogy was the obvious theoretical construct to exploit, and it is no coincidence that both de Broglie and Schrödinger gave it such a prominent place in their work.

40. Scott, *Schrödinger,* pp. 2–3.

BIRTH CRIES OF THE ELEMENTS: THEORY AND EXPERIMENT ALONG MILLIKAN'S ROUTE TO COSMIC RAYS

ROBERT KARGON

SCIENTISTS, especially in their ground-breaking papers, are usually more interested in lucid exposition than in accurately recounting their mode of discovery. Consequently their historical reconstructions are rather more logical than chronological. What they represent as the history of their path to discovery is really a brief, meant to convince their peers of the strength of their case. They show, alas, callous disregard for the time and energy of historians of future generations. Missing from such accounts are the numerous false starts, disappointments, and misconceptions from which few scientists are free. Even the context of the inquiry and the original motivation for research are often distorted beyond recognition. For the scientist, patient *post-facto* reconstruction of the original route is either fruitless or pointless. For the historian, however, such labor can be revealing and provocative.

An interesting and important example of a scientist's "reconstructed history" is Robert A. Millikan's version of his work in cosmic rays. Early in 1928, Millikan and his co-worker G. H. Cameron announced that on the basis of their experiments on the ionization of the upper atmosphere they had discovered "the first direct indications that all about us, either in the

The major manuscript resources referred to below are the George Ellery Hale Papers, the Robert A. Millikan Papers, and the Ira Bowen Papers at the Robert A. Millikan Library, the California Institute of Technology, Pasadena, California; the J. C. Merriam Papers, Library of Congress; the Henry Crew Papers, Niels Bohr Library, American Institute of Physics, New York City.

I should like to thank the officers and staff of the California Institute for their generous assistance; special thanks are due Dr. Judith Goodstein, Institute archivist. I should also like to thank Professor Lawrence Badash of the University of California, Santa Barbara, and Messrs. Michael Freedman and John Servos for helpful references and advice. I owe a debt of gratitude to Dr. Daniel Kevles of the California Institute of Technology for valuable discussions and for an opportunity to examine material from his studies on the history of physics in the United States.

stars, the nebulae or in the depths of space, the creative process is going on, and that the cosmic rays which have been studied for the past few years constitute the announcements broadcast through the heavens of the birth of the ordinary elements out of positive [protons] and negative electrons."[1] In October of that year, they published an elaborate account of a bold interpretation of their data in the form of an atom-building theory which they held to be of fundamental significance. They insisted that their "results were obtained from an empirical analysis of the ionization-depth curve and *entirely without the guidance of any theory*. They represent solely the general type of solution *demanded by the nature of the curve itself*."[2] In subsequent articles, Millikan repeated his contention that the theory was pressed upon him by empirical necessities.[3] Rather than theory-free, however, Millikan's route to cosmic rays was from the beginning theory-encumbered; his atom-building hypothesis, in turn, reflected a complex matrix of concerns which tended to disappear from his own sight as they receded from the research frontier. This paper reconstructs Millikan's original aims and concerns in order to provide a fuller understanding of his unquestionably important work in cosmic-ray physics, and to illuminate what I believe to be some fascinating dark areas in Millikan's intellectual biography.

ON THE TRAIL OF THE PHILOSOPHER'S STONE: THE SCIENTIFIC CONTEXT

At the beginning of 1921, George Ellery Hale, director of the Mount Wilson Observatory, was working with his usual nervous enthusiasm to bring Robert A. Millikan to the California Institute of Technology on a full-time basis. Hale had been a trustee of the institute and its predecessor, the Throop College of Technology, for almost a decade and a half. His scheme to establish in Pasadena major centers of research which would command world-wide attention hinged upon Millikan's participation. The Chicago physicist had already spent several winters at Caltech as its director of physics research; Hale's plan involved his appointment as president of the institute as well as its physics chief. Millikan felt strong ties of loyalty to the Ryerson Laboratory at Chicago; nonetheless the exciting possibilities outlined by Hale for a new world of co-operative research between the institute and the observatory were extraordinarily attractive. Millikan wrote to his

1. Millikan and G. H. Cameron, "Direct Evidence of Atom Building," *Science*, 67 (1928), 402. For a concise and effective review of Millikan's life and work see D. Kevles, "Robert Andrews Millikan," *Dictionary of Scientific Biography*, 9 (New York, 1974), 395–400.

2. Millikan and G. H. Cameron, "The Origin of Cosmic Rays," *Phys. Rev.*, 32 (1928), 534. Emphasis supplied.

3. Millikan, *Electrons (+ and −), Protons, Photons, Neutrons, Mesotrons and Cosmic Rays* (Chicago, 1947), pp. 304–313.

wife Greta that "I wish I could split myself up into parts. This is worse torture than courted damsels ever go through, for I could be perfectly happy with both suitors or with either of them."[4]

In order to break through the impasse created by this rare indecision on Millikan's part, Hale searched for a way to present an offer that Millikan would be unable to refuse. After long conversations with him, Hale turned to Henry M. Robinson and Arthur Fleming, fellow Caltech trustees who served at the same time as directors of a local private utility, the Southern California Edison Company. He proposed an interesting addition to the original offer. The Edison Company would build a laboratory for Millikan in addition to the already well-endowed Norman Bridge Laboratory for physics. This second building would house a tremendous new instrument which Millikan had wanted—a million-volt transformer—to be used for spectacular new forays into the secrets of the atom. Hale wrote, explaining his plan, to Robinson:

> The solution lies in making the opportunities here too attractive to be denied .... This is the Edison scheme.... The Institute could agree to run the plant if the Edison Company would give it, and I believe the Company would get its money back in the form of new information regarding insulation and other problems connected with high voltage lines, not to speak of the advertising value.... Millikan would also have the advantage of using enormous voltage to bust up some of his atoms. This possibility, which no other laboratory could match is what excites him.... When a man gets on the trail of the philosopher's stone, even if he isn't after gold, you can accomplish a great deal by offering it to him![5]

Was Hale's choice of words in explanation to Robinson deliberate and judicious? This paper supports the view that Hale was correct: Millikan was indeed on the trail, as Hale put it, of the philosopher's stone; that is, he was deeply concerned with the artificial transmutation of the elements, and the trail led (at least in part) not to gold but to his controversial theory of the origin of cosmic rays.

The efforts of Fleming and Robinson to build the High Voltage Laboratory were successful. So too was Hale's scheme: Robert Millikan agreed to come to the California Institute of Technology in 1921 as the chairman of its Executive Council, and as director of the Norman Bridge Laboratory, which was soon to be opened.

The decade prior to Millikan's arrival in Pasadena had been one of the most exciting and fruitful in the history of physics. The plethora of data and speculation opened by the exploitation of the spectroscope, by the discovery of X-rays, and by investigations into radioactivity, was beginning to come under some reasonable control. Ernest Rutherford's nuclear atom and the

4. Millikan to Greta Millikan, 17 May 1921, Millikan Papers, box 50, p. 4.
5. Hale to Robinson, 25 February 1921, Hale Papers, box 35.

physics of the quantum were joined in Niels Bohr's landmark atomic model. Bohr's atom was a spectroscopic atom: it was designed to cope with the enormous quantity of line-spectra data which physicists had long intuited would yield information about atomic structure.[6] Millikan later remarked that when "it was devised, spectroscopy was a veritable dark continent in physics. With the aid of the Bohr atom the dark continent in physics has become the best explored and best understood, the most civilized portion of the world of physics. It has been an exciting game of exploration."[7]

The "exciting game" was further stimulated by the discoveries of Henry G. J. Moseley, the British physicist who, working along lines first laid out by the X-ray crystallographers Max von Laue and the Braggs, William Henry and William Lawrence, accurately determined the frequencies of the X-rays characteristic of the elements. Moseley showed that the square roots of these frequencies are related in a simple way. They constitute a simple arithmetic progression: each member of the series is obtained from the previous member by adding the same quantity. Moseley saw at once that his formulas could be used to test the periodic table, for his series of X-ray frequencies (with a few exceptions) follows the chemists' series of atomic weights.[8] Furthermore, radioactivity research produced evidence which demonstrated that when a substance loses a doubly-charged positive particle (an alpha) it moves two places to the left in this periodic table, and when it loses a single negative particle it moves a place to the right. The chemical nature of a substance depends, therefore, upon its atomic number, or the number of positive charges in its nucleus. The periodic table, on the basis of Moseley's work, could now be reconstructed along physical lines, utilizing this notion of atomic number. "There is," Moseley stated, "in the atom a fundamental quantity which . . . increases by regular steps . . . from one element to the next. This quantity can only be the charge on the central positive nucleus."[9]

Moseley's work revived interest in William Prout's hypothesis which had been advanced a century earlier. Prout had argued that atomic weights were multiples of that of hydrogen and, as a consequence, hydrogen may be taken to be the primordial element. Prout's speculation foundered upon the rock of precise measurement: the multiple relationships are not exact. Twentieth-century physicists were confident of resolving all difficulties, and the assumption was widely held that the elements were built up out of hydrogen nuclei (protons) and electrons. Arnold Sommerfeld's magisterial *Atombau und spektrallinien* endorsed Prout in this way: "The atoms of the

6. A useful review is F. H. Loring, *Atomic Theories* (New York, 1921), pp. 41–68.

7. Millikan, "Holographic Autobiographical Notes," Millikan Papers, box 67, fol. 8.

8. On Moseley's work see *Phil. Mag.*, 26 (1912), 1024, and *Phil. Mag.*, 27 (1914), 703. See also J. Heilbron, *H. G. J. Moseley, The Life and Letters of an English Physicist, 1887–1915* (Berkeley, 1974).

9. Moseley, *Phil. Mag.*, 27 (1914), 703–713. See also Millikan, *The Electron* (Chicago, 1917), pp. 200–202.

various elements must be similarly constructed out of identical units."[10] Millikan himself put it a bit more dramatically in his book *The Electron* of 1917: "It looks as if the dream of Thales of Miletus had actually come true and that we have found not only a primordial element out of which all substances are made, but that that primordial element is hydrogen itself."[11]

The exact work of F. W. Aston with the mass spectograph served only to reinforce interest in atom-building. Aston's "whole number rule" that, with the exception of hydrogen, atomic weights of the elements are whole numbers, "removes the only serious objection to a unitary theory of matter."[12] Hydrogen was measured at very slightly more than unity; the "lost" mass in the building up of heavier elements from the proton was explained away by a "packing effect" which was originally viewed as an electromagnetic contraction.[13] Whatever the mechanism of packing, it was widely believed that in the construction of the heavier elements some mass remained to be accounted for. Aston's influential book *Isotopes* of 1922 was merely recording the commonly-held belief that "we may consider it absolutely certain that if hydrogen is transformed into helium a certain quantity of mass must be annihilated in the process."[14] "Should the research worker of the future," Aston continued, "discover some means of releasing this energy in a form which could be employed, the human race will have at its command powers beyond the dreams of science fiction."[15]

This revived concern with the construction of the elements from hydrogen nuclei and electrons was seen not only as possible in principle but as a process actually occurring in the heavens and possibly capable of realization on earth. Ernest Rutherford, invited by Hale to appear before the National Academy of Science, spoke on "The Constitution of Matter and the Evolution of the Elements" and splendidly summed up the views of many physicists and astrophysicists:

> It has been long thought probable that the elements are all built up of some fundamental substances, and Prout's well-known hypothesis that all atoms are composed of hydrogen is one of the best-known examples of this idea. . . . In the hottest stars the spectra of hydrogen and helium predominate, but with decreasing temperature the spectra become more complicated and the lines of the heavier elements appear. . . . [I]t is supposed that the light elements combine with decreasing temperature to form the heavier elements. There is no doubt that it will prove a difficult task to bring about the transmutation of matter under ordinary terrestrial conditions. The enormous evolution of energy which accompanies the transformation of radioactive matter affords

10. A. Sommerfeld, *Atombau und Spektrallinien* (Braunschweig, 1922), p. 75.
11. Millikan, *Electron,* pp. 202–203.
12. F. Aston, *Isotopes* (London, 1922), p. 90.
13. Ibid., p. 101; W. Harkins and E. Wilson, "Energy Relations Involved in the Formation of Complex Atoms," *Phil. Mag.,* 30 (1915), 723–734.
14. Aston, *Isotopes,* p. 104.
15. Ibid.

> some indication of the great intensity of the forces that will be required to build up lighter into heavier atoms.[16]

Millikan had long been deeply interested in radioactivity and the problem of the evolution of the elements. His first research program in the new physics concerned these subjects. In a review of the field for the *Popular Science Monthly* in 1904, he plainly revealed his enthusiasm: "the dreams of the ancient alchemists are true, for the radio-active elements all appear to be slowly but spontaneously transmuting themselves into other elements."[17] Millikan's noteworthy conclusion lays down a theme to which he would return many times in the succeeding two decades: "The studies of the last eight years upon radiation seem to indicate that in the atomic world also, at least *some* of the heaviest and most complex atomic structures are tending to disintegrate into simpler atoms. The analogy suggests the profoundly interesting question, as to whether or not there is any natural process which does, among the atoms, what the life process does among the molecules, i.e., which takes the simpler forms and builds them up into more complex ones."[18]

The "profoundly interesting question" of discovering a natural process of building complex nuclei from simpler ones was part of what by his own testimony was Millikan's first foray into artificial transmutation in 1912. Millikan and his student G. Winchester believed that they had produced hydrogen ions from aluminum by high-voltage discharges.[19] After Sir William Ramsay, Norman Collie and H. Patterson announced that they thought they had produced helium and neon using electrical discharges,[20] Winchester, at least, ventured forth into print, and suggested that while helium and neon were merely occluded gases, *hydrogen* can be liberated from aluminum "in somewhat the same manner as the α-particle is disintegrated from radium."[21]

The revolutionary work of Rutherford and James Chadwick in 1919–1921 on artificial nuclear transformations created fascinating new possibilities.[22] Deeper understanding of the atom-building process awaited further penetration into the mysteries of the nucleus. Studies of artificial transformations, it now seemed, offered clues to these mysteries. Millikan fol-

16. Rutherford, *Smith Inst. Ann. Rpt. 1915* (Washington, D.C., 1916), p. 201.

17. Millikan, "Recent Discoveries in Radiation and Their Significance," *Pop. Sci. Mon.*, 64 (1904), 498. Millikan had been one of the first in the U.S. to do research in radioactivity. See L. Badash, "Early Developments in Radioactivity with Emphasis on Contributions from the United States," Ph.D. dissertation, Yale University, 1964, pp. 220–221.

18. Millikan, "Recent Discoveries," p. 498.

19. Millikan, "Gulliver's Travels in Science," *Scribner's Magazine*, 74 (1923), 584.

20. See *Nature*, 90 (1913), 653–654.

21. *Phys. Rev.*, 3 (1914), 294. Winchester concluded that the helium and neon reported by Ramsay were not disintegration products, although hydrogen appeared to be such.

22. E. Rutherford, *The Newer Alchemy* (Cambridge, 1937), *passim.*

lowed Rutherford's work with his usual keen interest.[23] The possibilities which he foresaw were hinted at in an address given in Washington, D.C., shortly before his decision to come to Caltech on a permanent basis. Millikan's exuberance about the new world of physics just about to be opened was scarcely veiled:

> We have been forced to admit for the first time in history not only the possibility but the fact of the growth and decay of the elements of matter. With radium and with uranium we do not see anything but the decay. And yet somewhere, somehow, it is almost certain that these elements must be continually forming. They are probably being put together now in the laboratories of the stars. . . . Can we ever learn to control the process. Why not? Only research can tell. What is it worth to try it? A million dollars? A hundred million? A billion? It would be worth that much if it failed, for you could count on more than that amount in by-products. And if it succeeded, a new world for man![24]

It was this "new world for man" and the attempt to come to grips with it, which contributed to Millikan's interest in the construction of the High Voltage Laboratory and to which Hale referred when he wrote of Millikan's interest in "this possibility which no other laboratory could match," that is, a laboratory which would enable Millikan "to bust up some of his atoms."[25]

That Millikan's intended to use the high-voltage capacities of the proposed laboratory to probe the nucleus as well as for other problems is borne out by the prominence afforded these intentions in the major grant proposals constructed by him and by Hale for the Carnegie Corporation of New York in mid-1921. In a preliminary proposal submitted to the corporation's president, James Angell, on 4 June 1921 Hale referred to the High Voltage Laboratory more explicitly as "especially adapted for Dr. Millikan's researches on the breaking-down of atoms and the resolution of the chemical elements into simpler components."[26]

In the fuller grant proposal to the corporation submitted in the fall of the year by Millikan, Hale, and Noyes, director of Caltech's Gates Laboratory, Millikan revealed that nuclear transformations were to occupy a major place among the concerns of the California Institute of Technology, now emerging as an ambitious center for physical-science research. The heart of the proposal was a joint attack on problems of radiation and matter by the Gates Laboratory and by Hale's Mount Wilson Observatory as well as by

23. Millikan to Rutherford, 3 February 1920, Millikan Papers, box 42, fol. 9.

24. Millikan, "The Significance of Radium," *Science,* 54 (1921), 59–67.

25. Hale to Robinson, 25 February 1921, Hale Papers, box 35. Millikan would also require high voltages, although well under a million volts to prosecute his hot spark spectra work during the 1920's. See Millikan, "The Extension of the Ultra-Violet Spectrum," *Astrphs. Jnl.,* 52 (1920), 49.

26. Hale to Angell, 4 June 1921, Hale Papers, box 67, fol. "Carnegie Corp."

Millikan's soon-to-be-opened Norman Bridge Laboratory.[27] In their brief summary, Hale, Noyes, and Millikan pointed to several exciting new areas for exploration: radioactive transmutations and the existence of a basic element (hydrogen), direct attention to the possibility of artificial transmutation in the laboratory, and the probability "that in the stars the heavier elements are being built up from the lighter ones, a process not yet realized on earth." The projected physics program included work of high potential bearing "upon the nature of atoms and their possible transformations into one another."[28] It is clear that the problem of transmutations was to be approached by a projected research program which was to encompass both heavens and earth, that is, through a laboratory attack involving Millikan's plan to "bust up" atoms, and also through investigations of stellar processes by the staff of the Mount Wilson Observatory.

A press release which never saw the light of day was prepared by Sam Small, Jr., based upon interviews with Millikan in 1922. It reported that "it is the present task of the scientists at the Institute of Technology, by means of fundamental research working through several directions, to smash the nucleus of the atom and find out what it is made of."[29] Millikan's optimism along these lines continued on into 1923. In an article for *Scribner's Magazine* which appeared in that year, entitled "Gulliver's Travels in Science," Millikan reported upon the present state of knowledge of transmutable elements. Writing of such transformations, he asked rhetorically, "Does the process go on in both directions, heavier atoms being formed as well as continually disintegrating into ligher ones? Not on earth as far as we can see. Perhaps in God's laboratories, the stars."[30] The key question, however, remained: "Can we on earth artificially control the process? To a very slight degree we know already how to *disintegrate* artificially, but not as yet how to build up. As early as 1912, in the Ryerson Laboratory at Chicago, Doctor Winchester and I thought we had good evidence that we were knocking hydrogen out of aluminum and other metals by very powerful electrical discharges in vacuo. We still think our evidence to be good."[31] Millikan has here made an extraordinary claim. He believed, as of 1923, that he and Winchester had anticipated Rutherford's artificial transmutation, not as Rutherford had done, with alpha-particle bullets, but with high-voltage

27. On cooperative research at Caltech, see R. Kargon, "Temple to Science: Co-operative Research and the Birth of the California Institute of Technology," *Historical Studies in the Physical Sciences,* 8 (1977), 3–31.

28. "Memorandum relating to the application of the California Institute of Technology to the Carnegie Corporation of New York for aid in support of a project of research on the constitution of matter. . . ." p. 3, Hale Papers, box 6; CIT file.

29. California Institute of Technology publicity releases, Box 31-A, p. 12.

30. Millikan, "Gulliver's Travels," p. 584.

31. Ibid.

discharges.[32] No wonder then that Millikan was eager to erect and to use a high-voltage laboratory with its million-volt transformer. "How much farther," he continued, "can we go into this artificial transmutation of the elements? This is one of the supremely interesting problems of modern physics *upon which we are all assiduously working.*"[33]

The million-volt transformer was built for the High Voltage Laboratory by Royal Sorenson, professor of electrical engineering at Caltech. Four 250,000-volt, 50-cycle Westinghouse transformers were installed, cascade-fashion, on steel frames, supported by porcelain insulators.[34]

Whether the cascade transformer itself was used by Millikan for the study of nuclear transformations is doubtful. He more likely made trials similar to those published by Winchester in 1914. Winchester built vacuum tubes with an aluminum cathode and platinum anode, and placed high potentials (at that time about 100,000 volts) across them. He generated helium, neon, and hydrogen gases. After careful study he concluded that the neon and helium had been occluded at or near the surface of the electrodes, but that the hydrogen was possibly to be understood as a disintegration product of aluminum. It is reasonable to assume that given his stated interest, Millikan may have made further attempts along these lines in 1922 and 1923, but concluded that the hydrogen, previously identified spectroscopically, was not in fact a disintegration product.[35]

Millikan's plans to smash the nucleus with his high voltages were therefore never successfully prosecuted. In retrospect, with all the advantages of hindsight accrued, the attempt may seem bizarre. But as we have seen, in the context of the very fluid situation of the early 1920's, Millikan was by no means unjustified in risking some time and research capital. Transmutation was a popular enterprise in the 1920's. Successful attempts at turning base metals into gold using medium and high voltages were reported by A.

32. He repeated this claim in 1926: "The Last Fifteen Years of Physics," *Proc. Amer. Phil. Soc.*, 65 (1926), 74.

33. Millikan, "Gulliver's Travels," p. 584. Emphasis supplied.

34. Royal Sorensen, *Jour. Amer. Inst. of Elect. Eng.*, 44 (1925), 373–374. The fullest account of the High Voltage Laboratory is to be found in Rollin Olson, "Electrical Engineering Research at California Institute of Technology and Southern California Edison Company, 1915–1935," unpublished paper, The Johns Hopkins University (1965).

35. Winchester, *Phys. Rev.*, 3 (1914), 294. There is, at the present time, a gap in the historical record regarding Millikan's intentions to smash the nucleus through applications of high voltages. I have been unable to unearth among his remaining laboratory notebooks any which describe any such efforts. No detailed published accounts of such attempts reached the light of day. We may surmise from his strong public commitments and his references to such work in the grant proposal of 1921 that these efforts may have in fact been made but were unsuccessful. Millikan, like most others, preferred to bury quietly unfruitful enterprises; his position as fundraiser for an exciting new research center made such unpublicized interment a practical necessity.

Miethe and H. Stammreich,[36] and H. Nagaoka.[37] Lead was reportedly transmuted into mercury by A. Smits of Amsterdam.[38] These claims were opposed by F. Haber[39] and by F. Aston.[40] Millikan was perhaps too wise an experimentalist to move very far out on this limb.

Millikan turned his energies to a more promising subject which had intrigued him for the better part of a decade. Millikan's search for fuller understanding of the atomic nucleus and its transformations helped to concentrate his efforts in the area which came to be known by the name he gave it cosmic rays.

That a fuller understanding of atom-building and the study of astrophysical phenomena were widely seen during the 1920's to be closely related is reflected in Ernest Rutherford's presidential address before the Liverpool meeting of the British Association for the Advancement of Science in September 1923. In this address, Rutherford discussed "another method of attack" on the question of the meaning of artificial transformations. A close look at the atomic masses of hydrogen and helium shows that "in the synthesis of the helium nucleus from hydrogen nuclei a large amount of energy in the form of radiation has been released in the building of the helium nucleus from its components."[41] Arthur Eddington had suggested that this source of energy may account for the heat emission of sun and stars. Before the same forum in 1920, Eddington had speculated that the interior of stars is the place where the evolution of all the elements from hydrogen occurs, and, provocatively, if "the sub-atomic energy in the stars is being freely used to maintain their great furnaces, it seems to bring a little nearer to fulfillment our dream of controlling this latent power for the well-being of the human race—or for its suicide."[42]

Rutherford agreed that the evidence of stellar evolution "certainly indicates that the synthesis of helium, and perhaps other elements of higher atomic weight, may take place slowly in the interior of hot stars."[43] The facilities of the Mount Wilson Observatory, along with those of the Norman Bridge Laboratory in Pasadena, seemed to place Millikan in a strategic position to investigate these exciting problems.

COSMIC RAYS AND THE EVOLUTION OF THE ELEMENTS

There exists considerable evidence from 1922 and 1923 that Millikan was indeed actively concerned with the links between evidence from outer space

36. *Nature,* 114 (1924), 197–198.
37. *Nature,* 116 (1925), 95–96.
38. *Nature,* 117 (1926), 13, 621.
39. *Naturwissenschaften,* 14 (7 May 1926), 405–412.
40. *Nature,* 116 (1925), 902–904.
41. Rutherford, *BAAS Rpts.* (1923), p. 21.
42. Eddington, *BAAS Rpts.* (1920), p. 46.
43. Rutherford, *BAAS Rpts.* (1923), p. 21.

and the problems of radioactivity and nuclear transformation. He presented a paper before the Carnegie Institution of Washington entitled "Atomic Structure and Etherial Radiation," which made these links explicit. Writing to John C. Merriam of the Carnegie Institution on 3 May 1923, Millikan indicated that he chose this topic "because it illustrates the beauty, and the consistency too, of our vision, a vision which has been acquired within the past two decades into the structure of matter, and at the same time the abysmal depths of our ignorance when we attempt to go a little farther and relate these different fields of physical investigation sufficiently to a consistent whole."[44]

In the address Millikan noted that when "radioactivity was first discovered it was conjectured by some that the energy involved in this radioactive change might come from outside somewhere."[45] Focusing specifically on the energy for beta-decay, Millikan posed the important question, "Where does the energy come from which enables the negative electron to push itself out against the pull of the nucleus in which it certainly lies?" In reply he offered a possible "way out": "I have been interested in recent years because of that difficulty in attempting to find out whether there are any penetrating radiations that come in from the outside."[46] Hidden in this little known address we have an important clue to any reconstruction of Millikan's concern with cosmic rays. Millikan's cosmic ray researches were part of his overall program to probe the nature and structure of matter.

The notion to which he referred, that the energy for radioactivity may come from "outside somewhere" had first been raised by Marie Curie in 1898. Mme. Curie suggested that "all space is constantly traversed by rays analogous to Röntgen rays but which are much more penetrating and which can be absorbed only by certain elements of heavy atomic weight such as uranium and thorium."[47] The idea was revived with considerable *éclat* by Jean Perrin in 1919. Perrin pointed to the known existence of highly penetrating radiation widely viewed as extra-terrestrial in origin and suggested that these "rayons ultra X" were responsible for radioactive dissociation.[48] His views created, somewhat later, a flurry of experimental activity, especially after Millikan's work on cosmic rays renewed interest in this area.[49]

Without this background of developments in the physical conception of

44. Millikan to Merriam, 3 May 1923, Merriam Papers, Library of Congress, box 125.

45. Millikan, "Atomic Structure and Etherial Radiation," Millikan Papers, box 62, fol. 10, p. 15.

46. Millikan, "Etherial Radiations," p. 16. He added, "I am not trying to explain radioactivity that way, but it is an allied subject."

47. M. Curie, *Paris Acad. des Sci. C.R.*, 126 (1898), 1103.

48. J. Perrin, *Ann. de Phys.*, 11 (1919), 85–87.

49. See for example L. P. Maxwell, *J. Franklin Inst.*, 207 (1929), 619–628, for a late effort; see also N. Dobronravov, P. Lukirsky, and V. Pavlov, *Nature*, 123 (1929), 760, and L. N. Bogojavlensky, *Nature*, 123 (1929), 872.

the nature of matter and radiation it is impossible, I suggest, completely to understand Millikan's tenacious concern with what came to be known as cosmic rays. He had been acquainted with problems surrounding these phenomena for many years. It was known since the beginning of the twentieth century that the atmosphere is always slightly ionized. When Rutherford and Frederick Soddy announced their pioneering work on radioactive elements it was at first supposed that radiations from naturally radioactive elements were responsible for this ionization; Millikan had been among the earliest to attempt to shield laboratory experiments from them with lead.[50] The consensus which obtained for most of a decade held that these radiations were of terrestrial origin. In 1910, however, A. Gockel, in manned balloon flights, showed that the intensity at 4500 meters above the earth's surface was actually greater than at sea level.[51] Extensive balloon flights were made by Victor Hess and by W. Kolhörster which demonstrated the increase in radiation intensity with altitude.[52] Hess proposed in 1912 that "a radiation of very high penetrating power enters our atmosphere from above."[53] Still, the nature and origins of the penetrating radiation were matter of minor controversy, and according to Millikan's later testimony, he renewed his interest in the phenomena at Chicago shortly before World War I. He was beginning to design apparatus for unmanned sounding balloon flights at previously unreachable altitudes of 25 or 30 kilometers.[54] The war put an end to Millikan's work in this area. He became absorbed in National Research Council work, and the sounding balloons, manufactured in Warsaw, were unavailable for importation. After the war, however, he returned to the subject and requested support, ultimately granted, from the Carnegie Institution of Washington for it.[55]

The report which he made to the Carnegie Institution on his initial endeavors clarifies the eclectic nature of his concern with the problem. "These penetrating radiations must apparently have their origins," he wrote, "in nuclear changes going on in the atoms of the sun and stars, and their study is therefore a very fitting part of the program for the joint attack on the problem of the structure of matter from both the physical and astrophysical

50. Millikan, "Some Episodes in the Scientific Life of Robert A. Millikan," Millikan Papers, box 68, p. 10. See also J. McLennan and E. Burton, *Phys. Rev.*, 16 (1903), 184, and E. Rutherford and H. Cooke, *Phys. Rev.*, 16 (1903), 183.

51. A. Gockel, *Phys. Zeits.*, 11 (1910), 280.

52. V. Hess, *Wien. Ber.*, 120 (1911), 1575; 122 (1913), 1053, 1481; W. Kolhörster, *Verh. Deutsch. Phys. Gesell.*, 15 (1913), 1111. For an excellent review of the early literature see J. Stranathan, *The "Particles" of Modern Physics* (Philadelphia, 1942), pp. 460–463.

53. V. Hess, *Phys. Zeits.*, 13 (1912), 1084–1091. A translation of this article can be found in A. M. Hillas, *Cosmic Rays* (Oxford, 1972), pp. 139–147.

54. Millikan to H. Crew, 3 March 1920, Crew Papers, Niels Bohr Library, American Institute of Physics; Millikan, "Some Episodes," p. 11. J. B. Wright and L. J. Lassalle had worked with Millikan on cosmic ray problems. See L. J. Lassalle "Earth's Penetrating Radiation at Manila," *Phys. Rev.*, 5 (1915), 135–148.

55. Millikan to Woodward, 6 May 1919, Millikan Papers, box 20, fol. 14.

points of view."[56] Flights were made from Kelly Field, Texas, with instruments designed by Millikan and his assistant Ira Bowen which attained altitudes of over 15 kilometers.

Millikan and Bowen sent recording instruments aloft in sounding balloons, each with a barometer, thermometer, and electroscope which along with the accompanying photographic film weighed only seven ounces. Prevailing views would lead them to expect to find very high rates of discharge because only about 12 percent of the atmosphere was left to absorb the radiation. They did not, Millikan reported, "find anything like the computed rates of discharge."[57] They were able to show, however, that if the rays came from above they were far more penetrating than had been supposed.

Further flights were made in 1923 by Millikan and R. Otis from Pike's peak, but the results were generally agreed to be likewise inconclusive. In 1925, however, Millikan and G. H. Cameron traveled to Muir Lake, California (altitude 12,000 feet), and Lake Arrowhead (altitude 5000 feet). They found that the intensity of ionization demonstrated that the rays, coming exclusively from above had eighteen times the penetrating power of the hardest known gamma rays. Sinking their instruments in the upper lake to a depth of six feet (the water-equivalent in absorbing power of the layer of air between the surfaces of the two lakes) they showed that the readings of the lower lake were then identical with that of the upper. They concluded that the atmosphere contributed nothing to the intensity at the lower lake. The atmosphere acted only as an absorbing blanket. The rays came, they said, from space.[58]

Before the National Academy of Sciences at Madison, Wisconsin, on 9 November 1925, Millikan reported upon his findings and dubbed the highly-penetrating radiation "cosmic rays."[59] He was quick to draw cosmic conclusions. Since the most penetrating rays producible on earth are from radioactive transformations, he reasoned that "it is scarcely possible, then, to avoid the conclusion that these still more penetrating rays which we have been studying are produced similarly by nuclear transformations of some sort. . . . We can scarcely avoid the conclusion, then, that nuclear changes having an energy value perhaps fifty times as great as the energy changes involved in observed radioactive processes are taking place all through space, and that signals of these changes are being sent to us in these high-frequency rays."[60]

Millikan pointed out that the frequency of the hardest cosmic rays known

56. Millikan, *CIW Rpt.* (1922), pp. 385–386.
57. Millikan, "High Frequency Rays of Cosmic Origin," *Proc. Nat. Acad. Sci.*, 12 (1926), 149.
58. Millikan, *Electrons (+ and −)*, pp. 307–308.
59. Millikan, "High Frequency Waves of Cosmic Origin," *Proc. Nat. Acad. Sci.*, 12 (1926), 48–55.
60. Ibid., pp. 53–54.

at that time corresponds to the energy of formation of helium out of hydrogen, and corresponds closely also to the capture of an electron by a light nucleus.[61] Such nuclear captures may in fact be the likely explanation for the origin of such rays. He was very excited about this evidence for what he saw as the birth cries of infant atoms either by fusion or by capture.[62] Before the American Philosophical Society on 23 April 1926, he exclaimed that we can if we wish "call it the music of the spheres!"[63]

Almost immediately Millikan became what some might today call a "media figure." The *New York Times* chose to run a big spread on the newly christened "cosmic rays" in November 1925.[64] *Time* magazine featured Millikan on its cover, and with breathless urgency exclaimed that he had "detected the pulse of the universe."[65]

By February 1928, Millikan was prepared to announce another breakthrough. Before the physics seminar at Caltech, he analyzed the ionization-depth curves which he and his assistants had laboriously compiled. He indicated that the cosmic-ray absorption curve could be accounted for by summing the curves of three sets of presumed frequency bands: a low-frequency band responsible for most of the atmospheric absorption having a coefficient of 0.35, and two high-frequency bands having coefficients of 0.08 and 0.04. Using the precise work of Aston on atomic weights, Einstein's relation between mass and energy, and Paul Dirac's recently published formula for absorption through Compton scattering, he was able to compute energies for the cosmic rays (assumed to be photons) of about 26,110 and 220 million electron volts,[66] and to show that these energies corresponded to the building up, out of hydrogen, of helium, oxygen, and silicon. The cosmic ray photons were taken, then, to be "the announcements sent out through the ether of the birth of the elements."[67] "Creation continues, Millikan's Theory," exclaimed the *New York Times.* "Cosmic Rays Herald 'Birth of the Elements',"[68] "Super X-Rays Reveal the Secrets of Creation," beamed the *New York Times.*[69]

When the Klein-Nishina formula replaced Dirac's in 1928, Millikan was able successfully to hold fast to his atom-building hypothesis.[70] The revised energy calculations seemed in fact to bolster his position. Before the Royal

61. Millikan ruled out a solar origin for his rays because intensity did not vary by day or night or directionally.

62. Millikan, *Science and the New Civilization* (New York, 1930), p. 105.

63. Millikan, "The Last Fifteen Years of Physics," *Proc. Amer. Phil. Soc.,* 65 (1926), 78.

64. *New York Times,* 12 November 1925, p. 24.

65. *Time,* 6 (23 May 1925), 26–27; *Time,* 9 (25 April 1927), cover.

66. Millikan and Cameron, "The Origin of Cosmic Rays," *Phys. Rev.,* 32 (1928), 533–557; P. Dirac, *Proc. Roy. Soc.,* 111A (1926), 423; *Cosmic Rays* (New York, 1964), pp. 28–29.

67. Millikan and Cameron, "Direct Evidence of Atom-Building," *Science,* 67 (1928), 402.

68. *New York Times,* 18 March 1928, p. 1.

69. *New York Times,* 25 March 1928, sec. 10, p. 3.

70. See Millikan and Cameron, "Origin," pp. 533–557.

Society, however, Rutherford cautioned that although "the absorption coefficient of the most penetrating radiation deduced by Millikan and Cameron from their experiments is in excellent accord with that to be expected on the Klein-Nishina formula for a quantum of energy 940 million volts "we should be wary of relying on extrapolations of theories of absorption into the upper energy ranges."[71]

Millikan's atom building views, it should be noted, generated little controversy in the scientific community and even less support.[72] When during the course of the 1930's the work of A. H. Compton and others demonstrated that cosmic rays were not photons but were in fact composed of charged particles, the underpinnings of his views collapsed. Millikan and coworker Carl Anderson themselves directly measured cosmic ray energies in 1931 and found them to be far in excess of any which could be accounted for by packing fractions.[73] He began, at that point, a slow retreat to a fall-back position; from atom-building he and his coworkers ultimately passed to atom-annihilation.[74] He reviewed all the evidence for his last position on the origins of cosmic rays in an article for the *Reviews of Modern Physics* in 1949.[75] The *dénouement* came, however, only in Millikan's waning years. A note inked in upon a reprint of this article, and dated December 1950, confessed his final, reluctant abandonment of his cherished hypothesis: "New evidence," he wrote, "has appeared since this was written which is unfavorable to this hypothesis, but the experimental data contained herein is valid. The actual origin of the cosmic rays is still today an unsolved mystery."[76]

CONCLUSION: HISTORY MAKING AND HISTORY WRITING

The atom-building hypothesis represented a new turn in the scientific life of Robert Millikan. A man noted for extraordinary experimental intuitiveness and respected for his scientific caution and precision, Millikan was fully self-conscious during his excursion into dramatic theorizing. His ambivalence toward it is recorded in his own histories of his efforts in cosmic ray physics. His "history" is a reconstruction which suppresses the years of concern with the evolution of the elements and his interest in dissecting the

71. E. Rutherford, *Proc. Roy. Soc.*, 122A (1929), 15.

72. See, however, for a serious critique, E. C. Stoner, "Cosmic Rays and A Cyclic Universe," *Leeds Phil. and Lit. Soc. Proc.*, 1 (1929), 349.

73. Millikan and C. Anderson, "Cosmic Ray Energies and Their Bearing on the Photon and Neutron Hypotheses," *Phys. Rev.*, 40 (1932), p. 325; *Phys. Rev.*, 45 (1934), p. 352, and Millikan, *Electrons (+ and −)*, p. 554.

74. Millikan et al., "Hypothesis on the Origin of Cosmic Rays," *Phys. Rev.*, 61 (1942), 397–407; 63 (1943), 234–245; *Nature*, 151 (1943), 663.

75. Millikan, "The Present Status of the Evidence for the Atom-Annihilation," *Rev. of Modern Physics*, 21 (1949), 1–13.

76. Millikan Reprints, California Institute of Technology Archives.

nucleus in order better to rebuild it. The original context of his interest in cosmic rays was covered over, and a new, clean, theory-free path created. Millikan displayed his philosophy before the British Association for the Advancement of Science at Leeds in September 1927. "I told my audience," he later recounted, "that my subject presented 'a very beautiful illustration of the slow step-by-step process by building upon the past, but pushing on, if he is fortunate, a little beyond where his predecessors had gone.' "[77] As we have seen, the October 1928 paper for the *Physical Review* entitled "The Origin of Cosmic Rays" already insisted that his "results were obtained from an empirical analysis of the ionization-depth curve and entirely without the guidance of any theory. They represent solely the general type of solution demanded of the curve itself."[78] Just a paragraph later, while describing the physics seminar presented at Caltech in February 1928, he again insisted that "our minds being up to this time completely unbiased by any knowledge as to whether bands might be expected, or if so where they might occur, we set at the task of seeing whether we could find any theoretical justification for their existence, or for their energy values."[79]

The account of the discovery of cosmic rays and their interpretation as related in the 1947 edition of *Electrons (+ and −)* is similarly restricted to the purely empirical. His interest in cosmic rays derived, on this account, solely from his desire to resolve the controversy raised by Hess and others regarding the origins of the ionizing radiation. No explanation of *why* he believed the question to be critical is supplied. The atom-building hypothesis is shown, in addition, to be a strictly empirical generalization to which he was forced by the nature of his evidence and by the theoretical tools at his disposal at the time.[80] The same story is retold in his review article for the *Reviews of Modern Physics* in 1949.[81] His *Autobiography* published in 1950 makes no mention at all of nucleus smashing or of the atom-building hypothesis.[82]

It is important to recall, however, that rather than being forced by his data to each succeeding step, as he would have his readers believe, Millikan's work in cosmic rays was shaped by his theoretical concerns. As early as 1922 he published reports which indicated that he suspected the origins of cosmic rays lay in "nuclear changes going on in the atoms of the sun and stars."[83] His paper before the National Academy supported his earlier

77. Millikan, "History of Research in Cosmic Rays," *Nature,* 126 (1930), 15.

78. Millikan and G. H. Cameron, "The Origin of Cosmic Rays," *Phys. Rev.,* 32 (1928), 534.

79. Ibid.

80. Millikan, *Electrons (+ and −),* pp. 304–313.

81. Millikan, "Present Status," pp. 1–13.

82. Millikan, *The Autobiography of Robert A. Millikan* (New York, 1950), pp. 209–210; 246–247.

83. Millikan, *CIW Rpt.* (1922), pp. 385–386.

speculation concerning the origins of cosmic rays.[84] The atom-building hypothesis was not therefore forced upon a reluctant Millikan by his carefully collected data; it was on the contrary the primary reason for his initial interest in the problem. Millikan found what he was looking for.

Millikan's route to his theory of the origin of cosmic rays represents for the historian something more than the series of false starts which Millikan himself remembered and was concerned, for both pedagogical and practical reasons, to cover over. His scientific career demonstrates that he consistently sought after the critically important problems, the key ideas, and the fruitful techniques which would shed light upon them. Because of his zeal for the truly significant, Millikan was more successful than most of his contemporaries in opening up new areas for research. Even his false starts often bore fruit. In order to bolster his atom-building hypothesis Millikan required direct evidence of cosmic-ray energies. He urged his student Carl Anderson to design a cloud-chamber apparatus for the purposes of this study. Anderson discovered the positron.[85]

84. Millikan, "High Frequency Waves of Cosmic Origin," *Proc. Nat. Acad. Sci.*, 12 (1926), 48–55.

85. Carl Anderson, "The Positive Electron," *Phys. Rev.*, 43 (1933), 491–494.

VI

THE SUPERNOVA OF 1054

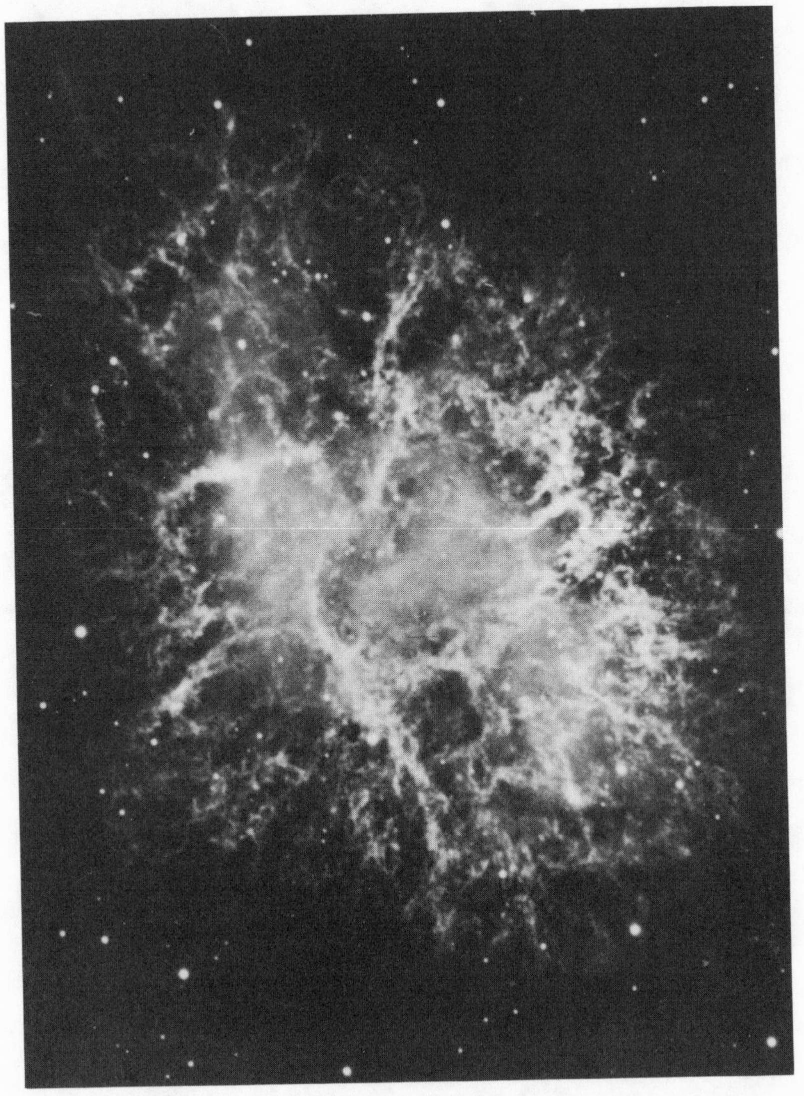

The Crab Nebula

THE SUPERNOVA OF 1054: A MEDIEVAL MYSTERY

L. PEARCE WILLIAMS

MODERN astronomers and historians of astronomy agree that the Crab Nebula is the visible remnant of a great stellar explosion.[1] The date which all modern authorities accept for this event is 4 July 1054, for on that date a "guest star" appeared in the heavens and was duly noted by the astronomers of China and Japan. The magnitude of the Crab Nebula indicates that the explosion that produced it must have been quite spectacular and the new star that suddenly waxed in the sky has been classified as one of the few supernovae on record. Just how spectacular the event must have been can be seen from the impact that it seems to have made on the Amerindians of the Southwest of the United States. Cave paintings there, it is claimed, record this event and are unique in Indian remains of the time.[2] All this is now such

Henry Guerlac is a scholar of brilliant insights and original ideas. No one can teach others to have either of these. But Henry Guerlac did teach his students that insights and ideas may have humble origins. He insisted, for example, that we never pass over a discrepancy in a historical account. If something looked odd or peculiar, he wanted us to go to the original sources where we might discover interesting things. This paper is offered as an example of what this method can lead to. I simply could not believe that an event as spectacular as the supernova that supposedly caused the Crab Nebula could go unnoticed in Western Europe, even in the eleventh century. Out of that disbelief came this critique of the whole Crab Nebula story.

1. The identification of the Crab Nebulae with a supernova in 1054 is to be found in both I. S. Shklovsky, *Supernovae* (New York, 1968), and *The Crab Nebula,* ed. R. D. Davies and F. G. Smith (New York and Dordrecht, 1971), in which the date is taken as proven. Joseph Needham in volume 3 of his *Science and Civilisation in China, Mathematics and the Sciences of the Heavens and the Earth* (Cambridge, 1959) also accepts the identification. With but one exception (to be noted later) every author who discusses the Crab Nebula and is cited in this paper considers it to be the result of a supernova in 1054. The photograph on the facing page is reproduced from "An Optical Atlas of Galactic Supernova Remnants," by S. van den Bergh, A. Marsher, and Yervant Terzian, *The Astrophysical Journal Supplement,* 26 (1973), 19.

2. In 1955, William C. Miller of the Mount Wilson and Palomar Observatories published a leaflet (No. 314) in the series sponsored by the Astronomical Society of the Pacific (ASP) entitled "Two Prehistoric Drawings of Possible Astronomical Significance." In this article,

common knowledge that it is repeated routinely in every account of the Crab Nebula and is taken for granted even in articles that appear in *The New York Times.*[3] It is the purpose of this article to examine this account critically and to show that serious doubt can be thrown on many aspects of it. It will be shown that there are a significant number of implausible inferences that need to be reassessed and that the identification of the Crab Nebula with the "guest star" of 1054 is not quite so clear-cut as has been supposed. To do this, it is necessary to look closely at the history of this identification.

The story begins in 1921 when John C. Duncan of the Mount Wilson Observatory published a short (one and one-half pages) article in the *Proceedings of the National Academy of Sciences* on "Changes Observed in the Crab Nebula in Taurus."[4] The figures that he gave indicated that the gases in the Crab Nebula were expanding at a measurable rate. In that same year, the Swedish astronomer, K. Lundmark, published a brief account of suspected novae that had been recorded in Chinese sources. Number 36 in Lundmark's list was an object whose date of appearance was 4 July 1054, whose location was "southeast of eta Tauri but near" and, in a footnote, "Near NGC 1952."[5] This location was later claimed to be an error and the new star was moved to zeta Tauri. In 1928, Dr. Edwin Hubble, the great astronomer also at Mount Wilson, having read both Duncan's and Lundmark's articles wrote a brief (two and one-half pages) paper which was published as Leaflet 15 of the Astronomical Society of the Pacific. His comments are worth quoting here. What follows is all that he said about the Crab Nebula. "The Crab Nebula, Messier No. 1, is possibly a third [nova near enough for its nebulosity to be seen and photographed] for it is expanding rapidly and at such a rate that it must have required about 900 years to reach its present dimensions. For, in the ancient accounts of celestial phenomena only one nova has been recorded in the region of the Crab Nebula. This account is found in the Chinese annals, the position fits as closely as it can be read, and the year was 1054!"[6] It was not until ten years later that Hubble's identification of the nova of 1054 with the Crab Nebula was taken up by astronomers.

In 1939, Duncan published a second article on the expansion of the Crab

---

Miller strongly suggested the identity of the supernova observed by the Chinese and Japanese and the astronomical object of the Amerindian drawings in Arizona. This thesis has since been supported by John C. Brandt et al., "Possible Rock Art Records of the Crab Nebula Supernova in the Western United States," in A. F. Aveni, ed., *Archaeoastronomy in Pre-Columbian America* (Austin, Texas, 1975).

3. See the article in *The New York Times,* 10 September 1975, p. 47.
4. *Proceedings of the National Academy of Sciences,* U.S., 7 (1921), 179.
5. *Publications of the Astronomical Society of the Pacific,* 33 (1921), 225.
6. Dr. Edwin Hubble, "Novae or Temporary Stars," Leaflet 14, January 1928, ASP, p. 58. I have used the later, collected edition of the leaflets in which the pagination is consecutive.

Nebula. From the new observations he assigned, "with considerable uncertainty—the year A.D. 1172 as the date of the outburst."[7] In this same year, another astronomer entered the picture, Nicholas U. Mayall of the Lick Observatory. Mayall had read Hubble's leaflet and fully adopted Hubble's suggestion that the Crab Nebula was identical to the nova reported in 1054.[8] What appears to have convinced Mayall was the translation in 1934 of the Japanese account of the "guest star" by the Japanese astronomer, Y. Iba.[9] That account stressed the brightness of the stellar object ("it was as bright as the planet Jupiter," Mayall reported)[10] and it was this fact, together with the calculated distance, that led Mayall to suggest that what the Chinese and Japanese had observed was not a nova but a supernova. Mayall's degree of commitment to this view should be noted: "It may be said that the identification of the Crab Nebula as a former supernova possesses a degree of probability sufficiently high to warrant its acceptance as a reasonable working hypothesis."[11]

With the publication of Mayall's leaflet in 1939, the first chapter in the story of the supernova of 1054 comes to an end. We may note a few points that will be of interest later. There is, first of all, no serious study of the original sources to be found in any of the accounts we have mentioned. Hubble noted that the Crab Nebula was roughly in a part of the sky where a nova was noted in 1054. Mayall merely added the comment on brightness to this. Duncan's admittedly imprecise conclusion from his data offered a date more than a century later than the one preferred by Hubble and Mayall. And even Mayall wished to go no further at this time than to suggest a "working hypothesis." One would assume from this that what must have happened in the years after 1939 was the discovery of further evidence that placed the identification of the Crab Nebula with the supernova of 1054 completely beyond doubt. Let us look at the new evidence that was forthcoming.

In 1941, J. J. L. Duyvendak, a Sinologist at the University of Leyden, published the Chinese and Japanese sources, with an English translation, in the journal *T'oung Pao*. These accounts are of such fundamental importance for my thesis that I include them here, with the permission of the publisher. All that has been omitted from Duyvendak's article are a few of his asides and explanatory notes that are not essential to the argument.[12] I have numbered the sources for convenience in later reference.

7. John C. Duncan, "Second Report on the Expansion of the Crab Nebula," *Astrophysical Journal* 89 (1939), 485.

8. Nicholas U. Mayall, "The Crab Nebula, A Probable Supernova," Leaflet 119, January 1939, ASP, p. 145.

9. *Popular Astronomy,* 42 (1934), 243.

10. Mayall, "The Crab Nebula," p. 146.

11. Ibid., p. 154.

12. J. J. L. Duyvendak, "The 'Guest Star' of 1054," *T'oung Pao,* 36 (1941), 174–178.

1.

The original text (quoted by Biot from Ma Tuan-lin's *Wenhsien T'ung-kao* ch. 294, p. 12a) is found in the *Sung-shih* 宋史 (Treatise on Astronomy, Paragraph on "guest-stars") ch. 56, p. 25a (Po-na ed.). It runs:

至和元年五月己丑出天關東南可數寸。歲餘稍沒。

"In the 1st year of the period Chih-ho (1054), the 5th moon, the day *chi-ch'ou* (July 4th) (a "guest-star") appeared approximately several inches south-east of T'ien-kuan (ζ Tauri). After more than a year it gradually became invisible."

2.

In the Annals of the *Sung-shih,* which do not report the phenomenon under the 1st year of Chih-ho, this information is supplemented in an entry under the 1st year of the period Chia-yu. There (ch. 12, p. 10b) we read:

辛未司天監言、自至和元年五月客星晨出東方、守天關、至是沒。

"On the day *hsin-wei* (of the 3rd moon of the 1st year of the period Chia-yu, i.e. April 17th 1056) the Chief of the Astronomical Bureau reported that from the 5th moon of the 1st year of the period Chih-ho (June 9th–July 7th 1054) a guest-star had appeared in the morning in the eastern heavens, remaining in T'ien-kuan (ζ Tauri), which only now had become invisible."

The duration of visibility of this "guest-star" was therefore from July 4th 1054–April 17th 1056.

3.

Further information is found in the *Sung Hui Yao* 宋會要, vol. 52, p. 2b (not numbered):

至和元年七月二十二日.....楊惟德言、伏覩客星出見、其星上微有光彩黃色。謹案皇帝掌握占云、客星不犯畢、明盛者主、國有大賢、乞付史館容。百官稱賀、詔送史館.

嘉祐元年三月司天監言、客星沒、客去之兆也。初至和元年五月晨出東方守天關、晝見如太白、芒角四出、色赤白。凡見二十三日。

"On the 22nd day of the 7th moon of the 1st year of the period Chih-ho (August 27th 1054) ...... Yang Wei-tê said:

"Prostrating myself, I have observed the appearance of a 'guest-star'; on the star there was slightly an irridescent yellow colour. Respectfully, according to the dispositions for Emperors, I have prognosticated, and the result said: The 'guest-star' does not infringe upon Aldebaran; this shows that a Plentiful One is Lord, and that the country has a Great Worthy! I request that this (prognostication) be given to the Bureau of Historiography to be preserved."

All the officials presented their congratulations and by Imperial Edict it was ordered that (the prognostication) should be sent to the Bureau of Historiography.

"In the 3rd moon of the 1st year of the period Chia-yu (March 19th–April 17th 1056) it was reported that the 'guest-star' had become invisible, which was an omen of the departure of guests.

Originally this star had become visible in the 5th moon of the 1st year of the period Chih-ho (June 9th–July 7th 1054) in the eastern heavens, in T'ien-kuan (ζ Tauri); it was visible by day, like Venus; pointed rays shot out from it on all sides; the colour was reddish-white. Altogether it was visible for 23 days."

The observations reported in the *Sung-shih* and the *Sung Hui Yao* were of course made at the capital K'ai-feng. Now it will be remembered that at that time the Liao dynasty had its capital at the site of the present Peking and it is very interesting that there, quite independently from the astronomical bureau at K'ai-feng, this "guest-star" was also observed. The official history, the *Liao-shih*, contains no report about it, but if we turn to the older *Ch'i-tan Kuo-chih* 契丹國志, we find it duly mentioned.

In ch. 8, p. 6b we read:

(重熙二十三年)。。。八月國主崩。。。。。先是日食、正陽客星出于昴。著作佐郎劉義叟曰、興宗其死乎、至是果驗。

4.

(In the 23rd year of the period Ch'ung-hsi [which is really the 24th year, or the 1st year of the period Ch'ing-ning 淸寗, i.e. 1055]) in the 8th moon the Lord of the country died ..... Previously there had been a sun-eclipse, and in the 1st moon (January 31st–Febr. 28th 1055) a "guest-star" had appeared in the Pleiades. Liu Yi-sou, Senior Vice-President of the Bureau of Historiography, said: "Now Hsing-tsung has died, (these omens) have indeed come true!"

5.

Not only at K'ai-feng-fu and at Peking, but also in Kyōto the "guest-star" was observed. The *Mei Getsuki* 明月記 and the *Ichidai-Yôki* 代要記 both contain texts which, though there are graphic variations, are virtually identical and are clearly derived from an earlier text written in running hand. The text in the *Mei Getsuki* runs as follows:

天喜二年四月中旬以後丑時客星出觜參度、見東方、孛天關星、大如歲星。

"In the middle ten-day period of the 4th moon of the 2nd year of the period Ten-ki (i.e. May 20th–29th 1054) and thereafter, between 1–3 a.m. a 'guest-star' appeared in the orbit of Orion; it was visible in the eastern heavens. It shone like a comet(?) in T'ien-kuan (ζ Tauri), and was as large as Jupiter."

This text claims that the "guest-star" appeared in the 4th moon, instead of in the 5th as reported by the other texts. This however is impossible, for, as Professor J. H. Oort tells me, at that time the sun was approximately in conjunction with this star, so that it must have been invisible. The 4th moon must therefore be an error for the "5th moon"; as happens so easily in chronological records, the notice has been shifted to a wrong moon.

The "middle ten-day period" would then correspond to June 19th–28th; this would date the Japanese observation from 16–6 days earlier than the earliest Chinese record (July 4th) of this "guest-star."

This account was followed by a short commentary by J. H. Oort, the astronomer who had asked Duyvendak to investigate the Chinese accounts.[13] In a footnote, Oort pointed out a difficulty.

It should be noted that a slight discrepancy exists between the position of the Crab nebula and that of the "guest star" reported in the History of the Sung dynasty. According to this report the star was situated south-east of [zeta] Tauri, whereas, if the identification is correct, it should have been situated a little over a degree to the north-west of [zeta] Tauri. Prof. Duyvendak has confirmed that it is unlikely that a mistake of copying was made in the Chinese sources. In view of the astronomical data it would seem that either an error has been made in the original Chinese record, or that T'ien kuan should not be identified with [zeta] Tauri; there is a possibility that T'ien kuan may have referred to a group of stars in that neighbourhood instead of to [zeta] Tauri alone.[14]

Oort was able to send Duyvendak's paper from Nazi-occupied Holland, through Sweden, to Mayall in the United States.[15] The translations of the Chinese and Japanese sources were immediately published by the Astronomical Society of the Pacific, followed by a discussion of the astronomical aspects by Mayall and Oort.[16] Mayall was now convinced that his "working hypothesis" of 1939 had been confirmed and that the identification of the supernova that produced the Crab Nebula with the "guest star" of 1054 had been established beyond doubt. What convinced Mayall was the evidence provided by the Chinese for the brightness of the "guest star". In Mayall's words: "The maximum apparent magnitude of the 1054 nova, as the second of the two quantities that determine its maximum absolute mag-

13. According to N. U. Mayall, "The Story of the Crab Nebula," *Science,* 137 (1962), 94.

14. Appendix to Duyvendak, "The 'Guest Star,'" n. 1.

15. Mayall, "The Story of the Crab Nebula," p. 94.

16. N. U. Mayall and J. H. Oort, "Further Data Bearing on the Identification of the Crab Nebula with the Supernova of 1054 A. D., Part II. The Astronomical Aspects," *Pub. ASP,* 54 (1942), 95.

nitude, may be estimated from the Chinese observation that 'it was visible by day, like Venus.' . . . A better way to estimate the maximum apparent magnitude appears to be to use the additional information: 'Altogether it was visible for 23 days.' "[17] From these remarks, Mayall then estimated the "limiting apparent magnitude m_1 of an object that can be seen with the unaided eye in daylight" by the following argument:

> In order to estimate m_1, we make general reference to the literature of planetary observations, in which there frequently occurs the statement that Venus may be seen in daylight with the unaided eye at nearly all points of its orbit. . . . Since the apparent magnitude of Venus, in this instance, was −3.3, it seems reasonable to infer that an object of −3½ vis. mag. could readily been [*sic*] seen farther from the sun, especially if the observer knew just where to look. . . . Under these circumstances, the Chinese probably knew precisely where to look for their guest-star in daytime, and may have been able to follow it in daylight to an apparent magnitude of −3½, which we shall therefore take as m_1.[18]

From observations of modern supernovae, Mayall then estimated that the fall off in brightness from maximum to bare visibility was of the order of 1½ magnitudes, and concluded that the "guest star" of 1054 had the visible magnitude of −5 at maximum brightness. This would make the "guest star" of 1054 one of the brightest heavenly objects ever observed (excluding, of course, the sun and the moon), second only to the great supernova of 1006 in Lupus whose estimated visual magnitude was −8.[19]

Mayall and Oort now concluded that "the case for the supernova character of the 1054 guest-star, and its identification with the Crab Nebula, seems to us to be so strong as to admit of no serious doubts."[20] These articles by Duyvendak, Oort, and Mayall are now referred to as the "proof" of this identification.[21] There may still, however, have been some who wondered if it was all that certain. Whatever doubts may have lingered in the minds of astronomers appear to have been erased by Miller's remarks on the cave paintings found in the American Southwest.

Photographs of these cave paintings are shown in Plate 1. What do they mean? The answer seemed clear to Miller.

> The key to the problem lay in the fact that both drawings showed the crescent moon in close association with the circle assumed to represent the supernova. The first step consisted of establishing the location of the moon at the time the supernova flared to maximum brilliance. The requirements for a favorable answer were that the phase of the moon be only a few days before new moon

17. Ibid., 96.
18. Ibid., 98.
19. Xi Ze-zong, Po Shu-jen, "Ancient Oriental Records of Novae and Supernovae," trans. K. S. Yang, *Science*, 154 (1966), table 2, p. 602.
20. Mayall and Oort, "Further Data," p. 100.
21. Xi Ze-zong et al., "Ancient Record," p. 598.

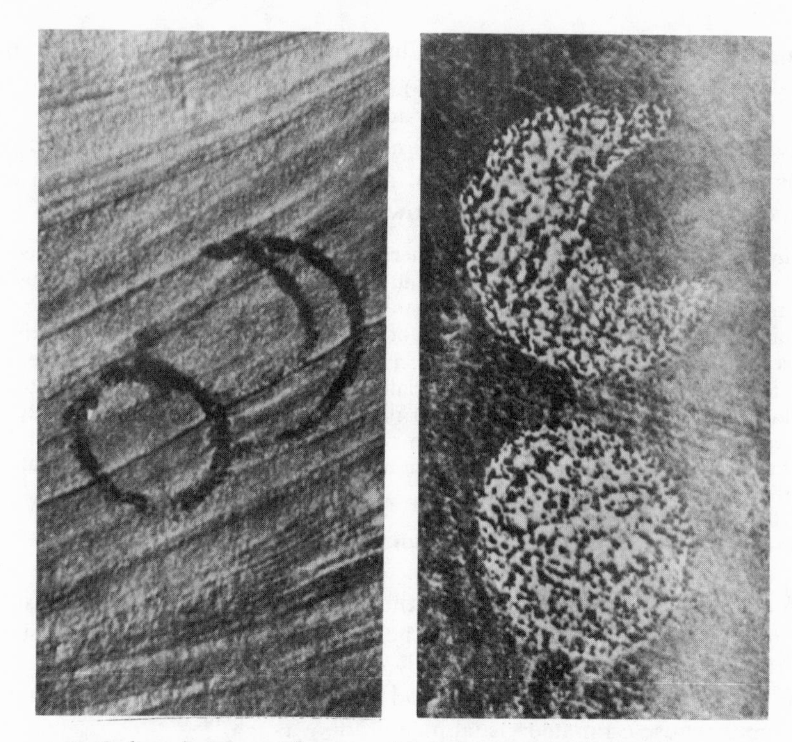

1. Indian drawings. These are the drawings that Miller claims support his interpretation that they represent the supernova of 1054.

> on or near July 4, 1054, and that it be located within a few degrees of the known position of the supernova which would have been a brilliant object near the eastern horizon about an hour before sunrise.[22]

When he did the computations, Miller discovered that "(1) the moon was at crescent phase in the morning sky on July 4 and 5, and (2) at about three o'clock on the morning of the 5th was directly north of the supernova and only two degrees distant. The rather stringent conditions for a favorable answer seem to be met and strongly suggest the possibility that the two drawings actually depict the event of the apparition of the supernova of 1054."[23]

Miller's account added powerful support to the belief that the Crab Nebula originated in the supernova of 1054. The crescent figure, Miller pointed out, is rare among the drawings of the Indians of the Southwest and, therefore, the paintings were not casual depictions of an ordinary astronomical event. Rather, they must be the result of something stupendous for the

22. Miller, "Two Prehistoric Drawings," p. 5.
23. Ibid., p. 6.

Indians to have bothered recording the event. And, as Miller claimed (erroneously), "this supernova was probably the brightest object other than the sun and moon ever to appear in the sky in the memory of man and would certainly attract attention."[24] Miller's article provided what seemed to be the necessary firm foundation for Mayall's and Oort's certainty, for it gave independent support for their estimation of the intrinsic brightness of the supernova on which the identification ultimately depended.

We are in possession of all the arguments and evidence in favor of the supposition that the Crab Nebula originated in a supernova of 1054. How well does the evidence hold up to critical scrutiny?

There are a number of ways to attack the problem. We shall first examine the astronomical evidence to see if it supports the identification. It is then possible to look more closely at the Chinese sources to discover if they are as unambiguous and accurate as those who have used them believe them to be. Miller's argument from the cave paintings can then be subjected to analysis to discover if it necessarily supports the identification. Finally, there is some evidence from Europe that can be introduced to cast doubt upon the nature of the "new star" of 1054.

In 1942, W. Baade at Mount Wilson published a paper on the Crab Nebula in which he discussed Duncan's measurements of the expansion of the nebula in the light of new information. Like Duncan, he found that a simple extrapolation backward of the expansion gave a date of origin some hundred years after 1054. It is safe, I think, to assume that if Mayall and Oort had not published their paper at the very time that Baade was doing his calculations, Baade would simply have let the matter rest there, but as he noted at the beginning of his paper: "The important new data about the nova of 1054 which Professor Duyvendak has recently made available leave hardly any doubt, as Oort and Mayall have shown, that this star is the parent of the Crab nebula and that at maximum it reached the absolute magnitude of −16.5 which means it was a supernova."[25] So, *knowing* that 1054 was the date of origin, Baade tried to account for the difference between his calculations and this date. The only possibility was an acceleration in the expansion but, as he frankly acknowledged, "difficulties arise, however, when we try to understand it in terms of known forces."[26] Radiation pressure, he showed, was insufficient and he was forced to conclude that:

> There remains the possibility that the acceleration took place during the months or perhaps years immediately following the outburst and that it had ceased by the time the star reached its present state. The growth of the radius of the nebula would then be of the type represented in Figure 1, with t_0 the

24. Ibid., p. 7.
25. W. Baade, "The Crab Nebula," *Astrophysical Journal*, 96 (1942), 188.
26. Ibid., p. 197.

> date of the outburst as inferred from the final rate of expansion. The difficulty in this case arises from the fact that, according to Duncan's measures, the time difference between t_0 and the true date of the outburst amounts to about 100 years. Since it is obvious from Figure 1 that the acceleration is still near its maximum value at the time t_0, the bolometric intensity of the star 100 years after the outburst must have been at least 10^3–10^4 times as high as at present to explain the observed acceleration. This is highly improbable because it implies that the Crab nebula, after its decline below the sixth magnitude in 1056, was again a naked-eye object of apparent magnitude 0 some 100 years later. There seems to be only one way out of these difficulties, namely, to assume that the measured rate of the angular expansion is too large and that the resulting acceleration of the expansion is spurious. To settle this question, as well as that of the proper motion of the nebula, we shall have to wait until the recent red photographs can be used for a redetermination of the motions.[27]

This passage is of fundamental importance for my argument. It baldly states that there is no way to account for the discrepancy in dates using forces known in 1942. Moreover, it makes explicit the fact that the mechanism of expansion of the Crab Nebula is both unique and extraordinary. Baade cannot suggest a model to explain it satisfactorily but can only hope that later observations will remove the difficulty. They did not. In 1968, Virginia Trimble of the Mount Wilson and Palomar Observatories reported once more on the motions of the filaments of the Crab Nebula and calculated a date of 1140 for the initial explosion. Her 1971 report does not modify this date significantly.[28] Hence we conclude that *the best and most accurate astronomical data available indicate that the Crab Nebula did not originate in 1054.* The point is worth making again that if the Chinese and Japanese sources for the "guest star" were not available, astronomers would be content to date the origins of the Crab Nebula from the middle of the twelfth century. This being so, it seems essential to ask how compelling the Chinese and Japanese evidence for a supernova really is.

Ho Peng-Yoke, the historian of astronomy, is the only scholar who has seriously challenged the attribution of the "guest star" to the origin of the Crab Nebula.[29] He focuses upon the discrepancy in the position noted by the Chinese and the actual location of the Crab Nebula. His argument is worth repeating in some detail.

Duyvendak's Chinese source 1 located the "guest star" "several inches south east of T'ien-kuan" and we have already noted Oort's note to the effect that the only possible error here could be in the identification of T'ien-kuan. After an exhaustive examination of all the references to T'ien-kuan, Ho Peng-Yoke and his co-workers conclude that there could be no

27. Ibid., p. 198.

28. Virginia Trimble, "Motions and Structure of the Filamentary Envelope of the Crab Nebula," *Astrophysical Journal,* 73 (1968) 540, and Davies and Smith, eds., *The Crab Nebula.*

29. Ho Peng-Yoke, F. W. Paar, and P. W. Parsons, "The Chinese Guest Star of A. D. 1054 and the Crab Nebula," *Vistas in Astronomy,* 13 (1972), 258.

mistake here. The eleventh century, they remind us, "happened to be a period renowned for its astronomical instruments and accurate observations .... It was between the years A.D. 1049 and A.D. 1053 that the positions of many stars were determined or checked using the armillary sphere made by Shu I-chien."[30] A modern check of these determinations has shown that "80 per cent of the observations had a margin of error of ± 0°!"[31] But the phrase "several inches south-east" is a troublesome one with which Ho Peng-Yoke wrestles. The Crab Nebula simply is not southeast of zeta Tau and Ho would move it even farther away. The passage can be translated another way to mean, "appeared to the south-east of T'ien-kuan, (measuring) about several inches"[32] which could be interpreted as meaning that the "guest star" was located several inches from T'ien-kuan on one of the armillary spheres for which Chinese astronomy has been justly praised by modern astronomers. This would put it some ten to fourteen degrees away from the Crab Nebula.[33] That is where Ho and his co-authors leave the matter. But it cannot be left there. We know that novae leave remnants that are detectable by modern astronomical means and if Ho is correct, there ought to be a pulsar or some other evidence of a nova southeast of T'ien-kuan. There is none.

We are now faced with a real predicament. Duyvendak, a noted Sinologist, doubts strongly that southeast can be a scribal error for northwest. Ho Peng-Yoke, a noted historian of astronomy, is convinced that T'ien-kuan must be zeta Tau. If both are right, then we have a nova for which there is no physical evidence and that would seem to be impossible. So, one of the two must be wrong. Let us suppose that Duyvendak is and that our earliest and probably most reliable source wrote northwest which, in copying, was changed to southeast.[34] That would leave us with the "several inches." "Several" must be more than two. Zeta Tau has a right ascension of $05^h\ 34^m$ and declination of 21° 06′. The Crab Nebula is at R.A. $05^h\ 31^m$, Dec., 21° 58′. I can find no way to get "several inches" out of this, even assuming that the measurement is made at arm's length with some kind of a scale held up to the sky. So even if the scribal error in direction had been made, the "guest star" would not be near the Crab Nebula. That leaves us with the only other alternative, namely, that zeta Tau is not T'ien-kuan. Ironically, a possible solution now appears from the very beginning of our story. Lundmark had used eta Tau as his reference point. Hubble had followed Lundmark in this. The Crab Nebula is, in fact, to the southeast of eta

30. Ibid., p. 9.
31. Ibid.
32. Ibid., p. 3.
33. Ibid.
34. It should be noted that this is a most unlikely error to make for the Chinese characters involved are quite dissimilar. Those such as Brandt et al. who use this argument have the responsibility of providing at least one such example if we are to take them seriously.

Tau and within the range suggested by Ho's translation and interpretation of the "several inches" phrase. If the Chinese source is to be taken as accurate, then the error must come in the identification of zeta Tau with T'ien-kuan. In spite of Ho Peng-Yoke, therefore, we are forced to conclude that, in this case, the Chinese astronomers did not equal their vaunted accuracy in observation or in description. Nor should this really come as a shock. Readers, by now, must share my impatience with the constant claim for accuracy of the Chinese when compared with the actual observation. Nothing could be fuzzier and less precise than an account in which "several inches" is offered without any reference point or any hint as to how the measurement should be carried out.

This "solution" still leaves a number of other difficulties with which we must deal. Among these is the question of the magnitude of the "guest star" of 1054.

Mayall based his classification of the "guest star" as a supernova on the descriptions of its brightness. We should note, first, that Mayall shifted his ground on this point between his "working hypothesis" of 1939 and his arrival at certainty in 1942. In 1939, it was the Japanese account that stated that the "guest star" was as bright as Jupiter that Mayall described as a "datum of crucial importance."[35] In the 1942 paper, however, Jupiter was eclipsed by the Chinese reports comparing the "guest star" to Venus. It is, nevertheless, interesting to follow Mayall in his rejection of that earlier datum of crucial importance.

> At the time of the apparition, Jupiter was not far from conjunction with the sun, and its apparent magnitude was close to −1.3. A brightness of this order for the nova, if it occurred about a week before a maximum observed brightness of −5, could be reconciled with the few pre-maximum observations of other supernovae. This view of the Japanese brightness-estimate does not, however, inspire much confidence, because at the time in question Jupiter was an evening object, while the nova was a morning object. Under these circumstances, a simultaneous comparison of the two objects was impossible, and the statement that the nova "was as large as Jupiter" probably is a gross underestimation of the actual apparent brightness of the nova.[36]

Mayall here looks suspiciously like a person picking and choosing only the evidence that will fit a preconceived notion. What is critical for Mayall's position is that the brightness of the "guest star" and Jupiter just would not do. So, Mayall in 1942 discredits the account that had gained his full credence in 1939. The reasons seem obvious: Jupiter will give him a magnitude of only −1.5 whereas the remarks by the Chinese about Venus will permit him to raise that to −5, which then makes the "guest star" into the supernova that he knew had to be there all along.

35. Mayall, "The Crab Nebula," (1939), p. 146.
36. Mayall and Oort, "Further Data" (1942), p. 100.

Do the Chinese sources support Mayall's conclusions on visible magnitude? If, as Miller claimed, the "guest star" in 1054 were truly the "brightest object other than the sun and moon ever to appear in the sky in the memory of man," the Chinese and Japanese were remarkably blasé about it. The earliest account comes from an author who died in 1084 and probably represents the description of an eye-witness. It reads simply, "On an i-ch'ou cyclical day (in the fifth lunar month) a guest star appeared to the south-east of T'ien-kuan, possibly several inches away" (Duyvendak, 1). There is no mention of its brightness, nor of its unusual size, nor of anything out of the ordinary. In the Annals of the Sung-Shih, the official history of the Sung dynasty completed in 1345 and apparently derived from an earlier work finished in 1280, we get the first hint of brightness in the phrase (Duyvendak, 2) that the "guest star" had appeared "in the morning in the eastern heavens." Now, at this time, the sun was more than two hours east of the nova[37] so this should be interpreted as meaning only that the "guest star" was visible in the early morning sky, that is, before sunrise. Only in the Sung Hui Yao, for which Ho et al. give no date, is there to be found the statement (Duyvendak, 3) that the "guest star" was "visible by day, like Venus." This is also the only source that implies that it was visible in daylight for twenty-three days. The context of this description is interesting. The first part of the account quotes the words of Yang Wei-te, Chief Astrologer, who makes no mention of anything out of the ordinary except to refer to the fact that the "guest star" was "slightly an iridescent yellow color." Yellow was the Imperial color and all that the Chief Astrologer could find to say about the appearance of the "brightest object . . . ever to appear in the sky in the memory of man" that had obvious imperial connections was, "This shows that a Plentiful One is Lord, and that the country has a Great Worthy!"(!) The section that compares the "guest star" to Venus is not a direct quotation and may well be a later interpolation. Indeed, it looks as if someone had seen the earlier account of the "guest star" appearing in the early morning (as Venus does as the Morning Star) and had embellished the account accordingly, exaggerating the brightness of the star for effect. In any case, it should be noted that this is the only source for the conclusion that the "guest star" was out of the ordinary. It seems a precarious perch from which to hang a firm conviction that the "guest star" was a supernova identical with the Crab Nebula.

We are left with one difficulty to remove before the hypothetical house of cards constructed by Mayall and Miller collapses entirely. What is to be done with Miller's cave paintings? Surely they do depict an unusual heavenly spectacle and it seems highly improbable that the position of the

37. Ibid., p. 98.

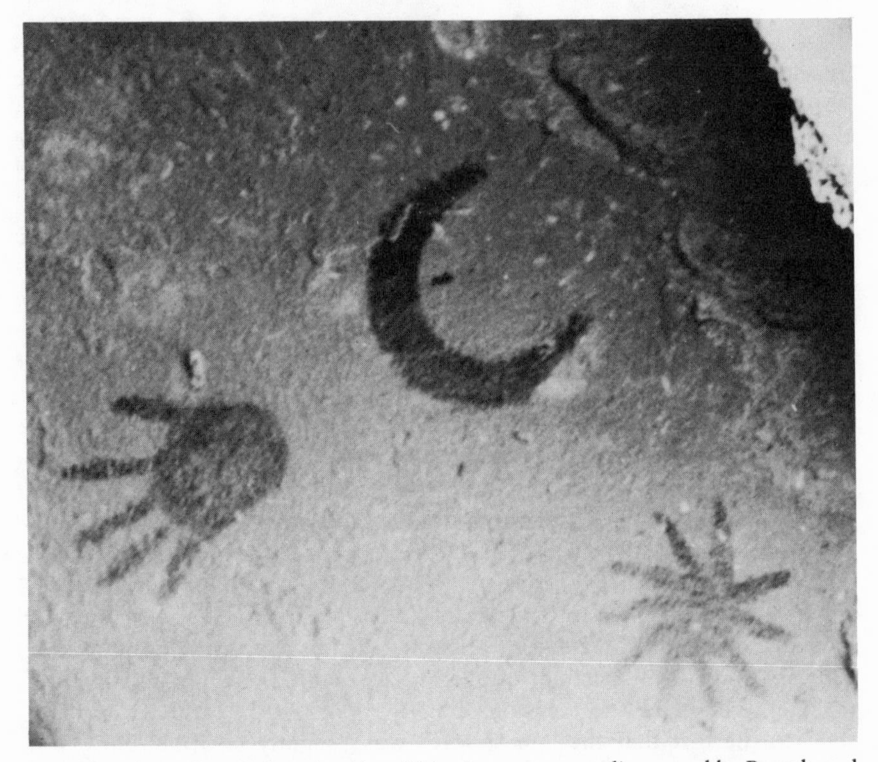

2. A more modern Indian drawing. This pictogram was discovered by Brandt and his team in the American Southwest. It is supposed to represent the supernova of 1054 since it looks very similar to Miller's examples. (Photograph courtesy of the National Aeronautics and Space Administration and Dr. Brandt.)

crescent moon and the "guest star" of 1054 is not reflected in the drawings themselves.

There are three possible explanations. First, Miller's claim that because the crescent figure is rare in Amerindian cave paintings, his figures must record an extraordinary event is not necessarily to be allowed. We now know that these figures are not quite so rare as Miller thought.[38] Hence they need not represent a unique event. Pictures that look very much like Miller's are rather common elsewhere (Plate 2). Babylonian cylinder seals abound with crescent moons and stars (see Plate 3) and the tomb of Antiochus I (d. 38 B.C.) at Nimrud Dagh shows the crescent moon and Regulus in precisely the same configuration.[39] It is entirely possible that the event recorded in the

38. Brandt et al., "Possible Records," and *The New York Times,* 10 September 1975.

39. I am deeply grateful to Professor A. Aaboe of Yale University for calling my attention to these representations. Ho Peng-Yoke et al. were conscious of this objection and tried to use it in their argument. Unfortunately, all the other cases of the crescent-star motif that they cited were probably later than 1054 and might, in fact, have been derived from the supposed supernova.

American Southwest is nothing more spectacular than a conjunction of the moon with Venus when Venus was the Morning Star. Since the "guest star" was specifically referred to Venus as the standard of brightness, such a conjunction would appear exactly as the "guest star" was supposed to have appeared in Mayall's view.

The second possibility, put forward by a respected anthropologist and student of the Indians of the Southwest, is that the pictograms have nothing whatsoever to do with either a conjunction of Venus or a nova.[40]

The third possibility is that the figures do represent an astronomical event of some magnitude which Miller identifies with the "guest star" of 1054. I am uneasy with this explanation, for the pictograms seem far more impressive than our accounts of the "guest star" warrant. In particular, the figure that shows the "star" as a sizable object accords ill with the appearance either of Venus or of the "guest star" as described in our sources. There is, however, an astronomical event that would nicely fit these pictures. In 1006, there was a supernova in Lupus which was described by an Islamic eyewitness.

> I will now describe for you a spectacle (*athar*) that I saw at the beginning of my education (*ta'alīm*). This spectacle (*athar*) appeared in the zodiacal sign Scorpio in opposition to the sun, at which time the sun was in the 15th degree of Taurus, and the spectacle (*nayzak*) in the 15th degree of Scorpio. It was a large *nayzak,* round in shape, and its size 2½ or 3 times the magnitude of Venus. Its light illuminated the horizon and it twinkled very much. The magnitude of its brightness was a little more than a quarter of the brightness of the moon.[41]

This supernova was, in fact, the brightest stellar object ever to appear in the sky. And it was so treated by the Chinese. A contemporary description states:

> On the first day of the fifth month, in the third year of the Ching-Te reign period [30 May 1006] the Director of the Astronomical Bureau said that at the first watch of the night, on the second day of the fourth month [1 May 1006] a large star, yellow in color, appeared in the east of *K'u Lou,* in the west of *Ch'i Kuan.* Its brightness had gradually increased. It was found in the third degree east of the *Ti*—thus it belongs to the geographical division of Cheng and the station of *Shou-Hsing*. . . . The star later increased in brightness. According to the star manuals there are four types of 'auspicious stars', one of which is called *Chou-po,* a yellow and brilliant [object] foreboding great prosperity for the State over which it appears.[42]

Carl Sagan of Cornell University, for example, thinks it possible that the Turkish flag was inspired by the supernova (personal communication).

40. F. H. Ellis, "A Thousand Years of the Pueblo Sun-Moon-Star Calendar," in Aveni, *Archaeoastronomy,* p. 59.

41. Bernard Goldstein, "Evidence for a Supernova of A.D. 1006," *The Astronomical Journal,* 70 (1965), 105. The article is incorrectly cited in Shklovsky, *Supernovae.*

42. Bernard R. Goldstein and Ho Peng-Yoke, "The 1006 Supernova in Far Eastern Sources," *The Astronomical Journal,* 70 (1965), 748.

3. Babylonian cylinder seals showing the crescent moon and a bright star near it.

A *Chou-po* star was quite unusual and, as the authors of the article from which the above description was taken remark, "*Chou-po* is not used anywhere else to describe an observed guest star."[43] We may just note, par-

43. Ibid., p. 749. This strikes me as a very odd thing. If the term *Chou-po* was used only for the nova of 1006, how did it get into the star manuals before the event?

enthetically, that the "guest star" of 1054 was also yellow and might reasonably have been expected to be described as a *Chou-po* if it had been, as claimed, the second most brilliant stellar appearance in history. It was, of course, only called a "guest star."

The supernova of 1006 is appealing as the source for the cave paintings in the American Southwest except for one fact. It was nowhere near the moon. That is, it was nowhere near the moon in space, but it occurred in time when the moon was new. Another Arabic account relates that it took place "in that year (A.H. 396) at the new moon. . . ."[44] We know that the Indians used a lunar calendar and therefore were accustomed to using the moon for time reckoning. Given that fact and the fact that they had to convey ideas by pictograms, it seems logical to assume that the verbal account offered by the Islamic author could be translated into pictograms to produce precisely what we have. These figures are not to be read "near the new moon in space" but "at the time of the new moon" there occurred a spectacular astronomical event. In this interpretation, we can salvage Miller's thesis that the figures do, in fact, represent a unique astronomical event and avoid the problem of reconciling a spectacular heavenly phenomenon with what appears to have been an ordinary astronomical observation in China and Japan.

There is one final piece of evidence to be presented that would reduce the magnitude of the nova of 1054. Since everyone in our story has assumed that the origin of the Crab Nebula must have been a spectacular event, the question was early raised as to why it was not observed in Europe. Mayall remarked that "Europe, of course, was at that time in the dark ages scientifically, and there are almost no records of astronomical events in the middle of the 11th century."[45] Joseph Needham goes even further. He implies that the medieval West was so blinkered by religious and philosophical prejudice that men and women there could not bear thinking about something as spectacular as the supernova of 1054. "Sarton has said, in another connection, that the failure of medieval Europeans and Arabs to recognize such phenomena was due, not to any difficulty in seeing them, but to prejudice and spiritual inertia connected with the groundless belief in celestial perfection. By this the Chinese were not handicapped."[46] Now on the face of it, we can discount this explanation. Aristotle's physics, with its insistence on the immutability of the heavens was not yet available to the West in 1054. Nor was the church particularly powerful at this time and able to enforce its will, even if that will had included a desire to see the heavens remain unperturbed by the sudden appearance of new stars. Moreover, the people of the medieval world had no difficulty observing and describing the supernova of

44. Goldstein, "Evidence," p. 107.
45. Mayall, "The Story of the Crab Nebula" (1962), 94.
46. Joseph Needham, *Science and Civilisation*, III, 428.

1006[47] or the great comet of 1066. The truth probably lies in the rather ordinary nature of the nova of 1054. Since, as we have seen, it was not the great spectacular event modern astronomers have tried to turn it into, there is less surprise in the fact that it was not particularly noted in 1054. But it was a nova and one might expect it to be picked up by someone. There is, in fact, a medieval account that would appear to be a report of the same nova that was observed by the Chinese and the Japanese.

In the *Rampona* chronicle, compiled in Bologna about the year 1476 from an earlier source that is now lost, the following passage occurs:

(1) Anno Christi M18 Henricus tertius imperavit annis x19. Hic primo venit Romam in mense maii.
(2) Cuius tempore fames et mortalitas fuit fere in universa terra.
(3) Et obscedit civitatem Tiburtinam diebus 3 mense iunii.
(4) Hic Henricus pater fuit matris comitisse Mathilde, ex qua Bonifacius marchio genuit ipsam Matheldam. Tempore ipsuius Henrici.
(5) Tempore huius stella clarissima in circuitu prime lune ingressa est, 13 Kalendas in nocte initio.[48]

The relevant parts may be translated as follows: 1. In the year of Christ M18 Henry the third had reigned nineteen years. He first came to Rome in the month of May. 2. At this time there was a famine and death throughout the whole land. 5. At this time a most clear star appeared in the first circuit of the moon, beginning in the night of the 13 Kalends. This is the complete entry for 1058, except for a long section on the activities of Hildebrand who later became Gregory VII.

There are some obvious difficulties with this text. In (1) the date is given as M18 for 1058. This mixture of Roman and Arabic numerals is quite common in the fifteenth century but almost impossible for the eleventh when Arabic numbers were just beginning to enter the West. Moreover, 1058 is an impossible date since Henry III died in 1056. We may, then, suspect an error in copying. If the original scribe had written mliiii, it could easily have been mistaken for mlviii. Precisely this kind of mistake is often made.[49] We cannot be certain that the correct date is 1054, but it is possible and the "stella clarissima" would then be the nova observed in China and Japan. Unfortunately, it is almost impossible to make astronomical sense

47. See, beside the Goldstein article mentioned above, N. A. Porter, "The Nova of A.D. 1006 in European and Arab Records," *Journal of the History of Astronomy,* 5 (1974), 99.

48. My attention was first called to this chronicle by a brief report in Robert R. Newton, *Medieval Chronicles and the Rotation of the Earth* (Baltimore, 1972), p. 690, which reported under the year 1058, "*Rampona,* a bright light within the circle of the new moon." The text of the chronicle may be consulted in Albano Sorbelli, ed., *Corpus Chronicorum Bononiensium,* vol. 1 (Bologna, 1939), 464.

49. Perhaps the most famous such mistake occurs in Michel de Montaigne's essay 34, "Observations on Julius Caesar's Methods of Making War," in which Montaigne wrote CIX thousand for IIX thousand thus multiplying Caesar's figure of 8000 to 109,000. Donald M. Frame, ed., *The Complete Works of Montaigne* (Stanford, 1957), p. 560, n. 1.

out of (5). The thirteenth Kalends is thirteen days before the first of the month to which the Kalends refer.[50] Thus, if we assume the month to be July, then the thirteenth Kalends would be June 18, only one day away from the date given by the Japanese for the first sighting of the nova. I simply have no idea what "in circuitu prime lune ingressa est" means except that it associates the bright star with the moon and Miller has shown that there was such an association between the moon and the nova of 1054. All in all, I would suggest that there is sufficient evidence for accepting this account as a sighting of the nova. It was not sighted elsewhere because, as (2) suggests, the climate of Europe that year had been very bad. Famine is the result of a shortage of wheat; the great enemy of wheat is wetness. If we assume a cloudy and rainy summer, the result would be both a famine and a failure to notice the nova.

CONCLUSION

It is time to assess the various arguments put forth in this paper and indicate just what has been proved. I think I can safely draw the following conclusions: The identification of the Crab Nebula with the "guest star" of 1054 is not so straightforward and as simple as it is presented in all the current literature. The "guest star" of 1054 was not a spectacular event, to be classed with the supernova of 1006. It is probable that the star of 1054 was observed in Western Europe. Beyond these, it is possible to offer problems for future research. These involve both the history of Chinese astronomy and modern astronomical research. If the nova of 1054 is where the Chinese sources place it, then historians of Chinese astronomy are faced with the problem of the identity of zeta Tau and T'ien-kuan. It seems essential to re-do Ho Peng-Yoke's work here and try to discover if an error has been made. If it is shown that there is absolutely no possibility of error, then there is only one way out. There is another pulsar at R.A. $05^h\ 25^m$, Dec. $21°\ 58'$ which might just fit the "several inches" description if a scribal error in direction were assumed. I am not persuaded of the truth of this suggestion, but I offer it for what it is worth.

If the Crab Nebula is, in fact, the "guest star" of 1054, then astronomers must still wrestle with the explosion data. All three of my astronomical colleagues at Cornell to whom I submitted this paper suggested that it would be possible to make an accurate physical and mathematical model to account for the discrepancy between extrapolated date and actual date of explosion. It should be emphasized that the model does not yet exist and it might be well worth some young astrophysicist's time to work it out. If

50. See R. T. Hampson, *Medii Aevi Kalendarium,* 2 vols. (London, 1841), I, 428, and II, 228.

nothing else, it would tie up one embarrassing loose filament in the Crab Nebula story.

There is another possibility worth mentioning. The nova of 1054 and the explosion that produced the Crab Nebula may not be the same. As mentioned earlier, if we did not have the Chinese and Japanese sources, we would date the origin of the Crab Nebula from the first half of the twelfth century. Let us suppose, for the sake of argument, that this is when the Crab Nebula was born. What would we have to assume to make this believable? The only thing against it is that no one reports a sighting of a nova at this time. Is that sufficient to make the hypothesis untenable? It might be if the explosion were as spectacular as everyone has assumed it to have been. Certainly an object of −5 visible magnitude could not have gone unnoticed. There are two important assumptions in this last statement that deserve close examination. The first is that whatever explosion produced the spectacular Crab Nebula must, itself, have been spectacular. Must that be so? We know that the converse is not. The supernova of 1006 did not create a modern astronomical object commensurate with that blow-up. Indeed, the origin of that supernova is rather difficult to find. May it not be foolish simply to assume that the birth of the Crab was such an event that it could not fail to be noticed? If this is so, then the visible magnitude of −5 is questionable. We have seen that the sources do not appear to reflect an event of such a nature. If we then scale the magnitude down, failure to see and report the nova of 1140 becomes at least possible. Faith in this hypothesis is strengthened by the fact that we do know of precisely such a case in which failure to observe the supernova seems inexplicable and impossible. Cassiopeia A is the location of a supernova that appeared around 1670.[51] It was seen by no one, anywhere in the world. This was after the invention of the telescope and in a period of the most intense astronomical investigation. Shklovsky writes of this: "There arises the interesting question, why was it not observed at this time. In the times of Newton and Bradley, European astronomy was at a sufficiently high level and the non-setting constellation Cassiopeia can be observed throughout the year. Below we shall discuss the problem in detail."[52] Significantly, no such dis-

51. Shklovsky, *Supernovae,* p. 86. Shklovsky determined the date by saying, "If this expansion has occurred all the time at a constant rate, it must have begun at about A.D. 1700 ± 14 and this estimate coincides with the outburst of the supernova." In other words, the date of Cassiopeia A is determined by precisely the kind of calculation that astronomers reject for the date of the Crab Nebula! I have used the date 1670, because that date, according to Carl Sagan and Yervant Terzian, seems to be the more modern estimate.

52. Ibid.

It is a pleasure to acknowledge my gratitude to the following people. Professor Carl Sagan whose lecture (following Needham) at Cornell Alumni University in the summer of 1975 first aroused my suspicions about the failure to observe the nova of 1054 in the west. Professor James J. John for invaluable assistance with the *Rampona* chronicle and medieval dating. Professor Brian Tierney for a critical reading of the paper and shrewd comments on the

cussion exists for the failure does seem to be inexplicable. But if men of the stature of Isaac Newton, James Bradley, Edmond Halley, and John Flamsteed missed a supernova in 1670, it should not be surprising that medieval men failed to see one in the twelfth century. This does not, to be sure, prove my case. But it should be noted that the argument is exactly analogous to Galileo's use of the moons of Jupiter as "proof" of the Copernican system. Galileo could not explain why the earth should move around the sun, but he could show that the moons of Jupiter went around that planet for an equally inexplicable reason. If you can believe the one, he argued, you can believe the other. I ask no more.

This, then, is where we end. If the "guest star" of 1054 was the origin of the Crab Nebula, then the sources of Chinese astronomy need to be examined carefully to discover how so glaring an error could be made by the most sophisticated and accurate astronomers of the time. And modern astrophysicists will have to devise a rather complicated and difficult model to explain the modern data on the expansion of the Crab so as to fit the origin of the expansion into the eleventh century. If, however, the Crab Nebula grew out of a nova in 1140 or thereabouts, then all we have to explain is why people failed to notice it, as professional astronomers were to miss the supernova of 1670. It will be interesting to see how future scholarship decides the issue.

---

medieval aspects. Professors Frank Drake and Yervant Terzian of the Cornell Department of Astronomy for a number of criticisms of an earlier version of this paper and for generally disbelieving my thesis. Professor A. Aaboe of Yale University who utterly destroyed my first attempt to make sense of the supernova-Crab Nebula story. The alumni of Cornell Alumni University who stimulated me to follow up the suspicions raised by Sagan.

CONTRIBUTORS

MARIE BOAS HALL, who received her Ph.D. at Cornell University in 1949 under Henry Guerlac, has since 1963 been Reader in History of Science and Technology at Imperial College, London. She is the author of, among other works, *The Establishment of the Mechanical Philosophy* (1952), *Robert Boyle and Seventeenth Century Chemistry* (1958), and *The Scientific Renaissance* (1962) and, with A. R. Hall, she edited *Unpublished Scientific Papers of Isaac Newton* (1962, 1979) and *The Correspondence of Henry Oldenburg* (1965–1980/81).

LESLIE J. BURLINGAME received her Ph.D. from Cornell University in 1973; she now teaches at Franklin and Marshall College in Lancaster, Pennsylvania. Since Henry Guerlac directed her dissertation on Lamarck, she has continued her research on Lamarck, published several articles on his work, and is now in the final stages of completing a book manuscript, *Natural History, Biology, and Evolution in Lamarck's Thought.*

J. B. GOUGH is Associate Professor of History at Washington State University. He specializes in the history of the experimental sciences, especially chemistry, in eighteenth-century France. He received his Ph.D. degree from Cornell in 1971.

French born and American educated, ROGER HAHN studied with Henry Guerlac from 1957 to 1960, completing his dissertation on the fall of the Paris Academy of Sciences during the French Revolution in 1962. He is presently Professor of History at the University of California, Berkeley.

THOMAS L. HANKINS is Professor of History at the University of Washington, Seattle. He is the author of *Jean d'Alembert: Science and the Enlightenment* (1970) and of *Sir William Rowan Hamilton,* a biography of the Irish mathematician (1980).

RIO HOWARD is currently an *attaché de recherche* with the Centre National de la Recherche Scientifique, Paris. She works in the history of medicine and biology in the sixteenth and seventeenth centures.

MARGARET JACOB is Associate Professor of History at the Graduate Center and at Baruch College of the City University of New York. Most recently she was a research fellow in the History of Science Department, Harvard University and at the Institute for Advanced Study, Princeton. Her latest book, *The Radical Enlightenment: Pantheists, Freemasons and Republicans,* was published in 1980.

DAVID KUBRIN was educated at the California Institute of Technology and Cornell University, obtaining his Ph.D. in 1968. He has taught at Dartmouth College, the University of Wisconsin, and the San Francisco Liberation School (where "Newton's Inside Out!" was originally given as a lecture), a Marxist education project he helped form in 1972. He has written on Newton, nineteenth-century health care in the United States, science, and ideology, and is currently writing a novel, *To Rear Cattle in the Evening.* He earns his living as a piping designer in the Bay Area.

GENEVIEVE MILLER, Ph.D., is Associate Professor Emeritus of the History of Science, Case Western Reserve University School of Medicine, and Retired Director of the Howard Dittrick Museum of Historical Medicine, Cleveland, Ohio.

CARLETON E. PERRIN is Associate Professor of Natural Science and Humanities at York University, Toronto. His current research interests include the chemical revolution, the international diffusion of Lavoisier's chemistry and the role of scientific networks in diffusion of ideas in early modern science.

RHODA RAPPAPORT is Professor of History at Vassar College. Her research interests center upon France in the eighteenth century, with emphasis on the history of the earth sciences.

CECIL J. SCHNEER teaches both crystallography and the history of science at the University of New Hampshire, where he has been studying relationships between the internal structure and the external plane polyhedral morphology of the kingdom of minerals since 1954.

MARTIN S. STAUM received his A.B. degree with Honors in history from Princeton University in 1964 and his Ph.D. from Cornell University in 1971. His monograph entitled *Cabanis: Enlightenment and Medical Philosophy in the French Revolution* was published in 1980.

L. PEARCE WILLIAMS is John Stambaugh Professor of the History of Science at Cornell University.

HARRY WOOLF, Ph.D. 1955, Cornell, was Willis K. Shepard Professor of the History of Science, at the Johns Hopkins University from 1961 to 1976. He has been Director of the Institute for Advanced Study, Princeton, New Jersey, since 1976. His latest publication is *Some Strangeness in the Proportion: A Centennial Symposium to Celebrate the Achievements of Albert Einstein,* 1980.

INDEX

THE ANALYTIC SPIRIT

Designed by G. T. Whipple, Jr.
Composed by The Composing Room of Michigan, Inc.
in 10 point VIP Sabon, 2 points leaded,
with display lines in Sabon Bold.
Printed offset by BookCrafters, Inc.
on Warren's Number 66 text, 50 pound basis.
Bound by BookCrafters, Inc.
in Holliston book cloth
and stamped in All Purpose foil.

Library of Congress Cataloging in Publication Data
Main entry under title:

The Analytic spirit.

Includes index.
CONTENTS: The chemical revolution: Gough, J. B. The origins of Lavoisier's theory of the gaseous state. Perrin, C. E. The triumph of the antiphlogistians. Burlingame, L. J. Lamarck's chemistry.—[etc.]
1. Science—History—Addresses, essays, lectures. I. Guerlac, Henry. II. Woolf, Harry.
Q126.8.A52 509 80-29083
ISBN 0-8014-1350-8